THE
NATURE *of* NATURE

THE
NATURE
OF
NATURE

**The Discovery of SuperWaves
and How It Changes *Everything***

IRV DARDIK

written with Estee Dardik Lichter

RODALE.

RODALE *wellness*

Live happy. Be healthy. Get inspired.

Sign up today to get exclusive access to our authors, exclusive bonuses,
and the most authoritative, useful, and cutting-edge information on
health, wellness, fitness, and living your life to the fullest.

Visit us online at RodaleWellness.com
Join us at RodaleWellness.com/Join

Rodale books may be purchased for business or promotional use or for special sales.
For information, please write to:
Special Markets Department, Rodale, Inc., 733 Third Avenue, New York, NY 10017

Printed in the United States of America
Rodale Inc. makes every effort to use acid-free ∞, recycled paper ♲.

Book design by Christina Gaugler

Library of Congress Cataloging-in-Publication Data is on file with the publisher.

ISBN-13 978–1–62336–935–4
Distributed to the trade by Macmillan

2 4 6 8 10 9 7 5 3 1 hardcover

We inspire health, healing, happiness, and love in the world.
Starting with you.

In honor of Planet Earth and Life

CONTENTS

NOTE FROM THE PUBLISHER ix

INTRODUCTION: THE DISCOVERY OF SUPERWAVES AND HOW
IT CHANGES *EVERYTHING*.. xi

PART 1: LIVING WITH NATURE

CHAPTER 1 • The Universality of Rhythms... 3

CHAPTER 2 • The Beginnings of Scientific Civilization.......................... 9

PART 2: HOW SCIENCE DEVELOPED FROM THE PUZZLE HYPOTHESIS

PRELUDE • The Origin of the Scientific Method 28

CHAPTER 3 • Matter—What Is It?... 35

CHAPTER 4 • Motion—How Does It Work? 49

CHAPTER 5 • Laws—What Holds Everything Together,
and Why Do Things Fall Apart? 75

PART 3: WHERE DO WE STAND TODAY?

PRELUDE • Do We Have a Theory of Everything? 90

CHAPTER 6 • Where Do We Stand Today with Regard to Matter? 92

CHAPTER 7 • Where Do We Stand Today with Regard to Motion?.... 100

CHAPTER 8 • Are the Laws Bringing Us Closer to a
Theory of Everything? 117

PART 4: THE DISCOVERY OF THE NATURE OF NATURE

PRELUDE • A Final Commonality—Waves 138

CHAPTER 9 • My Story 142

CHAPTER 10 • Tragedy Yields Discovery 148

CHAPTER 11 • The Three Principles of Waves Waving...................... 169

CHAPTER 12 • Cell Cycles 186

CHAPTER 13 • The Quantum Explained:
The Resolution of Wave-Particle Duality..................................... 197

CHAPTER 14 • Laws: The Quantum and Thermodynamics 224

CHAPTER 15 • The Origin of Health and Disease 241

CHAPTER 16 • SuperWaves, the Environment, and the
Origin and Survival of Life 269

CHAPTER 17 • The Solar System, the Galaxies, and the
Nature of Gravity 278

PART 5: THE FUTURE

CHAPTER 18 • A New Beginning 293

ACKNOWLEDGMENTS 301

SOURCE NOTES 303

INDEX 319

THE NATURE OF NATURE:
A History of Science, Thought, and Health Revolutionized and Reframed

Note from the Publisher

ONCE IN A LIFETIME, a publisher gets to publish a book that is so revolutionary that certain constraints must be set aside; when an author is both so brilliant and so frustrating that a book takes a whole lifetime to manifest—and when manifested is deeply imperfect and perfectly perfect.

This is not a book for everyone. It is a book about a theory that goes beyond what any theory has attempted to theorize. And it's very appropriate that it started with the human heart. The beating human heart is in motion, and healthiest when exhibiting variability. This variability is motion *in* motion. It is the hallmark of life. For motion and life are all waves.

It is only in the past century that we have asked the questions "Is something a particle or a wave? And how can it possibly be both?" Scientists from all different lineages have been unable to reconcile the differences, the gaps, the conflicting information, the mystery.

With this book, this theory, I believe it is resolved. It is a wave. Everything is waves. Or to be more specific, everything is SuperWaves. Everything is waves waving within waves. There are no particles; underneath it all, there are ultimately only waves.

Indigenous tribes have intuitively stated since the beginning that everything is connected and we are all one. Interestingly, without even awareness of or interest in these traditions, Irv Dardik has shown how this is not just possible but true.

That's not to say that the information science has gathered until now is not true. In fact, we couldn't truly believe in the oneness of everything without science inventing the tools to search for the pieces and try to solve the puzzle. This theory, this book, is a new lens to look through—the birth of a whole new kind of science, research, and exploration.

This is just the beginning. This is your invitation. Prepare to see the world, life, and science in a whole new way.

INTRODUCTION

The Discovery of SuperWaves and How It Changes *Everything*

IN THIS BOOK, I present a novel understanding of the nature of nature. I will share a discovery that reveals the nature of the "stuff" of the universe—what we know as stars and planets, living organisms and particles—ultimately, all that exists.

Our current understanding of the nature of nature, the product of generations of scientific endeavor, is that matter is the raw stuff of the universe and is composed of particles that interact in predictable ways. True, in the twilight world of quantum mechanics, strange and hard-to-believe phenomena do occur, including particles sometimes behaving as if they are waves and sometimes as if they are particles. Ultimately, however, our understanding of the nature of nature is that it is a whole made up of parts (particles) that are essentially independent of each other.

This view of nature has deep roots in our thinking. It seems to match our experience of the world around us, and it has allowed us to develop fields of science, technology, and medicine that appear to have served us well and continue to do so.

The understanding that I have developed over the past 30-some years and am presenting in this book is that nature is not a whole made of parts that can be disassembled and examined independently, as if it were a car engine. Instead, nature is the manifestation of an unbreakable continuum of waves waving within waves. This means that the reality underlying the world we know and experience is an inherent continuity of waves at all scales: natural waves that coalesce and spread out to be everything from the quantum world all the way up to the universe as a whole, including its living and nonliving elements. These are not the waves of mathematical equations, but rather are real, normal, observable waves in nature. Their novelty is that they relate in ways never before recognized. They possess a surprising interdependence across scales that lends nature a wholeness, different from the view of nature as a whole made of fully separate parts. And they offer unprecedented explanatory power for the entire natural universe.

As I developed this novel view of nature, I came to realize that waves nested within waves nested within waves are the stuff behind what we experience as matter, motion, space, time, order, and disorder. They are the deeper identity of all the natural phenomena we know, such as light and energy, forces, thermodynamics, emergent complexity, quantum physics, and gravity. They resolve the mysterious coexistence of waves and particles on the atomic and subatomic levels. They form a continuum whose discovery answers our questions of "what it is," "how it works," "what holds things together," and "why things fall apart." All of matter and the way it behaves, and the order and disorder we see in the world, is the outcome of this continuum of waves nested within waves nested within waves: what I call SuperWaves. Ultimately, SuperWaves underlies and *is* the universe.

Now, I am the first to admit that accepting that the pages on which you are reading these words, whether as a hard copy or on a screen, are the manifestation of waves—not particles—is counterintuitive at best, and may even seem impossible.

But if you can suspend your incredulity long enough to hear what is, in effect, one long argument that I present here, I believe you will come to see the reality and explanatory force of SuperWaves yourself.

SUPERWAVES IS A DISCOVERY—not a philosophy, not a suggestion, not a way of thinking about nature. It is the discovery of a wave phenomenon that exists. It is an independent, far-reaching wave behavior unlike any recognized before. The catch, as it were, is that it takes a fair amount of discussion to fully explain what it is and how it exists. Waves are not necessarily easily or intuitively understood, and to deal with them, we generally tack them on to the things or events through which we see them and dissect them mathematically. Understanding SuperWaves requires understanding waves in a different way. It takes time to see them as their own phenomenon, as they exist in nature, and to see them in the unique relationship that pointed me to the existence of SuperWaves.

If it seems strange to hear that any discovery could tell us what nature *is*, it may help to realize it's not the first time we've thought of such a thing. The ancient Greeks suggested the atom to fill that same role, millennia ago. And people thought that when we found it, the atom would be exactly that type of discovery: something that tells us what nature is. The atom would be something in nature, which we would see expressed throughout nature; and when we understood it and its characteristics, it would confirm what nature is. We thought the discovery of the atom would prove that nature is something like a jigsaw puzzle, made up of indivisible parts called atoms, and that understanding how atoms work together to be nature would answer all our questions.

The atom was supposed to be two things, in other words. It was supposed to be the fundamental part of nature, but—and this point usually goes under the radar but is most important here—it was also supposed to confirm that the design of nature is a whole made of parts, in the form of a puzzle.

The atom ended up being divisible into further parts after all. So it did not turn out to be the fundamental part of nature. That didn't make a dent in science's conviction that nature's design is that of a puzzle, however. And by using the word *design*, I do not mean to invoke the debate over intelligent design, but rather to turn our attention to an assumption we all make about the format, or design, of the physical stuff of nature.

We assume nature takes the design of a whole made of parts. It's almost impossible to imagine an alternative to it being a whole that is divisible into parts. Nevertheless, this is an assumption.

And though down in the quantum realm we found what we thought we would not find—supposedly indivisible atoms can be broken into protons, neutrons, and electrons, and those into further parts still, while a veritable zoo of particles flirt with wave characteristics unimaginable in our macro world—we kept, and continue to keep, the idea of building blocks.

I point this out not to veer into a transcendental idea of one versus many or a debate about mathematics and reality, but to question this assumption we make about the stuff of nature itself. The fact is, though it is committed to being unbiased, science incorporates an unconfirmed, a priori conviction about nature, namely that nature is a whole made of parts. When the first true scientists turned to the natural world, moving away from theology and toward earthly realities to understand the universe, they automatically began studying "parts of" nature. (I put *parts of* in quotes to call attention to that presumption.) "Nature is a whole made of parts" is a precondition built into our otherwise rational investigations.

The assumption runs so deep that we rely on it in our everyday thinking about the world. Indeed, the way our senses function makes it difficult to imagine that parts-and-wholes even *is* an assumption that casts a bias into our thinking and is not the way the world "is." This is why we proceeded from it as a starting point to consider nature rationally.

And as a precondition, it indicates two things. In addition to telling us the design of nature, it determines the steps we must take in order to understand it. It tells us that the way to understand the whole is to isolate parts and to test them to identify their properties and how they contribute to the whole. "Nature is a whole made of parts" was the original, unstated premise behind the scientific experiment.

Some scientists do recognize the existence of this preconception in our thinking, and its implications. For instance, renowned Cambridge physicist Stephen Hawking has said that "if everything in the universe depends on everything else in a funda-mental way, it might be impossible to get close to a full solution by investigating parts of the problem in isolation."[1]

Inherent in the SuperWaves view of nature is that "everything in the universe depends on everything else in a fundamental way." The nested, interconnected qual-ity of waves waving within waves means they cannot be isolated (in the way that individual atoms can appear isolated) and therefore, cannot be detected with the traditional scientific method. Think about what this means: *Real though it is, the SuperWaves continuum cannot be found by scientific experiment.*

That may seem shocking to you, but it also must be maintained as a possibility. It means that a discovery outside of the scientific method *could* happen and could show us what we've searched for all along: a better way to understand and work with nature. That discovery is SuperWaves, which I made by observing nature and by following the implications of what I observed, not through an isolated experiment.

NOW, I KNOW FULL well that saying an important discovery was made by a means other than scientific experiment sounds suspicious. We accept that certain discover-ies, like a new species or planet, may be discovered by searching the universe, but it seems important that something as fundamental as I am discussing be found and confirmed in a lab.

But which do we want—truth or science? Science is meant to be a finder of truth, but if it harbors a precondition that is getting in the way, we must explore that fact. The very reason you feel concerned is the reason you should continue. Anyone who wants an open-minded, honest, and rational investigation must be open to hear whether we have been misled by a precondition that exists in our scientific thinking.

You will eventually come to see that SuperWaves exhibits characteristic patterns and behaviors that match perfectly with the data science has accrued about the quantum-level and macro-level characteristics of matter, motion, space, time, order, and disorder. It offers an elegant explanation for the universe as we know it. The facts correspond with this explanation much more simply than they do with the parts-and-wholes connections proposed by science. By virtue of Occam's razor, it stands as truth.

And even beyond truth—for we care about our survival and success as well as about our understanding—the discovery grants us compelling, practical tools. Knowing nature's character means we can work with it to get what we want. Specifi-cally, we can tap into the very means by which order and organization come about. The discovery of SuperWaves shows that the cross-scale way waves wave is the mechanism behind order and disorder, and *we can influence nature's wave patterns*. For example, SuperWaves shows that we can cultivate health and renormalize chronic disease into health by shaping our rhythmic behavioral patterns of exercise-recovery, waking-sleeping, and other large-scale cycles that influence the cycles of our organ systems, cells, and biochemistry. You will learn how you can effectuate such change. And you will also learn how we can use our knowledge of SuperWaves to cultivate the order and organization we desire for practical technologies, for the environment, which is deeply connected with our personal health, and for the planet as a whole.

And as the true stuff of nature, SuperWaves offers a new understanding of the steps we must take to understand nature. The fact that there is an inherent connection between all scales of nature puts the technique of scientific isolation in a new perspective. Experiment still has merit, but inasmuch as nature is ultimately a continuum of waves, experimental data is placed in a new and solid context. The context explains why and how experiments come out as they do in closed systems, and why and how to understand them in the context of continuous waves that cannot be isolated in closed systems.

The discovery of SuperWaves does not negate science, in other words; rather, it shows that scientific discoveries, together with scientific mysteries, support this new understanding of what nature is.

Let me be clear: I am not arguing a philosophical point of one versus many, but rather am introducing an actual physical discovery that points us toward understanding nature as a one, and not many. Neither am I disrespecting science, as the facts of science are what validated the scope of SuperWaves once I'd discovered it. I am saying that there is an actual *thing in nature* that is continuous and that once you see its cross-scale continuity, you see it stretches in and through scales, unbroken and unbreakable, and that it makes up and is all of nature. I discovered it by seeing it, much the same way a person would discover a new species or planet. Except this discovery required no travel or specialized equipment, just a *different way of thinking* about nature. It required seeing nature happen without resorting to a parts-and-wholes framework.

AS THE AUTHOR OF THIS BOOK, my challenge has been to introduce the discovery while also addressing the way we think. The assumption that nature is a whole made of parts runs so deep in our thinking that in order to loosen its grip, I have to trace its rise and spread through the histories of both civilization and ideas about the universe. Otherwise, its influence might cause you to reject SuperWaves even as I explain what it is—for if an idea doesn't fit our parameters of what a proper idea looks like, we often reject it outright. That is why the story of the discovery, from my history as a vascular surgeon and founding chairman of the United States Olympic Sports Medicine Council to the series of events that led to my discovery of SuperWaves, begins in Part 4. It takes three sections to get to the point where the story can be told—three sections to build the argument that unravels the grip of our current worldview, such that the new one can be properly understood.

Some of the buildup may be familiar to you, some may be new, but all is necessary for showing why science never found this natural phenomenon that exists in, and is, nature.

I start by calling attention to the historical experiences of humanity and how they shaped the foundations of how we think. I then explain how those foundations set us up to understand nature through science and thereby generated the difficulty we have had in recognizing the SuperWaves phenomenon. Then I move to the discovery itself, through the tale of how I discovered it. Last, I discuss its practical applications.

Part 1 explores the underpinnings of the way we think about nature. It begins with a description of nature as it must have existed before we ever thought about it—in order to clarify what happened when we did begin to think about it. Specifically, we explore that nature was reality, that we lived with and within nature without absolute boundaries, and that rhythms were a critical aspect of daily, monthly, yearly, and generational life. The section then proceeds to address what happened when we began to observe nature. The functioning of our senses, combined with the advent of our thinking minds, told us that nature is designed as a whole made of parts, a puzzle. This was a practical assumption that seemed to be supported by experience. The other possibility, that the "stuff" of the natural universe was an ultimately indivisible whole with a single identity, appeared to be impractical and eventually seemed illogical. The section closes with a discussion showing that the idea that nature is a whole made of parts was our Original Theory of Everything. Our experience, as civilization developed, was a great experiment that seemed to support it. I call the theory the Puzzle Hypothesis.

Part 2 tells how science developed on the basis of the Puzzle Hypothesis. Going on the original assumption that nature is designed as a whole made of parts, we tried to make sense of those parts, which seemed best categorized as matter and motion, with varying degrees of order and disorder. The section travels from Thales of Miletus in ancient Greece, to the development of the atomic hypothesis, to the work of Galileo, to Planck's quantum physics, to Einstein's relativity theories, and to present-day work, such as string theory, to show that every so-called scientific revolution in history has been but an advance in how to apply the Puzzle Hypothesis to nature. We applied it to matter, we applied it to motion, we applied it to patterns of order and disorder, and we brought ourselves to science as we know it today.

Part 3 discusses where we stand today. Has the Puzzle Hypothesis brought us to an ultimate understanding of everything? If not, are we at least close to answering what it is, how it works, what holds it together, and why things also fall apart? This is an important test: We adopted the Puzzle Hypothesis because we thought it accurately reflects nature and that it will bring us to a simple understanding of nature. If nature has become difficult and complicated to understand through its maneuverings, an equally reasonable possibility that is simpler merits consideration.

The prelude to Part 4 introduces a commonality found throughout nature and the sciences: waves. Waves have always existed without absolute boundaries in nature, long preceding our tendency to break things into parts for understanding. They possess unique qualities that, because science has not brought us all the answers and also leaves us with mysteries and paradoxes, demand we reconsider their power and what they are.

Part 4 then begins with my story. I briefly describe my background in medicine and with the Olympics and the unexpected loss of a friend that made me question our assumptions about health and disease. I describe how I came to recognize the heartwave, a phenomenon of waves changing together, simultaneously, across scales. Simultaneity is a type of motion disallowed by the reductionist approach of science, but when I looked at waves, which are pure motion, alone—instead of starting with material objects as if they are primary and seeing waves through "their" motion—I saw that such simultaneity is real. This was an important discovery.

From there, the section goes through the principles of wave behavior that such simultaneity entails. I then trace these principles in action, down through cellular, biochemical, and molecular behavior, until I reach the quantum realm. There, in the deepest levels of matter, the inherently continuous nested wave phenomenon I'd discovered on the level of the organism explains the mysteries of quantum physics. It in fact solves and explains all the mysteries and paradoxes described in Parts 1, 2, and 3. The section includes the ultimate explanation for emergent complexity (order) and decomplexification (disorder), with a focus on health and disease, the environment, gravity, and the true identity of energy. It spells out what we must do to protect our personal health and the health of the environment, and how we can apply our new knowledge to create what we would today call clean energy.

Part 5 closes the book with a reflection on the discovery of SuperWaves and a call to action. Now that we know the nature of nature, we know what we must do.

A word of caution: If you are a scientist, you may find much of the discussion in this book a departure from the way you usually treat nature. There is little involvement with the intricate details that scientists daily consider. But the early chapters are not about science; their point is to identify the *reasoning behind* science. This should be taken seriously by everyone, even if the subject matter is familiar. We have to get down to the basics, going even further than arguing for reductionism or holism, to identify the thread that runs through both approaches. We have to go down to the a priori approach that generates the dispute between them, in fact. Only then can we see that there is an alternative, a unity that is neither reducible nor holistic in the way we understand those terms now.

Let me restate what this book is *not*. It is not:

- A culmination of scientific thinking
- An attack on scientific thinking
- A transcendental explanation of nature
- A philosophical argument for the nature of nature
- A holistic view of nature

But it is:

- A discovery of something existent *in* nature
- An explanation of how that discovery accounts for—essentially *is*—nature
- An introduction, at the same time, to a new understanding of the steps we must take from now on to understand the natural universe
- An introduction to how to approach science in this new context

Even with all this said, I anticipate that some people will still misunderstand—they will say that this has been said before, that scientists already know the problems I present, that Eastern religions have long spoken of the rhythmic breath of the universe, that reductionism and holism are an ancient pair, and that my arguments present nothing new.

So here I offer a word of guidance for this book: It's easy to skim a book to see how it expands the breadth of what you already know, but that won't work here. Read carefully, read deeply; know that some ideas may seem elemental and others surprising. It will be different things to different people, depending on what you already know. I have to cover all the bases to properly introduce SuperWaves. Of course, you may—and I hope you do—grasp it quickly and delight in the new ideas. As long as you can let go of the old framework, you can see that the new ideas are simple, natural, and logical—even elegant.

I hope that by reading this, you will open your mind to possibility: that instead of judging from the get-go, you will accept the possibility that you may be about to swing around a bend and confront something totally new. That experience, of course, can bring on the same emotions as any unexpected turn of events that forces a new perspective: thrilling and upsetting in equal measure, with a fair amount of discomfort and even denial cropping up from time to time. New is not necessarily easy. But ultimately, when correct, it rewards as nothing else does.

Though new to us, SuperWaves is as old as the universe itself. Nature is what it is and always has been: SuperWaves.

AS WE GO FORWARD through the observations and arguments I present in the following chapters, I would like you to hold in your mind the two views of nature I presented at the beginning of this introduction. On one hand is the view of nature as a whole composed of individual, independent parts—the Puzzle Hypothesis of nature. On the other is the possibility that nature is the manifestation of an inherent nested continuum of waves, an interdependent oneness—the SuperWaves Principle that I will now introduce. My hope—my expectation, really—is that as you come to think differently about nature and science, your view of nature will shift from the former to the latter, and you will experience a gestalt flip in your perception of the nature of nature.

PART 1

Living with Nature

In the following two chapters, I tell the story of how the assumption that nature takes the design of a whole made of parts* entered our worldview with force and permanence. It is the story of how this hypothesis first arose and was then "proven" by experience—how it was a natural guess based on the design of our sensory perceptions, sank into unconscious acceptance because it was so practical, became a worldview, and eventually seemed so fully confirmed by the success of civilization that it became our theory of the universe. As we go through this story, however, we will never lose sight of the fact that "nature is a whole made of parts" is still an assumption.

Unacknowledged assumptions and preconditions to understanding are the bane of rational thinking. Nobel laureate physicist Louis de Broglie, the man who discovered that all matter has wave characteristics, has said that "history shows clearly that the advances of science have always been frustrated by the tyrannical influences of certain preconceived notions which were turned into unassailable dogmas. For that reason alone, every serious scientist should periodically make a profound reexamination of his basic principles."[1]

I am suggesting that the notion of nature as a whole made of parts is and has been a powerful preconception that has shaped scientific thinking for millennia. Although some have been aware of it (Albert Einstein said that "it is an outcome of faith that nature—as she is perceptible to our five senses—takes the character of such a well formulated puzzle"[2]), rarely do we recognize, to use de Broglie's words, the "tyrannical influence" it has. The precondition that nature is a whole made of parts dictates that we process information about the world through one specific perspective to the exclusion of any other possibility.

---

* In this case and for the duration of the book, I use the word *design* to refer to the pattern of organization of nature. I am only talking about nature itself, as we perceive it to be. The word *design* in this case should not be mistaken for design in a supernatural sense, as is used in arguments over "intelligent design."

CHAPTER 1

The Universality of Rhythms

THE STORY OF OUR PRECONDITION to thinking about nature—that it is a whole made of parts—begins not with us but with the natural world onto which we placed it. We have to recognize how nature must have been in the time before the earliest hints of civilization. It gives us a frame of reference through which the story of science becomes the story of human beings attempting to understand nature through a singular point of view. It lays the groundwork against which we can identify the birth of the human perspective.

By knowing what nature looked like then, before we began thinking about it, we will come to appreciate how our perspective shaped our thoughts on the reality that existed before we ever did.

Our starting point is that nature exists. This has been the starting point, throughout history, of all people who earnestly attempt to understand the world around us. We trust that nature's character, whatever it may be, is true to itself.

The human quest to understand nature, on the other hand, began some thousands of years ago.

Therefore our journey begins not with people but in the time before human understanding. Things then were not the way they are now. They were not the way we believed them to have been in the earliest days of civilization. They were purely natural: Nature in its truest form.

I am going to describe nature as it surely was in that period of time. I will tell it as if recounting a scene so you can get a good sense of what it must have been like. It is a description anyone could give with a small measure of reflection.

Nature was vibrant, and very importantly, it was all there was to reality. In the countless years before the first fire was created, the first agricultural seed was planted, or the first complex tool was used, people had to live in a way that is best described as "with and within" nature. We lived in nature "as is."

When nature itself was our home, no walls or other artificial boundaries separated us from the natural world. Even as we sheltered ourselves, we took shelter in nature just as other wild animals did. We did not divide time into segments or schedule activities to be carried out no matter the environment. Living meant going with the flow.

Let's think about what that means. It is rarely noted that we were living, and had to live, with and within nature at that time, but I think it is fair to say that it was key for our basic survival. Specifically, we had to change as nature changed. So if we want to understand early human survival, we have to understand the changing world we changed with.

To picture this world, the world we lived in naturally, I'll ask you to backtrack through the changes we have wrought. Erase the boundaries that we live in today. The walls of our homes, the paved streets we travel, city lines and national borders—they are a far distant future. A thoroughly vibrant world in motion predated those human constructs.

Restore the natural flow from one ecosystem to the next. Picture open plains settling into teeming swamps, vast forests and rivers giving way to high mountain ranges and, elsewhere, to the sea. Swell the animal and plant populations far beyond that which exists today. And add more species altogether, such that the full magnitude of natural life on earth, all around the globe, is throbbing with vitality.

That was our world—a world without artificial boundaries. There were no absolute divisions in the world before *we* began to build walls. Before we ever began to think about it, nature enjoyed unfettered, unrestrained dynamics. Of course there was variability, but starkly absolute borders did not exist. This is a natural realization that a moment's consideration will confirm. We did not live in a box, nor was anything else definitively boxed in. The boundaries of the material objects we encountered, just like ourselves, subsisted in a larger context.

In this variable, unsegmented world, rhythms guided our lives. I have seen little discussion on the role of nature's rhythms in early humans' survival. But it is not an overstatement to say that rhythms guided and pervaded everything. In fact, on reflection, you cannot escape the recognition that our lives then are best characterized as having been rhythmic.

For every living being, including ourselves, "living in nature" specifically meant living with and within the rhythms of nature. We had no choice about this. Our home, the natural world, and we, ourselves, were rhythmic. Countless rhythms concurred across scales. If you think about it, it becomes obvious that we experienced and had to live in accordance with all of them, including, to name but a few, the day-night cycle, shifting tides, seasonal changes, periodic droughts and storms, cycles of human fertility, waking and sleeping, the growth and dormancy of fruits and seeds, and the movements and migrations of various animals. And all the while, temperatures rose and fell through the days and seasons. If you read that list again, you see that cycles were all-pervasive.

These cycles happened, as they continue to happen, all at once—that is, simultaneously. They naturally carried forward together. All the cycles we lived with and within—from the skies to the world around us to the smallest creatures—happened, if you will allow a grand term for a fundamentally normal occurrence, in symphony. They were a universality—perhaps, *the* universality—of our world.

Think of how it must have been: People rose with the sun and slept in the dark. They gathered food and hunted in the day and took shelter at night. They hunted heavily when game was plentiful, whereas when prey was scarce, they tended to other tasks. They followed seasonal migrations. They gathered and ate what food was available; when the land was lean, they were lean. This is a fairly simple fact to recognize and will become relevant later on. The rhythms of nature informed and determined our own rhythms. And all of our basic behaviors shared the commonality of being rhythmic. There was no other way.

Here I must point out that the rhythms of nature, or really any rhythm, though pervasive, exhibit no sharp edges and no fixed boundaries like other "things" in nature. Nor do they appear to have physical substance or tangible permanence. This too will be relevant later on. Rhythms and cycles somehow evade the category of what we deem substantive in our world. But make no mistake about their reality. Their existence is undeniable. Rhythms are and were omnipresent forms or patterns of regularity that move up and down or back and forth, changing in space and over time. They just don't have natural edges and boundaries like other things do.

Nowadays, we identify rhythms, waves, and cycles as separate from one another and break them up in order to study them, but my goal here is to show how nature existed before we started analyzing it that way. The fact is that rhythms, waves, and cycles are real and ubiquitous. They exist in nature, not as things, and without clear boundaries—yet were, and are, critical for our very existence. When we lived with and within nature, our behavioral cycles were rhythmic.

And they were more than just rhythmic: They were simultaneously rhythmic. They happened together with one another. People hunted with bursts of multiple rhythmic behaviors at once, like cheetahs in the wild do. Cycles within cycles played out at the same time.

An ancient hunter stalking an antelope, for example, experienced many cycles changing together. He was awake, and being awake is part of a cycle—an upswing of the daily wake-sleep cycle. During that cycle, he was also hungry, which is an aspect of another repetitive cycle, a hunger-satiation cycle, where hungriness is an upswing and eating is recovery. (I call these *cycles* because that's what they are: behavior patterns that repeat in a cyclic rhythm.) This hunter, while awake and hungry, experienced other cycles simultaneously. He physically exerted himself and experienced heightened emotional arousal—his exercise-recovery and anxiety-relaxation cycles rose and came to a peak. His heart beat faster with exertion and with anxiety. And his breathing, also a cycle, sped up.

After the hunt, all these cycles came down together. His heartbeat and breathing slowed down into recovery. He experienced relief. He relaxed. He ate. He slept. The cycles, true to their unbounداried nature, had peaked and fallen together, as well as within one another.

It's worth taking a minute to flesh out how real this confluence of cycles was. I think it is best seen in the heart. During the hunt and then after it, this hunter's heartbeat sped up and slowed down—that is to say, his heartbeat rose, peaked, and fell. This was a cycle. At the same time, every heartbeat itself was an even smaller scale cycle of contraction and relaxation. This means that each heartbeat, as a cycle, occurred at the same time as—and within—the larger single cycle of the heart beating faster during the excitement of the hunt and beating more slowly when the hunt concluded.

It's not a hard reality to see, though it may seem a strange idea to bother with. Nevertheless, it will become outstandingly important later on. So, I restate my point. Many cycles happened at once. They naturally, inherently, rose and fell together. And because rhythms have no fixed boundaries, there was nothing to stop them from having this unusual, natural, inherent, and often nested relationship.

A further example is the act of breathing. It too, like the heart, showcased cycles within cycles. It sped up and slowed down with activity and rest and also during different stages of sleep. But it also went up in the day, overall, and down, overall, at night. These were nested cycles—cycles of accelerated and decelerated breathing within overall cycles of breathing faster during the day and more slowly at night. Body temperatures likewise rose with the heat of the day and fell with the cooling of the night, with cycles within those cycles. These cycles happened together and often nested within one another.

All behavioral cycles rolled forward in this way, weaving together, rising and falling, throughout a human life. None was localized or finite. And all were equally essential.

There is really no limit to the reach of natural rhythms. Even the most primal phenomenon of the emergence of human life is possible only through a confluence of cycles. Fertilization occurs thanks to cycles of fertility concurring for both sexes. Then during the powerful rhythms of birth, the mother's uterine contractions are deeply experienced by the infant, whose head is cyclically compressed and decompressed as it is pushed through the birth canal. The infant's heart rate also exhibits a well-documented phenomenon of speeding up and slowing down.[1]

At the time we lived in nature, birth was but one of the innumerable rhythmic phenomena that we experienced. It was a rhythmic launch *into* the world of rhythms.

There's another fact worth noting about the rhythmicity of nature, which was surely salient to those living at that time. As rhythmic wave patterns moved forward on all scales, they repeated on the whole, but none were ever precisely the same. And somehow, an inherent regularity set a pace for nature's rhythms even as they changed. It was, as it still is, an ordered regularity. Even after disruptions, rhythms returned. Indeed, even disturbances themselves were rhythmic—storms and earthquakes would come and go, come and go.

There was no shelter, no escape, from the cycles of nature at that time. Not to say that people would have wanted one—rhythms and cycles were simply the nature of nature. In the realms of the heavens and the environment, and in our very own bodies, rhythms *were*. They were a powerful common denominator of all that existed.

That is why, to survive in the purest sense, a person had to be in sync with them. Rhythms were as real and critical as a tree or river, even though they exhibited no distinct boundaries or qualities that would allow them to be confined. If a person *didn't* cycle his behaviors in sync—if he did not take cover at night and act in the day or hunt game when it was available—he would perish.

I think it's fair to say that the original survival was the survival of the rhythmic. The test was not how well your *parts* fit into nature but whether your rhythmic behaviors were well nested within the cycles around you. You clinched survival by fitting in with nature's cycles. If your rhythms did not mesh with nature's, you would die. There was no other way to survive.

Before the rise of civilization, before we began to build artificial boundaries, rhythmic waves were the universal language of nature, ourselves included. This was nature—the natural world—which we would later try to understand.

We have been talking about the rhythmic world as experienced by early humans, but I'd like to add that the significance of rhythms obviously extends beyond the behavior of the individual. There has also always been the order of the collective: the extraordinarily synchronous rhythmic interplay throughout the living world. Fireflies flash in sync, crickets sound in harmony, ants cluster to form bridges and then disperse, and schools of fish and flocks of birds cooperate in incredible and intricate rhythmic patterns.

Animals to this day survive on the basis of cyclic behavior. "In nature, animals that stop and start win the race," notes science writer Elizabeth Pennisi in an article for the journal *Science*. It has been found that "intermittent locomotion offers [animals in the wild] a surprising array of advantages over keeping a steady pace."[2] Most, if not all, hunting and foraging is some sort of stop-start behavior. Hawks glide around steadily and then swoop in on prey; crocodiles lie quietly, waiting to strike. Cyclic behaviors are "regular in foraging birds, lizards, and insects. It seems likely, in fact, that all search behavior can be placed on a 'stop and go' continuum."[3]

Marine mammals in the wild also exhibit cyclic behavior. Dolphins begin their descent to the deep with strokes of the tail, but then glide, motionless, for a minute or two. The pattern continues as the dolphin extends its dive, swimming and gliding intermittently, and is repeated as the dolphin ascends to the surface.[4] This swim-glide, swim-glide cycle is exhibited by *all* marine mammals, regardless of size or swimming style.[5] Marine mammals swim more efficiently and even further when they stop and go in cycles. Natural behavior is cyclic behavior. It always has been.

Just how, precisely, people lived in nature in the time before history is subject to conjecture. That we did, at one time, live rhythmically within the rhythms of nature is not. It is a self-evident truth. In order to survive, we had to be dynamically and responsively rhythmic because life and nature were equally dynamic and responsive. Our survival in nature depended on the universality of rhythms. Making waves with the environment was not only necessary for survival in nature but, in many ways, the essential character of survival itself.

CHAPTER 2

The Beginnings of Scientific Civilization

THEN THE WORLD CHANGED. Nature, as it simply exists, is no longer the domain in which we function. And no longer do we live with nature's rhythms as the primary guide for our own rhythms.

The catalyst came, of all places, from within ourselves. At some point, whenever it happened and however it happened, we developed a capacity to think. And our thinking changed everything.

We think about nature through a combination of sensory experience and using our minds. On the surface, this is an obvious truth. It is as true today as it was when we first began thinking. We began to see, touch, taste, smell, and hear not only for instinctive survival but also to learn about nature—about rocks and trees, fruit and fish, rain and sun.

I want to emphasize a turn of events that subtly accompanied this new ability. A second change took place, too: We formed a deep conclusion about nature itself from our experiences.

Look at a rock next to a river: What does it tell you? Common sense about the experience indicates that the rock and river are two separate things. This was what I would call a commonsense experience—an experience we interpreted and accepted based on native intuition. Our commonsense experience was that the world is a *whole made of parts.*

While telling us about specific objects, in other words, the combination of our sense experience and our thinking minds was also informing us that the world has a design, the design of a jigsaw puzzle. It was our first guess about the nature of nature—a guess so appealing we retain it today.

The main reason for this development was that our senses actually work by detecting what we perceive as edges and boundaries. They work this way because our physiology cannot cope with seeing all aspects of everything, all at once. We also cannot notice and process entire contexts. "Sensory perception is a matter of selectively throwing away information," write researchers Terry Bossomaier and David Green in their book *Patterns in the Sand*. "The process of sight . . . involves selective ignoring of most of the details transmitted, so that only vital features remain."[1]

And what are those vital features? In large part, they are the edges and boundaries of an object. "The eye is so wired up among the rods and cones that it actually looks for straight edges," writes mathematician Jacob Bronowski in his book *The Origins of Knowledge and Imagination*. "The very odd thing about the eye . . . is that the signaling of boundaries to the brain is the main thing that [it is] wired up to do."[2] That means that when faced with huge blurs of color, our eyes will signal sharp changes, which we construe as the boundary of a lion or tree.

It is reasonable to assume that at some time early on, the way human vision works gave us the idea that edges and boundaries naturally exist in nature. There was no reason for anyone to think that edges and boundaries were not inherent to nature itself and something we ourselves apply to nature as we sense it. As the saying goes, seeing is believing: The colloquial expression "I see" is equivalent to "I understand." "No real distinction can be made between seeing with the mind, and seeing through the eyes; the two are inextricably intertwined,"[3] says physicist and author F. David Peat in his book *From Certainty to Uncertainty*.

In the same way, touch and taste are also stimulated by encounters with what seem to be boundaries: the boundary of another person, of a flower petal, or of a piece of meat. And we experience the beginning and end of sounds and smells as we encounter them—of a song, of a river's roar as we approach and leave it, or of a fire. Each seems a distinct "thing."

We think boundaries because we sense boundaries and vice versa.

Said another way: Shifts in sensory input indicate boundaries to us, and we take those boundaries to mean that there is some "thing" there to think about. The beginning and end of a thunderclap bracket the object we think of as "thunder."

In fact, it is through boundaries that we understand something to exist in nature altogether. Without edges and boundaries to encapsulate whatever it is we think about, we drift into a shapeless infinity; there is no "thing" for us to comprehend. The word *existence* itself derives from the Latin word *existere*, which means "to step forth, appear." To exist is to be real. And our notion of something being real encompasses, at its root, the quality of standing out or apart from the environment—which is possible only by way of perceived edges and boundaries.

This idea about nature had a practical effect. If edges and boundaries are an innate feature of the world, which common sense told us was the case, it was natural for us to conclude that we can—and, in fact, can only—learn about nature piece by piece.

I want to point out the enormity of these assumptions. They were an acceptance of two things. One, we believed we knew the design of nature: a whole made of parts. Two, that belief—that nature takes the design of a jigsaw puzzle—indicated *how* to go about understanding it: piece by piece. After all, as we sensed it in segments, nature appeared too complex to be understood all at once. On top of that, our physical experience of the world was not everything all at once. Rather, we experienced it item by item, as we sensed it.

This was our ancient starting point. Understanding equaled knowing nature's parts.

I make this point because it is not generally discussed, despite its incalculable influence on our history. It is crucial to understand that there was such a launch, into "understanding how to go about understanding." Our philosophy was launched the moment we formed the concept that we retain today: that nature is patterned as a puzzle. Another term for this characteristic is *discontinuity*. We believed and believe that discontinuity, a genuine ability to be divided into utterly separate parts, is a natural and fundamental property of the world.

We use this understanding as we go about understanding. Nature appears to be built from the bottom up—a whole made of parts. The process of acquiring knowledge itself set sail with that original understanding impelling it.

We will now scope out how this understanding, that nature is by its nature discontinuous and that we must therefore go about understanding it part by part, directed us as we made sense of nature. We will see how it shaped our thinking. And we will see that its schema was so powerful that as it filtered our every experience, it eventually became invisible to us.

This exploration will help you see the all-powerful grip this assumption has on our thinking. It will help you see how, eventually, it comes to clash with the discovery that is SuperWaves.

WHEN WE BEGAN to think in terms of discontinuous wholes and parts, we became, for the first time, observers of nature. To be an observer implicitly meant to accept, and to rely upon, the commonsense experience that the observer is separate from the observed.

Our senses drove home the plain intuition that the different parts of nature are separate from us. After all, it is easy to feel grass beneath your feet, but the grass itself does not appear to be "one" with your body at all.

I think it is fair to posit that this experience, of ourselves and nature, propelled the discontinuity we sensed in nature to the status of fundamental truth. We observed that nature was a whole made of parts, where we ourselves were part of nature but separate from other distinct parts. We assumed that all parts were as separate from one another as we ourselves seemed from them.

Observation, as a means to approach and understand all of nature, commenced on that basis. *

As we observed the world around us, it appeared to feature a number of apparently distinct phenomena that were its elements: matter, space, motion, time, and order and disorder. That is to say, from our early history through today, people have taken for granted that nature, as a puzzle, has these certain basic parts. And to this day, all of our explanations of nature are attempts to account for them.

I can only speak of this cataclysmic shift, which must have happened at some time in our earliest history, in the most general terms. But its enduring impact leaves no doubt that it, at some point, occurred. It is evident from the way all cultures throughout history have discussed nature. Here, I'll set out the categories as they were likely first reasoned by the people who thought of them.

Nature seemed to be matter more than anything else. In a very immediate sense, to early humans, "nature" (or "the world") was likely the water they drank, the roots they ate, the game they hunted—the objects of pressing importance in their daily lives. Material objects also easily fit their scheme of understanding, for as people handled and used them, they seemed to be as thoroughly separate from one another as people's senses suggested they were. Likewise, between an observer (me) and something observed (that tree) there seemed nothing less than total and absolute boundaries. So too could all "things" out in the world be localized.

It seems that people came to understand that space exists because of their experiences with matter. Space was the arena that matter occupied. Experience showed that matter and space were different from one another: Since that deer moves through space, matter and space cannot be the same.

People also observed that matter coexists with motion, which experience proved was another fundamental part of nature. Motion was noticeable any time there was a change *of* matter or its position in space and time. Motion and change are ultimately the same thing, but apparently, as people made the basic categories of nature, *motion* seemed the more comprehensive term.

Last, people had a sense of time, which was that through which motion unfolded. Motion occurred in time (and through space) the same way matter occurred in space.

Thus, at some point in our history, people surmised that matter, space,

---

* I italicize "observation" here because it is the first instance of a way we treated nature that seemed commonsensical if nature is designed as a puzzle, and that would later be incorporated into the scientific method. I will do the same for the first mention of *hypothesis, experiment,* and *theory* as commonsense steps in our quest to understand nature.

motion, and time were parts of nature. They experienced and treated them, as we continue to experience and treat them, as if they were separate. It is how we experienced and thought of nature when we applied the hypothesis that nature is a whole made of parts.

I have to point out, however, that in reality, motion was and is impossible to separate from space and time. Motion must happen through space and through time: It cannot transpire otherwise. So too must matter be in space and through time in order to simply exist. Nevertheless, in the way we categorized them in our minds, and continue to categorize them, these things were (and are) separate. It was an outcome of our seeking to understand nature through parts and wholes.

Experience did seem to prove that they were separate. If you didn't know about the internal motion of particles, for example, a boulder at rest would seem to simply be matter occupying space. Neither motion nor time would appear involved in any inherent way. Give the boulder a push, however, and motion—through space and through time—would seem to bring it to the bottom of the hill. It would appear that an "independent" instance of motion had sent the boulder downhill, through space and over time—which themselves seemed to exist whether or not the boulder moves.

All these were different aspects of nature—this much must have seemed a simple truth to humanity in its early days. Matter, motion, space, and time accounted for what people experienced in nature.

But as categories, they did not cover everything in the world.

Also essential was the experience of order and chaos, or stability and instability.

Nature surely seemed, as it still does today, to be, in general, organized. Such organization and stability were very important. They indicated a future that people could anticipate and rely upon. Organization and stability remain, to this day, important qualities for this very reason.

Of all the parts of nature, matter was surely the paragon of stability. An apple in hand or the earth beneath one's feet did not spontaneously vanish or contort into something else. This was an important part of life.

Motion, though it was change and therefore might seem the antithesis of stability, had its own stability, too. Stable motion was motion whose changes were consistent. A person could expect that the sun will rise and a river will flow or that a stone will fly and then land when thrown.

Similarly, regularities in nature, from the changing tides to the changing seasons, were ordered as they repeated themselves. People depended on these stable changes.

But matter could fall apart, and motion could be wild: These were chaos. Chaos, what we describe as disorganization or instability, appeared in diseases, hurricanes, landslides, stampedes.

These were the parts of the puzzle nature. Matter and motion, space and time, order and chaos. At some time in history, humanity made a shift to thinking about nature in this way.

The reality is, of course, that though people regarded them as different parts, as we still do today, matter, motion, space, time, and order and chaos are inseparable. None exist or can exist without the others. That is the way nature is. Nonetheless, we treated them, and understood them to be, different from one another. That is the way we understood. They were, to us, pairs of opposites that comprise nature.

Divisions thereupon rippled through nature.

Nature was not only broken into wholes and parts but also stratified into different scales. All of nature was the whole. Its parts were land, sky, water, trees, animals, and so on. But it did not stop there. A tree, for example, itself had parts: its trunk, roots, branches, and leaves. Each of those parts had smaller parts within, too. And the tree could itself be a part of a larger forest or meadow, which in turn was a part of an entire region. Each part *had* parts from one angle and *was* a part of something larger, too. Nature, as people saw it, was parts made of parts. And it presented a mystery, of how they connected. There was a lot we had to understand.

What happened to the waves, rhythms, and cycles in which we had always, seamlessly lived? The ones that naturally unfolded together, spanning scales in their unique way? It seems that the overwhelming ubiquity of waves did not capture particularly special attention. I also think that their lack of hard edges and clear boundaries probably removed them from easy sensory evaluation. They were less "thinglike" than other important things we had to consider, like water and shelter. Thinking about nature by focusing on local, boundaried objects was a process far removed from naturally going with the flow of the rhythms in which we lived.

Of course, we still lived with and within the ubiquitous rhythms of nature. But to build our knowledge of the puzzle we believed nature to be, we looked at rhythms as if they were different from one another and proceeded to divide them also. The unspoken assumption of discontinuity was our guide.

In the most general sense, waves themselves became a part of the puzzle. To isolate them as objects with edges and boundaries, we conceptually turned the cycles back on themselves. Instead of unfolding forward, as they naturally do, we thought of them as circles and spheres. Day-night was its own "thing," an entity separate from seasonal cycles, and day was the opposite of night. So too a chasm emerged between exercise and recovery as two separate things, as well between anxiety and relaxation, and hunger and eating. In the natural hunt, these cycles truly concurred, but in our thinking, they were separate things.

Inasmuch as nature was a puzzle, tampering with rhythms did not seem unnatural. A flock of migratory birds, and all the waves of clustering and spreading they undergo as they travel then rest, were broken into pieces of the puzzle: a flock *of* birds. Wholes and parts. The birds were separate from one another. The flock was distinct from the birds themselves—a "thing" that is composed *of* them. We firmly separated their rhythmic motion from the material of the birds, too.

What was the result? The rhythmic fireflies that synchronously light up the banks of Malaysia without a leader or cue from the environment became a mystery, as did countless other coordinated events in nature. We saw these things as discontinuous—as separate events and entities, as if they too had edges and boundaries, but which were "somehow" organized, somehow synchronized. Nature showcased all kinds of matter and motion, organization and chaos, and rhythmic regularities and irregularities yet somehow retained its overall character even as change happened. Just because we knew it had wholes and parts did not mean we knew how it worked. And just because today we have partial explanations and computer models to account for how organization comes about piece by piece, the point still stands: We needed this type of explanation altogether because we viewed nature as a discontinuous puzzle.

Such was our human understanding. Observation seemed to inform us that the nature of nature is a puzzle.

EVEN THOUGH THE IDEA that nature is a puzzle, a whole made of parts, seems fundamentally—if not a priori—true, it is extremely important to realize that there was another choice. "Nature is a puzzle" was not a fact but a *hypothesis*. It was something we presupposed about nature and not nature itself. I call this guess about nature the *Puzzle Hypothesis*.

The Puzzle Hypothesis is not a scientific hypothesis—not in the way we usually mean it. This is a guess about nature, albeit not a conscious one. Humanity has entertained it since our earliest thinking days. And it has colored all earnest scientific inquiry, as we will see ahead. Nonetheless, as it is a guess about nature, it leaves open another choice.

What was the other choice, the choice we did not make? It was that nature might be an inherent continuum: a "one." There was and is a possibility that some sort of "ultimately indivisible unity," which we are as yet unable to recognize, *is* nature. The natural world may be something whose nature renders it ultimately indivisible even though it has characteristics that our senses perceive to be local boundaries. It was and is possible that our idea of a whole made of parts—the one and the many—is wrong. It may not be the true nature of nature.

This is the reality that I will eventually show you when I introduce SuperWaves. It is a phenomenon that is "one," that ultimately cannot be broken into parts, whose characteristics allow us to experience and work with it, on our scale, as if it is divisible. But such a way of understanding is far removed from our current thinking.

And it has been far removed from our thinking for all of our history.

When we first looked at nature itself—nothing supernatural—the idea of an inherent continuity, an only "one," simply did not make sense.

We had no idea what the pattern or design of nature would look like as an only "one." There was nothing obvious in nature to suggest it. It was all but impossible to imagine what nature could even be if not a whole made of parts. You might say that it did not seem natural to nature.

Even if we tried to see nature as one, our perception of what nature *is*—matter, motion, space, time, order and disorder—necessitated an impossible choice. One of those parts would have to generate the others. And that was difficult. You hit problems each way.

Let's say matter were the "one." But if there were only matter, there would be no motion or change—not a leaf falling, a child growing, or the sun rising. Everything would be an entangled mess, a blob. There would be no cause and effect as we know it. No motion would also remove the passage of time, which means that everything would either happen all at once or be frozen—another entangled mess.

But motion as the "one" seemed no better a choice. If nature were only motion, how could there be matter? In our experience, there must be some*thing* moving for motion to exist at all. Matter certainly (seems to) make motion, but motion could not be or become matter in any way we could fathom. Waves presented a unique type of motion, not contained like other instances of motion, but if motion made nature through waves—as prevalent as waves are—it would seem that there must be nothing at all: Waves were immaterial and, in our experience, could not make matter.

Nor could there be only space or only time, nor only order or only chaos—these too were incomprehensible. Our way of thinking told us nature *must* be a puzzle. Oneness as a normal quality of the physical world seemed unfeasible.

And yet, though we did not see it, there was a way that the other choice—of oneness—could have worked.

Instead of assuming a "one" would lie within the pieces of nature—instead of wholeheartedly accepting discontinuity—there might be a continuity to nature that is whole and unbroken. It could be: not a unity *of* pieces, but a unity that sidesteps pieces altogether.

However, we encounter a problem before even beginning to grapple with this possibility of how nature might be continuity. The problem is that we actually cannot conceive of continuity while also retaining our belief in discontinuity.* "Continuity and discontinuity are two totally opposite qualities; a thing cannot be both discontinuous and continuous, or change from a continuous to a discontinuous

---

* It is hard to let go of our first guess. We retain our conviction that nature is discontinuous even while we try to imagine continuity. But we must let it go to have a shot at understanding. True continuity—the "other choice" with regard to nature's character—and discontinuity, what we assume is true of nature, are mutually exclusive.

state," states physics professor Françoise Balibar in her book *Einstein: Decoding the Universe*, implicitly accepting discontinuity as her initial premise. "At the most it is possible to make false continuity from discontinuity, just as sand may appear continuous when viewed from above, although it is not."[4]

That is to say, the way "continuity" is generally understood—and I put *continuity* in quotes because it is not a concept of true continuity but rather based on discontinuity—is as continuous *divisibility*. Physicist and mathematician John D. Barrow illustrates this point in his book *Pi in the Sky* in the extreme.

The physicists' search for a "Theory of Everything" brings us face to face with a deep question concerning the bedrock of space and time in which we move and have our being. The physicist takes this to be a smooth continuum, infinitely divisible, which acts as the cradle of the laws of nature. . . . Yet in opposition to this picture of the world stands a new paradigm, as yet immature and naked of the clothing of its full consequences, in which the world is not at root a continuum. It is discrete and bitty.[5]

Though Barrow seems to make a dramatic contrast between continuity and discontinuity, the truth is, he is really comparing two types of discontinuity. Something is considered continuous by science today if it can be continuously *divided*. Obviously this is based on the fundamental idea of discontinuity, derived from the Puzzle Hypothesis, that nature is a whole made of parts that can then be continually divided. For science, already steeped in the puzzle perspective, the choices are one sort of discontinuity or another—divisible forever or only divisible to a certain degree.

If one has nothing to replace the choice of discontinuity, then there is no way to move forward with a belief in continuity. As mathematician E. T. Bell writes in *Men of Mathematics*, "From the earliest times two opposing tendencies . . . have governed the whole involved development of mathematics. Roughly these are the discrete and the continuous. The discrete struggles to describe all nature . . . in terms of distinct, recognizable individual elements, like the bricks in a wall." Bell recognizes that the other way to comprehend nature is to see continuity, such as "the course of a planet in its orbit, the flow of a current of electricity, the rise and fall of the tides." These seem familiar and truly continuous, as Bell points out, and although "intuitively we feel" we know what continuity is, "today . . . 'flow,' or its equivalent, 'continuity,' is so unclear as to be almost devoid of meaning."[6]

What, then, would true continuity look like? How can continuity be not "devoid of meaning," to use Bell's words, but a meaningful reality?

A true continuity is not *of* soldered pieces. A true continuity is a unity that precedes pieces. It is whole and more than integrated (for integration is, again, *of* pieces). It would be neither matter as we recognize it nor motion as we identify it nor space nor time nor order nor chaos, but rather an unbroken single phenomenon that exhibits multitudinous features that we are inclined to perceive separately. It would be "one."

It would be ultimately unbroken and unbreakable. We could recognize that even though our senses construe edges and boundaries from whatever it might be, that is our *perception* of it but not its true nature. It would exhibit no edges or boundaries within it. As a continuity, it would be whole and inherently indivisible.

We cannot relinquish this possibility as a true choice even though it is a huge challenge to imagine. If our desire is to try to understand with as little prejudice as possible, we must—absolutely must—acknowledge this choice exists. It of course goes without saying that I eventually will show you that this is the underlying reality of nature. But for now, as we trace the history of how our thinking about nature developed as it did, we must realize that there was, from the beginning, a choice to see nature as a "one."

And yet we were, as we continue to be, trapped. Even when earnestly trying to conjure an image of true continuity in nature, we cannot help but do it with discontinuous pieces. Stephen Hawking made a profound point when he said, "If everything in the universe depends on everything else in a fundamental way, it might be impossible to get close to a full solution by investigating parts of the problem in isolation."[7] Because we treat nature discontinuously as we seek to understand it, it is nearly impossible, if not impossible, to find continuity at all. The possibility that nature is "one" is *barred* from consideration. We do not look at—or for—continuous facts. We look *only* at discontinuous pieces. The only "continuity" that we entertain is of things we can divide infinitely, like a "continuous" line (which, even while we profess its continuity, we treat as a procession of discrete points).

That is how we landed ourselves in an ironic situation. Open-minded though we are, inherent continuity—which is a legitimate possibility—ends up, in fact, not making sense. We have unwittingly faltered into a fallacy of negation. We looked only at the pieces and thereby have "disproved" oneness in nature itself. Choosing discontinuity makes inherent continuity impossible. We discounted a logical possibility, and in doing so, closed our minds to it forever.

Even if presented with a sound reason to believe nature is essentially an indivisible whole, we would be deeply challenged to understand the concept because of the very way we consider nature to exist. I will later introduce SuperWaves and show how it exists, but we first must come to terms with the truth. Coming from the Puzzle Hypothesis as we do, the other possibility is nearly beyond the limits of our understanding. The original combination of thinking and sensory experience of parts meant to us that nature *is* parts, and if our mind-senses combination has misled us about the nature of nature—well, we can scarcely comprehend the alternative.

That is how the other choice slipped out of the realm of possibility. In fact, instead of retaining any sort of tentative quality, "nature is a puzzle" seemed so self-evident that it became axiomatic. In his book *The World within the World*, Barrow, in describing nature as discontinuous, calls it an "axiom of faith . . . that we can study the world locally in small pieces, or in small regions of space or time, and infer the global structure by putting the small-scale features together. If the world were of an intrinsically holistic character this would be false."[8] Barrow calls it an "axiom of faith" because even though there did not appear to be an alternative to discontinuity, there was and still is another choice.

Einstein recognized the same issue, as I mentioned earlier: "It is an outcome of faith that nature—as she is perceptible to our five senses—takes the character of such a well formulated puzzle."[9] It is a guess, an assumption, something we simply believe—a hypothesis. What I want to emphasize here is that we can never say with 100 percent certainty that because we make sense of nature that way, nature *is* that way. The distinctions given by our observations that led to the Puzzle Hypothesis *could* be an interpretation that is practical but misrepresents reality.

At the time we embraced the Puzzle Hypothesis eons ago, it did not appear to be an "outcome of faith." From a practical perspective, it was a huge leap forward, a tremendous shift to understanding nature. We did not see that we might be mistaken about the nature of nature on a very foundational level, that there was another choice.

THE ENTIRE GROWTH of civilization, right from its beginning, was rooted in the idea that nature is designed as a puzzle. The Puzzle Hypothesis was in fact a mighty tool for survival. Everything in the natural universe seemed ripe for cutting into wholes and parts. Each piece of nature begged for the question, how can it be taken apart and how can it be put together with the other pieces? How does this piece *fit*?

The only thing to do was to check it out. We began to treat nature itself, and our developing civilization, in the only way we could: piece by piece, by trial and error. The *experiment* was born.

A rock hit just so with another rock cleaved along a plane that made a sharp spear point. A different rock didn't work. Two other rocks together created sparks. We tried it again and again. In the domain of the Puzzle Hypothesis, the experiment was the natural way to understand the universe. It allowed us to predict, control, and make progress based on *local cause-and-effect* relationships. When there was change, it seemed clear that a certain piece of nature must have effectuated that change. We wanted to know what and how. We tested. And the beauty of the experiment was that it not only seemed natural and simple—perfectly logical—but also could be put to use.

Language also strongly reinforced the hypothesis that nature is a puzzle. Language deepened our sense of boundaries between one object and the next: "the river" and "the delta" separated in our minds when we referred to them separately. Words themselves are vehicles for capturing differences and distinguishing one entity from another.* A *wolf,* a *bird,* a *cave,* a *pit*—these words clarified which separate object they referred to. Words delivered nature in pieces. Together, the pieces made up the puzzle of nature.

As astronomer and science writer David Darling writes in *Equations of Eternity,* "It is no accident that the first handmade tools and the first glimmerings of a spoken tongue should evolve at roughly the same time. . . . Once you can see the world piecemeal, and consciously isolate and focus on its components, the next step . . . is to pair off each object (or action or emotion) with a symbol—a symbol that can be vocalized in some reasonably predictable, agreed-upon way. From that stems immense survival value, especially for hunter-gatherers who rely upon close collaboration with their fellows."[10]

The Puzzle Hypothesis gave rise to all technological progress from the earliest days of civilization to the present time. When the nature of nature is a puzzle made of parts, one can experiment and use those parts as building blocks from which to invent tools and technology. We are, in fact, the only species who use tools to make tools, who use parts to make wholes, which are themselves parts of greater wholes. This was all made possible by experiment, which stemmed from the hypothesis that nature is a puzzle, which, in turn, stemmed from the original observation that nature is a whole made of parts. While these processes and experiences were not the formal scientific method as we know it today, they were the everyday mode of human operation and lay the groundwork for the development of science, as we will see in the next section.

Mastering fire—which was otherwise a natural part of nature—was a momentous achievement. We tamed it raw and eventually created steam engines and the lightbulb, burned wood and gas and coal and oil, and even developed nuclear reactors in attempts to create ever more advanced sources of light and heat. Fire enabled us to stay awake at night and to migrate to the darker and colder northern and southern latitudes. We learned how to control fire to cook and preserve food, to gain warmth and protection from the environment, and to make further tools and weapons. Fire was a tool that progressively separated us from the natural environment and its rhythms throughout the course of civilization.

Our lives as hunters and gatherers drifted away as we divided nature into parts.

---

* It is worth noting that in contrast, words are ill-suited to address issues of unity. It is difficult to put to words anything that fails to exhibit distinct edges. If one wants to describe a situation in which one item flows seamlessly into the next—a situation of continuity—it can be confounding. How can you refer to a boundless—? It can't be a boundless *thing* because a thing is some*thing* with boundaries. Words cannot easily deliver the concept of a genuine continuity, but rather only of the sort of continuity in which two separate entities, which can be named with two separate words, are joined. Linguistically, true continuity avoids capture. Things must be separate in order to be directly referred to.

We cultivated plants on fertile "pieces" of land, separating from "wild" nature—undergoing what we, in retrospect, call an agricultural revolution. We built walls to protect ourselves from scavengers and predators, as well as from rough weather. We were a piece of nature, and we separated ourselves from other pieces that threatened us.

We built homes, roads, and cities and diverted rivers. We stopped moving fluidly across the landscape in tune with the rhythms of the seasons, the rhythmic migrations of animals, and countless other natural rhythms, and we began to stay in one place.

These experiments were all deeply transformative experiences. The positive results we had when developing civilization validated the Puzzle Hypothesis. And this trend travels straight through to today.

Modern technology—from antibiotics to electricity to airplanes to computers—provides continuous testimony that our approach to nature is on target. We trust practical advances to verify the Puzzle Hypothesis. Our advances have certainly been impressive. What began with language, tools, and fire has continued for millennia, and still today—with electric lights, the refrigerator, the automobile, and even space travel—we continue to become ever more independent from the rhythms of nature in which we had originally simply and naturally lived.

We have shifted from being *with and within* nature to progressively separating ourselves *from* nature, to where we are defending ourselves *against* nature. And perhaps that shift has been unavoidable. For it is consistent with the way nature works—if nature is a puzzle. Whereas cooperation *with* nature and its rhythms had originally been crucial for survival, civilization shifted in the direction of protection *from* nature in order to survive. As Sigmund Freud said, "The principle task of civilization, its actual raison d'être, is to defend us against nature."[11] We seemed to have had no choice and seemed to be doing no harm.

In a puzzle, because pieces are discontinuous from one another, it is possible for them to be in a state of incompatible opposition. The Darwinian idea of survival of the fittest sprang from this concept. It supported the notion that not only are we in combat against nature but everything is competing against everything else. In the somber picture cast by the words of the English poet Alfred, Lord Tennyson, ours is no symphonic environment, but "nature, red in tooth and claw."

Natural rhythms continued to lurk in nature, but we treated them as if they were pieces of the puzzle. We modified and separated many of them so that they fit into the puzzle-style world of civilization. Waves came to be perceived as an unnecessary "side effect" of nature that needed to be straightened out and controlled. In fact, technological progress was partially measured as successful to the extent that it walled off rhythms—as if they were secondary—and freed us from our apparent dependence on them. Think of the usefulness of artificial light and what it allows us to do at night. It seemed liberating to be freed from the "constraints" of natural rhythms.

Based upon our progress, we began to perceive ourselves as the most successful part of nature. Nature took on the feel of something that needed to be mastered. We human beings are the only organisms on the planet, of course, with the intent to survive artificially separated from the natural environment. It is a matter of human pride that unlike other animals, we do not need to depend completely on nature. Indeed, we have advanced technology to the point that we can literally leave our natural environment. We have ventured, in recent years, into the depths of the oceans and into outer space. Nature smacks of wilderness, and we are civilized. Few people familiar with modern technology would be comfortable living with and within nature. We do not consider the natural environment to be our home.

All along, our successes in progressively separating ourselves from the natural world and in dividing the natural environment itself—we have done it, and survived—quietly continued to validate our most fundamental idea: that discontinuity, as represented by the Puzzle Hypothesis, is an inherent quality of nature.

We have come a long way. The commonality of experience was once synchronized rhythms. The experiment of civilization flourished on the basis of a new commonality of experience—a world of edges and boundaries—in which each piece could be put to use for the human piece: for us.

IN THIS CHAPTER, I have introduced a tacitly accepted but generally unrecognized fact: that the basis for our earliest thoughts about nature is that the world is patterned as a jigsaw puzzle. It is we who declared that nature takes the pattern of "a well formulated puzzle," to use Einstein's phrase. And it is we who built civilization on that assumption. These three factors—our tendency to observe edges and boundaries, our hypothesis that nature is designed as a puzzle, and our local experiments based on that hypothesis—were so essentially useful during the development of civilization that they catapulted our puzzle worldview into the category of the unquestionable. Even today, inherent continuity seems preposterous because we rely so heavily on the axiom that the puzzle pattern is the nature of nature.

There was one final step through which the Puzzle Hypothesis took its ultimate role in civilization.

That step was the development of mathematics. Language (i.e., words) reinforced our sense that nature is a puzzle. But we invented another language, mathematics, to formalize it.

Mathematics is in fact a natural language for describing a world of discrete units. Mathematics began with numbers, and numbers began when people counted objects. What could be more absolutely distinct than the number one? Is there not a total, even infinite, distinction between one apple and two apples? A number itself signifies a bordered, finite entity. It epitomizes the quality of divisibility and of wholes made of parts. The ability to break something into discrete pieces is indeed the heart of the very idea of numbers. Each number is absolutely and unequivocally separate from the next.

As Darling writes in *Equations of Eternity*, "The same object, as we perceive it, may coincidentally be tasty . . . tell good jokes, or, through its exquisite sounds, stir powerful emotions within us. But whatever the object is, by virtue of appearing to us as a separate entity, it has 'oneness.' That is the clinical point of origin for mathematical thought, and it means that from its very beginnings mathematics has built into it the idea that the world is divided into objects."[12]

Numbers are useful for simple counting, of course, but their stellar usefulness becomes apparent when they leave the realm of the objects they describe. This happens when relationships between sets of numbers emerge. That a field produces 50 bushels of grain when sowed one way versus 30 from another method, or that the stars change in a certain pattern that can be used to indicate the best time to plant, is more than just valuable information. It is a way of discussing relationships between parts in an abstract way. It is a way of identifying patterns free from the specifics of the parts.

Mathematics is the language through which people craft those messages. It describes how numbers—which are themselves discrete and represent discrete entities—change. It lays out patterns of change in the numbers we use. Those numeric patterns, being patterns (which tend to repeat), can be used to predict and to control.

It should go without saying that this was of enormous importance to people. If numbers are parts, mathematics is the way to unify them: to describe relationships between the parts we see, and to tell us what to expect to happen, in a language all its own. It introduced a way to model the reality of nature, what seemed to us a whole made of parts, that we longed to describe and explain.

I want to emphasize the point that I am making here: Whereas verbal language validated our sense that nature can rightly be described as parts and wholes, mathematics validated our sense that one can understand abstract relationships between parts without a tethered reference to the parts themselves. It was a liberating achievement for our species. It freed us from being tied to concrete realities and sent us soaring into the realm of the transcendental. Relationships between parts could be understood in an abstract way.

Thus the compelling elixir of sensory observation, the hypothesis that nature is patterned as a puzzle, and the development of civilization through local experiments all blended together upon the development of mathematics into an irrefutable answer. It answered the question people throughout history have asked. We have always wanted to know, what is nature? The development of mathematics secured the Puzzle Hypothesis as our answer. Nature is a puzzle whose parts relate to one another in abstract, mathematically describable patterns.

I call this our Original Theory of Everything.

I want to point your attention to this unrecognized fact. Incredibly, we have been looking for an all-encompassing understanding of reality—a *theory* of everything—without realizing that we already have one. This Original Theory of Everything—the final form of the Puzzle Hypothesis—has firmly guided our understanding of the natural universe and how to treat it. Even today, the Original Theory of Everything sits at the forefront of our reality: It dictates that all new information, all incoming data, must be construed as discrete pieces of nature. It dictates that the world *is* a whole made of parts that can be aptly represented by mathematics. Our perception that nature is matter, moving through space and time, has riveted us to the idea that nature must be a puzzle. The fact that we, as a civilized world, already operate with a Theory of Everything is something that science has not grasped.

What makes this particularly fascinating is that we have always been looking for an ultimate understanding of nature. And yet, it is our own understanding, the one with which we have been working from the very beginning, that paints the nature of nature as remote and complicated—that nature is designed as a puzzle.

We are so sure that this is the way it is that it is hard to comprehend that an explanation could be anything other than superimposed upon reality—or what that would even mean—though this is what I will eventually come to show. A fluid, self-evident understanding that emerges from nature itself and therefore generates no mysteries is currently outside of our scope. There has been no hope of—or strategy for—understanding nature as a whole: "We can only do it piece by piece," says Nobel laureate physicist Richard P. Feynman about the way to do physics in *Six Easy Pieces*.[13] This means any answer we compose will be a patchwork of pieces. And to the extent that they extend us a bit of victory over local areas of nature, every discontinuous explanation of each "part," every step we take, further reinforces the veracity of a discontinuous worldview.

An indivisible unity, as the reality underlying nature, would engulf those local victories in a different sort of explanation, but we are too busy building victory upon victory to understand that. We have our eye exclusively on an explanation of a single type: one that will overlay and reveal the secret behind the parts of nature we perceive.

The Original Theory of Everything is not, of course, the final theory for which scientists hunt today. It is not the ultimate secret of the universe. It only proposes that nature has a secret that we might solve if we study it piece by piece. It is the theory that nature *is* a puzzle of every*thing*. But it does not solve it.

That is pretty astounding. We have set up nature as a puzzle and determined we will properly describe it with mathematics, but nature, as a puzzle, is so vast that we can only chip away at it through the trials and errors of local experiments. We are left with a technique to use on the puzzle but no obvious direction for solving it.

The goal has been to solve the puzzle: to discover the secret of the universe. What is hidden in this universal puzzle that will reveal the ultimate reality that is existence? It seemed that the puzzle's solution would be the nature of nature.

PART 2

How Science Developed from the Puzzle Hypothesis

Prelude: The Origin of the Scientific Method

The development of human thought grew from the original understanding that nature can be understood and treated as a discontinuous whole made of parts.

It began with trying to understand the world as if nature is shaped and put together like a puzzle. That created the goal that we still have today: to *solve* the puzzle. We seek a simple solution to the complicated world. For though we had the design—a parts-and-wholes pattern—we had not a whisper as to what the ultimate reality of nature itself *is*.

In Part 2, I will lay out the four great questions civilizations have asked based on the presumption that nature takes the design of a puzzle. The chapters that follow in this section delve deeper into them. The questions are as follows, and as you will see, they are necessarily paired.

1. What is it?
2. How does it work?
3. What holds everything together?
4. Why do things fall apart?

I will show that the approach of these questions entails that there is only one way to answer them. I will explain how that approach leads to one final question about the universe as a whole. I will also take you on a tour through history, through the development of the scientific method and science itself, to see it all in action: to see how our puzzle-derived questions compelled us to develop science in the way that we have.

By the end of Part 2, you will come to see that science is not a pure look at nature but rather a study based on the presumption that nature takes the form of a puzzle.

You will see why science, as it has developed, will not and cannot fully recognize a discovery that shows nature's ultimate character to be indivisible: specifically, the discovery of SuperWaves.

The first of the four great questions—What is it?—has been asked throughout history. It is likely the first question you, like most every child and adult, will ask when you encounter something unknown.

Our observations have told us that we exist in a physical universe—that is, a material universe.* Therefore, the instant we see any "part" of nature, we are confronted with this question: What is this part?

The question itself is asked in a variety of ways. "What is this thing," "what is everything," "what is the material universe," and "what is it made of" are all variations of "what is it?" They imply that we are dealing with a parts-and-wholes medley

---

* The word *physical* comes from the Greek word *physis*, which means "nature"; physical used to mean natural. But long ago, physical came to mean material because that is what seems natural.

whose overall solution is not immediately available to us; we must first ask to identify each localized piece.

"If we want to begin to comprehend the universe, our starting point must be an understanding of matter—its forms, its organization, its movement, and its transformations," writes professor of physical chemistry Brian L. Silver in his book *The Ascent of Science*.[1] Whether applied to a rock or a quark, identifying "it" is the first step people take when working on the puzzle.

On the heels of this question about *matter*, there is always a question about *motion*. The parts of nature, after all, are not just matter but matter that *moves*. We always ask of whatever matter we are considering, how does it work? Whatever "it" is, we want to know how "it" moves and changes through space and time.

Throughout our history, we have made this double inquiry—What is it and how does it work?—of every item we have ever wanted to learn about, from stars to rain to sprouting wheat.

Any answer to this double inquiry prompts a sense that we have understood a part of nature—on the unstated assumption that nature is composed of parts. This is how we believe one acquires knowledge.

"What is it and how does it work?" is a set of regular, everyday questions that have been around for millennia and are still going strong, but I want to point out their unique relationship. They are more than simply a pair of questions. In many people's minds, they present a split: a dichotomy. We treat them as necessarily paired, but as antonyms. More than just matter *and* motion, they smack of being opposites—it seems to be the matter *versus* the motion of the thing we are trying to understand. The realms both relate to what exists but seem so utterly different they pass for being incompatible.

Beyond the two important questions of "what is it?" and "how does it work?" matter in motion originally brought out further questions meant to help us understand the puzzle-designed world, as they still do today. Matter in motion pointed to a phenomenon: stability, whose opposite is instability or chaos. Our experiences of stability and chaos generated the second set of great questions that stay with us to this day. To repeat, they are:

What holds everything together?

Why do things fall apart?

Whether we realize it consciously or not, we all count on the persistence of stability to make every part of nature stay the way it is—for without stability, the world as we know it would slip into undifferentiated chaos.

In the natural world, matter has always epitomized stability. The earth beneath our feet has seemed overwhelmingly stable and secure since our earliest days, as do material objects in general. Experience has told us that of everything in the world, the most stable stuff is matter. Stability also firms up the regular movements of nature that we depend upon, and that we can depend upon inasmuch as they do not change (i.e., they are predictable). Most plainly stable is relatively smooth motion: an apple falling straight from a tree rather than jittering erratically on the way down, or an arrow completing its arc rather than randomly shifting course. Also stable, and essential, is repetitive motion such as day and night cycles, waking and sleeping, the flow of seasons. A more fluid and complex stability has always shone through as living things, such as plants and migrating animals, repeatedly cycle by the seasons.

Stability is not a trivial quality to be found in matter and motion. It is an anchor. Because stable things don't change, we realized early on that we could refine our knowledge of them to use them for survival. At the same time, again because they don't change, we also realized we could use our knowledge of them to build an understanding of nature. Each stable thing is a reliable part of the puzzle.

Moreover, stability stands in opposition to a frightening part of nature: *in*stability. While stability suggests permanence, longevity, and survival—qualities we crave for ourselves—things can and do become unstable, chaotic, and fall apart. Instability in matter causes trees to crash down, mountainsides to crumble, and the human body to weaken with age and disease. Similarly, instability affects motion. Things can and do change erratically without warning. Healthy people sometimes die unexpectedly, storms strike, and clear, normal migrations turn suddenly to stampedes.

The difference between stability and instability is the difference between order and chaos. Our instinct to survive in a potentially chaotic world, as well as our thirst to comprehend our world as a sum of stable parts, has translated into a deep human yearning for order and stability throughout the ages.

Our questions about stability expose an inherent shortcoming of the Puzzle Hypothesis. They reveal our ignorance about the coherence of nature—a gnawing problem insofar as we presume the parts of nature are all separate. What *is* it that binds parts and wholes into their unique dual relationship? How do the parts of nature acquire and maintain their stable identities? Without this secret, we cannot explain nature as we perceive it to be.

Even the most basic natural objects present the problem. Though we are sure that nature is such that parts make wholes, it is not, and has never been, immediately clear how parts make wholes and stay together—how leaves, branches, and a trunk meld together to *be* a tree, or how the parts within a leaf make what we recognize as a leaf. Nor is it clear why those parts do not crumble to soil with every passing moment or what changes transpire so they do sometimes fall apart that way. Nature may be a puzzle, but the glue that maintains it—an essential, even elemental, aspect of our viewpoint—eludes us. This glue is the missing ingredient that would explain why and how each and every item in nature exhibits and maintains its particular character.

Across each level or scale, the levels themselves also appear somehow bound together as a whole. All the leaves and their parts, all the branches and their parts, and all the roots and their parts make up a tree, and that tree joins together with other trees to make an entire, long-lived forest. Meanwhile everything is changing, somehow staying together as it changes. Again, the missing ingredient that infuses stability escapes us. How does our world stay together as a whole as all its parts whirl about in an epic dance of both harmony and chaos?

It is no small question. The source of stability in nature, going on the assumption that nature is a puzzle, would be, bottom line, the secret of the universe. It would tell us what makes nature be and stay the way it is. Our longing for this enduring core of it all encapsulates a desire-to-know that is perhaps without parallel as an unspoken guide for what we look for in nature.

Again, these are the four basic questions that have arisen and arise in every human society about each item in the natural world, as well as about the entire world itself, because of the Puzzle Hypothesis.

1. What is it?
2. How does it work?
3. What holds things together?
4. Why do things fall apart?

As I move on to recount the way these questions were answered in the past, we must remember here that, while our questions arose from our observations of matter and motion, boundaries between matter and motion as they happen in nature have never been really clear. We wondered about stability in matter and in motion as two separate things, but in the real world, matter and motion were not (and are not) separate. Motion has always been seen as motion *of* something material, never just motion. And changes of matter—a huge issue of stability—has had to do with how the parts of matter move. The appeal of treating matter and motion separately stems from the Puzzle Hypothesis. So does the appeal of treating stability separately from change and also separately from instability. The Puzzle Hypothesis has told us that since nature is designed as a puzzle, our confusion can be mitigated by addressing each part separately.

Now we turn our attention to how we deal with these fundamental questions. What do we look for when we try to answer them?

This query invites us to reveal more than might be supposed. What we search for to answer our questions—the type of answer we seek and expect to find—exposes a secret belief about what holds nature together.

To explain nature, we sought, and continue to seek, that which does not change. We have sought that which is *invariant*. Unvarying, invariable, constant, unalterable: The invariant would fulfill our sense of what explains the universe.

It is no accident that the answers we have sought assume this solitary guise. We have had no practical way to think of stability and order as coming from nature as a "one" without parts. The only choice we have had is that there exists *some*thing—some "thing"—in nature that is responsible for stability and permanence in both matter and motion. Therefore, throughout our history, people have intuited that something must exist that does not change despite the changing universe. Across cultures, people have always assumed that something invariant keeps parts together as a stable whole.

"There is something attractive about permanence," John D. Barrow writes in his book *New Theories of Everything*. "We feel instinctively that things that have remained unchanged for centuries must possess some attribute that is intrinsically good. They have stood the test of time. . . . And, despite the constant flux of changing events, we feel that the world possesses some invariant bedrock whose general aspect remains the same."[2]

This elementary argument has orchestrated our entire search for knowledge.

It appears that something separate, hidden from our sight, is the secret ingredient that gives nature its stability. Whatever it is, it makes nature as we know it *stay* the way we know it.

The promise of this invariant even goes beyond ultimate knowledge. The invariant might be useful—for ourselves and for civilization. We, as one part of nature, might control the other parts, in hopes of bettering human life, if we could tap into what keeps nature together.

Thus, the Puzzle Hypothesis wordlessly set us on a mission we follow to the present day. It formed an intuition deep and incontrovertible. It gave an unnamed but totalitarian sense of what constitutes proper knowledge. Invariance, it seemed, determines the character of each part of the puzzle; by the same logic, an ultimate invariant must make the whole of nature be the way it is.

The unnoticed assumption that nature was a puzzle directly led to the galvanizing conclusion, untold generations ago, that finding an ultimate invariant is the only way to solve the puzzle of nature. We aim to use our mounting stores of information to answer the final question we have always wondered about nature:

Is there an ultimate reality—an ultimate invariant—that explains everything?

This is the final question the puzzle worldview generates.

NATURE APPEARED TO HARBOR a grand dichotomy. On one hand, there was the puzzle: that nature is patterned as a whole made of parts. On the other hand, there was the stuff of nature itself: matter, motion, space and time, and order and chaos. There was no apparent connection between the intangible design of nature and the natural world itself.

This dichotomy offered two choices of where the invariant we sought might be hidden. One could view nature with the understanding that the whole governs the parts. This idea meant we should look for an invariant above and beyond: somewhere outside of the fragmented parts of nature. Coming from the other direction, with a focus on the parts making the whole, necessarily meant looking *inside*, to the parts of the puzzle themselves.

The first possibility—that the invariant cause of stability lay, somehow, hidden beyond the physical parts of the puzzle—called for a supernatural cause, meaning, quite literally, beyond nature. This possibility indicated that something top-down, outside-in, some all-encompassing entity, transcended the parts of nature while organizing them all.

This was what we might call a mystical, spiritual, transcendent, or religious origin of order in the universe. It began with polytheism, in which stability came from separate gods controlling each separate piece of nature. Eventually, the perspective resolved into more unitary explanations of how nature arose: from a single cause. Here was an ultimate invariant, a "one," that gave order to nature. One god would govern everything from above.

This approach did not negate the worldview that nature itself takes the pattern of a puzzle; rather, it was the view that there is something beyond the puzzle. Those who ascribed to a transcendent way of thinking looked to the "one" ultimate invariant that granted order and coherence to the natural world.

The other choice for finding the ultimate invariant—that something hidden *inside* the puzzle generates the order and stability of nature—placed the mystery squarely in the heart of nature itself. This was the choice that set us on the hunt that we engage in yet today. The natural universe seemed fundamentally material. Therefore the answers to our questions must come from that physical reality. Put in terms of the puzzle, *the answer must lie somewhere, somehow within the parts and somehow ultimately connects those parts as a whole.*

This perspective led us to the foundation of modern science: reductionism and unification. It is the notion that we identify what the component parts of nature are, discover how they work, and, from this, expect to understand how the whole works.

Although we have called many advances in science "revolutions," there has actually only ever been one true scientific revolution. It is the one that created science itself. When we looked inside nature and started treating it as if it were a puzzle, taking it apart and putting it together, science was born. It is an astounding truth that every "revolution" since then *has been but a new way of applying the same, original approach* to new areas of the puzzle. This is an extraordinary way to look back at our history. We blazed that trail only once.

In fact, this method is the only method we have ever used when looking to nature itself for answers to our great questions. It has always been: take it apart and put it together. And that is because once science determined that nature is a puzzle, science had no choice but to *treat* nature as a puzzle.

In this way, science formalized the way we had been treating nature since the beginning of civilization, by standardizing the puzzle-derived steps of observe, hypothesize, experiment, and theorize. Albert Einstein said, "The whole of science is nothing more than a refinement of everyday thinking."[3] It created an algorithm— a directive for how to go about knowing—that today is called the scientific method.

The scientific method has become indispensable for finding invariants, as anyone familiar with modern science will know. Finding invariants is what the scientific experiment is designed to do: Isolate an entity, take a stab at an invariant property it might possess or display, repeatedly test to determine if the hypothesized invariant indeed does not change, and then—if it does not change—grant it the status of a puzzle piece and theorize about its role and relationship with other invariants. We place enormous stock in controlled experiments—experiments that hinge upon observation and hypothesis, which themselves draw credibility from the assumption that nature can and must be understood as a puzzle.

The primal appeal of success through experiment has only grown in strength. All our eggs are in that basket. As Richard Feynman has said: "The principle of science, the definition, almost, is the following: *the test of all knowledge is experiment.* Experiment is the *sole judge* of scientific 'truth.'"[4] And like all acts of gaining scientific knowledge, experiment involves dividing and isolating the thing in question from its environment.*

It is extremely important to remember that the scientific method bars us, without us knowing, from looking at nature in any other way. But as the Stephen Hawking quote you saw earlier implied, if "everything in the universe depends on everything else in a fundamental way"—if nature is *not* a puzzle—*science will never find that out.*

The fact that there is another logical, if difficult, choice, that nature itself might be a realistic "only one," was excluded from the scope of the scientific method. And though it may have originally been excluded because it was impractical, as time passed, it was silently rejected as a possibility.

It is this intellectual conviction that pits the scientific method against us ever recognizing SuperWaves. Like a giant treasure hunt for jigsaw pieces scattered across the cosmos, the ultimate goal of science is not just to collect data of individual experiments and theories but also to, one day in the future, put the entire puzzle together as the ultimate reality of nature itself. There is no option for discovering that nature is ultimately indivisible in the first place.

Feynman said, "What is necessary 'for the very existence of science,' and what the characteristics of nature are, are not to be determined by pompous preconditions, they are determined always by the material with which we work, by nature herself. We look, and we see what we find, and we cannot say ahead of time successfully what it is going to look like. . . . In fact it is necessary for the very existence of science that minds exist which do not allow that nature must satisfy some preconceived conditions."[5]

Yet all along, ironically enough, science has held fast to a precondition that does say ahead of time what the characteristics of nature will look like: the Puzzle Hypothesis.

---

* In fact, the word *science* derives from the Latin word *scindere*, "to cut" or "to split"—similar to "scissors," "schism," and "schizophrenia."

CHAPTER 3

Matter—What Is It?

GREEK PHILOSOPHER THALES OF MILETUS (624–545 BCE) is often called the first scientist because he looked only to nature itself for answers. Thales "left the gods out" of his explanations, to use the oft-repeated phrase. In terms of advancing the Puzzle Hypothesis, this was big news. Yet while Thales is seen as a revolutionary, the actual revolution he advanced was not a new idea but rather a refined commitment to the Puzzle Hypothesis that we had embraced for all time. He looked at nature—a puzzle abundant with varying pieces—and turned to the pieces themselves.

Thales sought a single, material invariant to account for all of nature even as it changes, often phrased as "a unique substratum that, itself unchanging, underlay all change." He thus starkly advanced the hypothesis that everything is a whole made of parts, which was our Original Theory of Everything. And significantly, he secured the idea that solving the riddle of this conglomerate's changes depends on finding the right invariant. His particular proposal was pulled from everyday experience: that all of nature originates from, and is, water.

Science as we know it had begun.

A number of other thinkers proceeded to do as Thales did: to isolate different materials in nature and propose ways they fit together to make nature. The Greek philosopher Empedocles (493–433 BCE), for example, theorized that all matter is composed of earth, air, fire, and water and that nothing comes into being or is destroyed, but recombines according to attractive and repulsive forces. This is science at heart, if in an unrefined form: Isolate invariant parts of nature and propose reconnections (i.e., reduce and unify) to explain the world as we know it.

The trend climaxed in the 5th century BCE with the pivotal theory of Greek philosophers Leucippus and Democritus. Leucippus and Democritus took the approach of their Greek predecessors further than anyone had before. They followed it all the way down to its logical end. This approach so refined and clarified the old Puzzle Hypothesis that, in its crystallized form, it seemed revolutionary and new.

Whereas Thales and others had highlighted particular parts of nature as we experience it and then postulated their ultimate invariance, Leucippus and Democritus left everyday experience behind—and ventured into the realm of something *hidden*. The Puzzle Hypothesis implicitly permits this, for when the invariant we seek is supposed to come from within the puzzle, logic allows for taking it a step further: to say it is hidden more deeply, as an invisible piece within, rather than something that is visible like water or fire.

It was a logically consistent step whose impact far exceeded the creativity of the idea. The hidden item they proposed was the atom.

Taking the commonsense approach of splitting nature into parts and extending it all the way down, Leucippus and Democritus postulated that if you repeatedly split any particular material item in nature—cut a piece of material in half again and again—you eventually reach a point where you cannot cut any further. That ultimate part, the indivisible block, is the *atomos*—*a* (not) *tomos* (cuttable)—that which cannot be further divided. The genius of Leucippus and Democritus was to propose that this is the nature of the puzzle of nature. This unit—the atom—does not change: It cannot be split, which means it must be ultimately stable. Its stability would grant stability to all of nature.

The atomic hypothesis, as developed by Democritus, posited that huge numbers of "atoms" combine to form everything in nature. Atoms are the invariant parts, and nature is the whole. It is the arrangement of this invariant—the way the atoms fit in the puzzle—and the changing ways atoms relate to one another that give us the incredible, spectacular, unimaginably fantastic diversity and complexity of the universe.

LET US PAUSE HERE to consolidate how the atomic hypothesis—which we retain today—developed from concepts that are implied by the Puzzle Hypothesis.

The first is the idea of absolute boundaries. The individual atom was perceived as a *closed system*. Whether in a void or in space, the atom was conceptualized as closed, isolated, and separate from its environment and its surroundings. This model borrows directly from our original perceptions of how we believe nature to be.

A closed system entails *locality*. When every item has edges and boundaries, you can find it in a discrete position. It is localized, so its influence is limited, extending only to things with which it comes into contact. This concept pairs off directly with our earliest perceptions of matter occupying space.

Partner to these ideas is the concept of building blocks. Atoms were imagined as tiny bricks from which the universe is built.

Then there is the concept of invariance. The Puzzle Hypothesis led us to believe that some invariant grants stability to every isolated part of nature, and Democritus "found" it in his atom. Invariance is the defining characteristic of the atom.

The confluence of these ideas confronts us with a remarkable realization about the Puzzle Hypothesis. While it itself is a single theory, that nature has the design of a puzzle, it comes with a number of attendant ideas that bolster and reinforce it. Together, the ideas we already possessed had pointed to the atomic hypothesis, which in turn seemed to confirm them all. The Puzzle Hypothesis offers what seems like a variety of angles for understanding nature itself, which in reality are all off-shoots of that one, original preconception.

This group of ideas also played a corollary role of monumental importance: They characterized atoms, like all of nature, as the sort of thing on which you could use numbers and mathematics. Just as the Puzzle Hypothesis indicated that the separateness of the basic parts of nature meant that those parts were fit for numbers, numbers seemed a fitting way to label, manipulate, and understand atoms. I mentioned in the last chapter that numbers have total boundaries in the same way we believe all parts of nature do, and numbers add up together, just as we believe the parts do to create the whole of nature. Physicist F. David Peat, in *From Certainty to Uncertainty*, recounts the scientists' perspective.

> *And so we turn to mathematics for a final certainty and begin with one of the simplest and purest operations—the act of counting. Of all things, common sense tells us that counting should be totally certain and free from all ambiguity and confusion. . . . Of all certainties counting seems to head the list. No matter what we may wish, no matter what society as a whole chooses to believe, counting and arithmetic remain objective certainties. . . . No matter how hard we try, we cannot "believe" that two plus two will ever equal five. . . . Scientists and astronomers are convinced that mathematics is the universal language of the cosmos.*[1]

If "no matter what we may wish . . . counting and arithmetic remain objective certainties," then numbers themselves stand out as the most objective certainty of all. It seemed obvious that there was an innate relationship between nature and numbers. This certainty was based on a single principle: that invariant, discontinuous parts combine in different ways to make a variety of different wholes. Parts and wholes were "the way it is," whether particular parts of nature or numbers; either way, they were understood to add up to make a greater entirety. Numbers—mathematics— merely embodied the Puzzle Hypothesis itself in an abstract form.

Like a self-fulfilling prophecy, the effectiveness of numbers and mathematics seemed to confirm the discontinuous design we believed—and still believe—characterizes nature. If there was another way to understand nature, a way that draws on an indivisible "oneness" instead of discontinuous parts, we were directed away from it when we used numbers. Common thought continues to be so totally reliant on this idea that it is hard to conceptualize that there is any option *other* than that numbers are the natural language of the material universe. The Puzzle Hypothesis is remarkably stealthy and potent—and self-validating—in action.

WITH THIS IN MIND, we return to the atomic hypothesis itself. After it was proposed, it lasted almost 2,500 years without any verification. Throughout history, many thinkers—such as Epicurus in ancient Greece, Lucretius in the Roman Empire, and scientist John Dalton in the 1800s—resurrected it. Civilizations changed, but this idea kept on coming back. No doubt, its durability is tribute to the fact that it is a perfect, distilled version of our original Puzzle Hypothesis. It tells us that nature is a puzzle and exactly what kind of puzzle it is: built up from the bottom, with tiny, invariant building blocks, whose unchangingness accounts for stability even as things change.

Democritus's ancient theory may be antiquated in some respects, but in the most critical way, it is no different from every modern theory of matter. It presupposes that nature is a puzzle and suggests an ultimate part *of* nature that might explain its order and stability. It embraces as well as cements the idea of a puzzle, the precondition to our understanding of nature that excludes the possibility that nature is "one."

It was not until 1905 that Albert Einstein mathematically validated the existence of something akin to Democritus's atoms for the scientific world. Einstein deduced that the jiggling of pollen grains or dust particles in a drop of water visible under a microscope, known as Brownian motion, was the product of the pollen or dust being bombarded by invisible atoms.

It is critical to understand that Einstein framed his calculations in terms permitted by the Puzzle Hypothesis. Physicist Mark Haw succinctly explains the requirements behind Einstein's ideas in *Middle World: The Restless Heart of Matter and Life*: "The particles would have to be as spherical as possible. Pollen particles are more often tiny tubes or cylinders. Einstein's theory demanded perfect spheres—only that way were the calculations straightforward."[2] Though pollen grains are not identical perfect spheres, Einstein cast them that way anyway, in order to model them with math. The way we perceive nature allows for that stretch: Everything is a whole made of parts that may be cast as idealized closed systems with absolute boundaries, and we make that leap because that is the only way to describe the parts with numbers. It is how Einstein crafted the calculations that seemed to validate the atom.

The old Greek atomic hypothesis, itself born of the Puzzle Hypothesis, survived to modernity. And then it thrived in the face of challenge.

The tenacious atomic hypothesis holds on despite every bump in the road of modern discovery, in fact. The atom—which had been conceived of as the smallest, indivisible part of nature—turned out to have smaller parts yet: *part*icles. For example, J. J. Thomson identified the electron in 1897, and in 1898, Marie and Pierre

Curie determined that radioactive atoms eject parts of themselves as they break down. Ernest Rutherford then declared, in 1911, that the atom has a centralized, positively charged nucleus with Thomson's negatively charged electrons surrounding it. Still we did not let go of the idea of a fundamental invariant that composes nature. We just shifted our expectations scales downward.

That is how we arrived where we are today, in the world of the quantum. Quantum physics is the science that deals with the atom and subatomic component particles that make up the atom. Particles are treated here as the new building blocks of the universe, and our hope and expectation today is that dividing them further will lead us to the ultimate invisible invariant.

Today, the Standard Model of particle physics is the map physicists use to chart the particles of nature. You might call it a modern incarnation of Empedocles's suggestion, that an assortment of building blocks combine and recombine to form nature. The Puzzle Hypothesis has led us to this particular model, in which we rely on the idea behind material entities—that there exist truly boundaried, discrete items—to account for not only matter but also forces and fields. Of course, though the model currently relies on an assortment of particles, scientists would prefer if there were fewer or, even better, only one underlying it all.

In short, the particles of the Standard Model are quarks and leptons, known as fermions, with a number of bosons, such as photons, which are said to "carry" force. In other words, we account for matter *and* motion with a host of assorted particles—every one of which is treated as a discrete object.

It is not my purpose here to explain how the Standard Model works, but rather to highlight the framework physicists use to explain how it works. For example, Nobel laureate physicist Leon Lederman states in *The God Particle*: "There are two manifestations of particles: real and virtual. Real particles can travel from point A to point B. . . . Messenger particles—force carriers—can be real particles, but more frequently they appear in the theory as virtual particles, so the two terms are often synonymous. It is virtual particles that carry the force message from particle to particle."[3] In Lederman's words, there is a wholehearted, even glaring, but unstated dependence on the nature-is-built-from-discrete-parts model.

With this framework to guide us, scientists continue to search for particles, such as the graviton, thought to be responsible for the force of gravity—again evidencing that we always think from a particle perspective. We are comfortable with this entire description because it fits our puzzle. And in step with the ultimate goal that bloomed from the Puzzle Hypothesis, we do not content ourselves with the idea that nature is composed of a smorgasbord of particles. From protons and neutrons to the quarks that compose *them*, from neutrinos and photons to virtual and other subatomic particles, we persist in looking for the fundamental, invariant piece of nature. We have not let up on our goal.

The Large Hadron Collider, buried beneath the French-Swiss border northwest of Geneva, illustrated this point in the extreme. It was an over 7 billion euro attempt to, among other things, find a "Higgs boson"—that, it was proposed, is responsible for mass. Lederman has called the Higgs boson the God Particle. If it gives mass, it makes matter, and matter, in our thinking, gives it substance and stability, just like a god would. The excitement surrounding its potential confirmation, in March 2013, was astronomical because it smacks of a fundamental invariant—even though it has not shown the existence of other elementary particles, such as the graviton supposedly responsible for gravity or the dark matter supposedly responsible for the stability of galaxies. It was nevertheless reminiscent of what we truly want to find: the ultimate invariant piece of physical nature. It is the same thing 2,500 years ago and today.

The workings of the quantum world, as you may know, extend far beyond identifying the particles of the Standard Model. The quantum realm has confronted scientists with numerous findings that spotlight the role of the Puzzle Hypothesis and the search for invariants. The findings are, to say the least, very unusual, and we will discuss them further in Part 3. For now, while we are examining how the Puzzle Hypothesis determined every "revolution" in science, let's take a moment to examine the developments that led to quantum physics.

The physics term *quantum* was coined by German physicist Max Planck in 1900 in response to a phenomenon known as blackbody radiation.* Most people are familiar with the way heated iron will begin to glow red, then become orange-yellow as its temperature increases, and eventually turn white-hot and even blue. This progressive change in color with an increase of temperature is blackbody radiation.

In Planck's day, current physics understood light energy (including, of course, the light emitted in blackbody radiation) to be electromagnetic waves. Those light waves assumed different frequencies; lower frequency waves were interpreted by the human eye as red, and as frequency increased, color moved from orange to yellow and, through the spectrum, up to violet. In blackbody radiation, it was a problem for light to be electromagnetic waves. The equations dealing with those waves predicted what was called the ultraviolet catastrophe, that the light emitted by a heated black object should immediately shoot right into extremely high frequency, the ultraviolet range, and infinitely beyond. Yet, of course, anyone who had ever heated coals or witnessed a blacksmith at work knew this was not the case.

Planck figured out how to make the equations work. Into the equations for electromagnetic waves, he introduced a new concept: the quantum. The quantum—from the Latin *quantus*, meaning "how much"—was a proposal to work with electromagnetic wave energy as if it were coming in discrete units that could not be

* A blackbody is any opaque object that absorbs (not reflects) all incoming light and emits thermal radiation. Such an object appears black at room temperature.

divided. This technique departed from the way waves were treated at that time. Rather than granting waves the flexibility to assume any wavelength, which by calculation would allow them to shoot right up to and through the ultraviolet range, Planck's idea was to treat electromagnetic waves as if they could only be emitted in discrete bundles. The first stop was at the red spectrum. Add more energy, and the energy emitted by the blackbody held at red until the right quantity had been reached to bump it up to the orange-yellow spectrum. With the right number of additional energy quanta, the energy emitted kept jumping and eventually became blue to ultraviolet.

There was a necessary theoretical implication to Planck's successful formula. Since the formula worked, Planck had no choice but to propose that we understand energy as a granular substance with the same divisible character as matter, which eventually stopped at a smallest indivisible unit similar to the uncuttable material *atomos*. In that sweep, he transformed the idea of waves of energy. It went from what was thought of as a continually flowing "substance" into a phenomenon with fixed, discrete units.

Planck, though respectful of its practical success, disliked the idea that energy comes in quanta so much that he later recalled the proposal as an "act of desperation."[4] The thought of discrete quanta of light waves, however well it worked in the formula, was abhorrent in the natural world. There was a marked preference for the idea of continuity. Note, however, that the continuity Planck preferred was the sort of continuity that we imagine in a puzzle universe: the continuity of infinitely small units (i.e., continuous divisibility).

That is to say, electromagnetic waves were supposed to be infinitely divisible. They should be an endless parade of light that we can always divide further. To Planck's mind, and to all of scientific thinking, this infinite divisibility was a smooth representation of nature (i.e., continuous), rather than the quanta he had proposed, which portrayed energy as fundamentally emitted in lumps. But, of course, the "continuity" of continuous divisibility was as mired in the puzzle worldview as the concept of the quantum was. Planck's distaste for the idea of the quantum was based on his preference for smooth divisibility over divisibility punctuated by ultimately discrete chunks. True continuity, of the sort discussed in Chapter 2, was incomprehensible and, if acknowledged at all, relegated to the supernatural.

The practical success of Planck's quantum sparked the quantum revolution. It led to the evolution of quantum mechanics, believed to be the fundamental mechanism of matter.

Einstein took hold of Planck's idea of a discrete quantum, and with it, he solved for science the problem known as the photoelectric effect, for which he won the 1921 Nobel Prize. Taking the idea further than Planck had, Einstein concluded that light itself is not only emitted in discrete portions, as Planck said, but turns out to actually *be* discrete particles. Light existed as a bombardment of granules rather than a flowing river to Einstein, and the photon was the elementary particle of light. This idea was a logical extension of Planck's concept of the quantum, and as a new application, it would resound in the field of physics.

The photon was considered a radical idea. It demolished the previous understanding of light and also suggested another sort of elementary particle (the photon). Science journalist Albrecht Fölsing called it "the most 'revolutionary' sentence written by a physicist in the twentieth century" when Einstein said, "On the assumption to be here considered, energy during the propagation of a ray of light is not distributed continuously over steadily increasing spaces, but consists of a finite number of energy quanta localized in points in space, moving without dividing and capable of being absorbed or generated only as [whole] entities."[5] Einstein made the following analogy to explain how he extended Planck's idea: Planck suggested that energy may be distributed only in quanta, much like beer that may be sold in only pint bottles (and not, as one might expect, any size glass you choose); Einstein proposed, going further, that the beer actually *exists* in indivisible pint portions.

Yet what seemed to be major advances were, here again, but further applications of the old techniques of the Puzzle Hypothesis. To explain blackbody radiation and the photoelectric effect, Planck and Einstein were backed by the preconception that light (like all phenomena in nature) can be divided. They had merely to choose between continuous divisibility and fundamental discreteness. Both were choices provided by the Puzzle Hypothesis and, in that way, I'd like to point out, not revolutionary at all. It perhaps goes without saying that neither Planck nor Einstein offered a perspective in which light energy could be seen as *truly* continuous, an "only one" that ultimately cannot be divided.

Einstein set the playing field for advances in the quantum perspective. Of the choices he had, he chose in favor of discreteness alone. Einstein embraced the act of circumscribing, and with the photon, he suggested that there was a particle of light that was whole and boundaried in space. It was a bold act, but one drawn directly from the ancient idea that nature has the design of a puzzle. It confirmed our sense that "what everything is made of" is discrete parts like a puzzle. The Puzzle Hypothesis had edged its way ever more deeply into our explanations.

And the science of the discrete went even further. Louis de Broglie borrowed Planck and Einstein's concept and applied it in the other direction: He postulated that if waves can be particles, all material objects—electrons and even atoms, for example— can be waves. In suit, physicists Max Born and Erwin Schrödinger developed a method (Schrödinger's wave equation) to statistically calculate how those electrons would behave—treating the waves as discrete packets so entirely that they were originally called wavicles or wave packets.

With that practical success, science was due for a new theoretical framework, as particles and waves are not easily resolved. Danish physicist Niels Bohr described it with the complementarity principle. Complementarity suggests a duality in which the particle simply *is* both a particle and a wave.

While the purpose here is not to get into a discussion of quantum theory and what physicists call its "weirdness," there is great relevance to be found underlying Bohr's conceptualization, as it epitomizes the premise behind the entire line of quantum thinking. Bohr began with the ideas of discreteness that we have accepted and applied to the material world since the earliest days of the Puzzle Hypothesis. While the electron and photon, whatever they may be, exhibit both wave and particle characteristics, all ideas about them are, first and foremost, derived from our original perception of matter existing in space—the discrete, local particle—and ascribe *to it*, to the particle, both the material and wave qualities. That is an incredible truth. It demonstrates once again that we think of nature fundamentally in terms of local objects with defined boundaries, no matter what we find. In this case, the material objects of the atomic world exhibit a duality of particle and wave behaviors and are accepted as such.

Indeed, it is testimony to the astounding depth to which our commitment to the Puzzle Hypothesis is buried in our thinking that the quantum was considered revolutionary. Yes, it is true that in those days, waves of energy had seemed to flow as a continuum and it was shocking to think of energy as emitted only in units, however tiny they were, that could not be split at all. But now that we know we have always treated nature as a puzzle, we can ask: How is it different from the approach that the ancient Greeks used, conceptually? Planck's approach to the problems he faced understanding nature, and Einstein's extension of them, was the same technique we have used for all time. They beat back the appearance of continuity with new proposals of underlying, finite, discrete units. The indivisibility of the quantum is conceptually no different from the uncuttability of the *atomos* suggested by Leucippus and Democritus. It is the same thing over again: We look for, propose, and proceed to use discrete pieces to solve the puzzle of nature.

The quantum powerfully reinforced the Puzzle Hypothesis. When we found indivisible discreteness deep in the subatomic world, it confirmed that the world *is* a sum of separate parts. On the subatomic level, it made the idea of nature being "only one," an inherent continuum throughout, seem downright impossible. Moreover, it lent powerful confirmation for the Puzzle Hypothesis just as the earliest tools did when we used them to successfully develop civilization. The quantum was a useful tool—it worked when we tried to work with the parts of nature. Just as an axe used to split a tree confirmed that nature is by its nature divisible, so too the effectiveness of quantum technologies confirmed that the subatomic world is discrete. It works. British science writer John Gribbin articulates the scientific acceptance of this situation in his book *Almost Everyone's Guide to Science*. The bold text below is mine for emphasis:

*The model which treats molecules of air as little hard balls is a good one, because it works when you use it to calculate. . . . None of the models is the ultimate Deep Truth, but they all have their part to play. They are tools which we use to help our imaginations to get a picture of what is going on, and to calculate things which we can test directly by measurement. . . . Everything in science is about models and predictions, about finding ways to get a picture in your head of how the Universe works. . . . **They are all true, in that they agree with experiment; they all fit together, like pieces in a jigsaw puzzle, to give a coherent picture of how the Universe, and everything in it, works.**[6]*

Science is about what works when coming from a puzzle worldview, and what works ever more deeply confirms the worldview from which it originated. The quantum revolution again demonstrated that, because it is so effective from a technological perspective, the Puzzle Hypothesis and its outcomes seem correct.

People continue working on the parts of nature's puzzle from different directions. One well-known proposal is string theory, which construes the fundamental stuff of nature as one-dimensional strings vibrating in 10-dimensional space to make up pointlike particles, such as quarks. According to this über-reductionist theory, the strings vibrate in different ways to give us different particles. According to Gribbin, in the physicist's conceptualization, "a loop of string is as much smaller than an atom as an atom is smaller than the Solar System."[7] Without getting into the theory, note that even in this conception, the string is a loop: It is a closed system, which is by definition a discrete entity. So too the waves of the vibrations are discrete. Thus, even when we try to depart from the particularity of matter and to conceptualize it as a vibration, we are still bound to apply the Puzzle Hypothesis. It is a particulate vibration. We conceptualize even the strings whose existence we only imagine as being discrete pieces.

Similarly, our methods lead us to understand cosmology in terms of discrete, individual entities, known as singularities. At the far reaches of our imaginations and calculations—from black holes to the big bang theory—everything disappears into a discrete point. Whether at the beginning of time or in a climax of gravity, these ultimate incarnations resolve into but a single point—a singularity—that is by definition discontinuous and discrete. And in the other direction, we see the universe itself expanding, which necessarily casts it as a closed system. Just as we do in the micro world, in the macro world we treat nature as local parts with absolute boundaries. The edges and boundaries of the puzzle, rife with entirely divisible separate pieces, permeate the character of nature as we strive to find the source of its invariance.

At one extreme, we have no particular ideas to guide us about the structure of a particle or wave—but still treat them as discrete. And in the other direction, our furthest abstractions, such as the big bang and black holes, are conceptualized as discrete. What does this tell you? Nothing in our approach has changed since Thales. Never do we see our findings as a suggestion that there *is* no fundamental hidden invariant. As British astrophysicist Sir Robert Wilson says in his book *Astronomy through the Ages,* "The universe is therefore composed mainly of some kind of matter, whose nature and properties are completely unknown, but we have no choice but to base our theories of cosmology on that part which we can see and understand."[8] Our findings only propel us to seek it further. It is amazing that we persist in this one strategy for understanding, but it is also to be expected, coming from the Puzzle Hypothesis as we do. No matter the discovery, we always ask, "What is it, what are its parts, what does not change within it?" The process has not changed. We need invariants to solve the puzzle. And we have not found the ultimate invariant.

THE HUMAN BODY is arguably the most important object in the material world as far as human beings are concerned. In the human body, we see the precious phenomenon of life. And in addition to the countless inert objects with which we interact and which sustain us, the world teems with life. We care about life—our lives, the lives of multitudes of creatures, and the life of the planet. It has been of deep practical import to us that we have sustained ourselves through the puzzle worldview. The technique of seeing the nonliving material universe as pieces was so practically successful that it granted credibility to following the same route for living organisms.

Our permission to treat life as we treat the nonliving world stems from our view that the vitality of life, like all motion, belongs to material objects and that the material world itself is a whole built from particles, as Richard Feynman explains in his book *The Character of Physical Law.* "First of all there is matter—and, remarkably enough, all matter is the same. . . . The same kinds of atoms appear to be in living creatures as in non-living creatures. . . . So that makes our problem simpler; we have nothing but atoms, all the same, everywhere."[9]

We see life as a material whole made of parts, from centuries ago through today—a spin-off of the original atomic hypothesis and its forebear, the Puzzle Hypothesis. To this day, it persists as the dominant method for understanding all life, including our own.

The approach became possible in 1665, when Robert Hooke discovered the walls of plant cells. It was a perfect match with how we like to think about nature: It suggested a puzzlelike quality to *organic* material. Hooke coined the term *cell* for what he saw—testimony to the fact that, in step with the Puzzle Hypothesis, the individual, discrete unit (the boundaried "cell") was salient to him. This discovery paved the way for cell theory, which is the well-accepted theory that cells are the fundamental unit of the structure of living things. Living organisms, says cell theory, are a puzzle, too, one made of pieces called cells. This is how we proceeded to explain the incredible phenomenon of life: based on matter (cells). Cells are perceived to be building blocks: building blocks of life.

It should come as no surprise that Charles Darwin also focused on matter, in pieces, and so took us further down into what we would later identify as genes. Darwin's theory of evolution is a theory of how change happens, and yet it revolves around matter—the material stuff that is the living creature's body. In Darwin's view, life itself is a material object with no unique force propelling it. The inextricable relationship of matter and motion was irrelevant or, perhaps, invisible to him. In the evolutionary perspective, the motivating factor of change is material.

Like all revolutionary scientists before and after, Darwin had found a new way to treat nature as a puzzle. His technique for explaining evolution is the selfsame one we have been seeing. He expertly divided the living body into constituent pieces that correspond with pieces of the environment. The entire theory of evolution by natural selection, replete with the drama of survival as it is, ultimately is a theory about how matter fits in with other matter. The physical parts of the body are what do or don't allow an organism to survive to reproduce in its environment. The materials of its body and the materials it encounters in the world must complement each other as puzzle parts fit together—or the living creature will perish.

The very complexities of the processes of life are attributed to its material, parts-and-wholes construction. Science writer K. C. Cole sums up the Darwinian perspective in *Mind over Matter: Conversations with the Cosmos*: "Life is complicated in part because it has been pieced together by evolution, borrowing whatever worked from whatever ingredients were handy (off-the-shelf, as it were) at the time."[10] Indeed, the phenomenon of life is nothing more, in Darwin's eyes, than a composite of different material layers, or scales. Today we call these scales the whole organism—the organs, tissues, cells, DNA, and genes, which themselves are made of atoms and particles.

James D. Watson and Francis H. C. Crick, who in 1953 announced their discovery of the double helix of DNA, cemented the apparently causative connection between the changes of life and DNA. It was implicitly understood that the sequence of molecules on the double helix is determinant. In other words, the material structure of the molecule, and not the means or motions by which life unfolds, determines the fate of life. Indeed, Watson and Crick believed that in biology, material objects, such as the cell and DNA, are the way to understand life. The double helix banished the notions of élan vital and intelligent design. "That is why the double helix was so important. It brought the Enlightenment's revolution in materialistic thinking into the cell," says Watson in his book *DNA: The Secret of Life*. He continues: "The intellectual journey that had begun with Copernicus displacing humans from the center of the universe and continued with Darwin's insistence that humans are merely modified monkeys had finally focused in on the very essence of life. And there was nothing special about it. The double helix is an elegant structure, but its message is downright prosaic: life is simply a matter of chemistry."[11]

The intellectual journey in which Watson envisioned his work was in fact a deeper one, begun well before Copernicus. It was the tradition of looking to the parts of nature for the answers to everything and to singling out matter in particular as the vaulting point for change.

Genes are now the vogue invariant, the thing to isolate and test to see how it affects and effectuates other pieces of matter. Indeed, the discovery of the gene has empowered scientists to full-throttle pursue the idea that life itself results from material interactions (i.e., local cause and effect), as in physics. Biology has, like physics, taken matter (or rather, the material objects that exhibit life) and come up with a legion of parts: cells, nuclei, proteins, molecules, DNA, and genes. You might say that a gene is the equivalent of an atom. It is a basic, universal structure made of particles called nucleotides, which in turn have their own parts (a phosphate group; one of the bases adenine, cytosine, guanine, or thymine; and a sugar). Scientists are determined to find the tiniest bit of matter that causes everything to happen. Like the atom in the atomic hypothesis, the gene is a building block that interacts through local, linear, cause-and-effect mechanisms to cause health and disease.

If you come at it from the other way, keeping in mind that there is another possibility, it is quite remarkable how these theories attribute so much motion, so much activity, and so much coordinated change that characterizes life to discrete material objects. But then again, coming from the Puzzle Hypothesis, there really is no choice.

So deep is our commitment to discrete units of matter as the essential puzzle piece that Feynman has said the following:

> *If, in some cataclysm, all of scientific knowledge were to be destroyed, and only one sentence passed on to the next generation of creatures, what statement would contain the most information in the fewest words? I believe it is the* **atomic hypothesis** *(or atomic* **fact***, or whatever you wish to call it) that* **all things are made of atoms—little particles that move around in perpetual motion, attracting each other when they are a little distance apart, but repelling upon being squeezed into one another.** *In that one sentence, you will see, there is an enormous amount of information about the world, if just a little imagination and thinking are applied.[12]*

The atomic hypothesis is, according to Feynman, the ultimate fact of reality. Of equal, if not greater, significance is its pronouncement that nature is designed as a puzzle, built from the bottom up. Note that Feynman's atoms are replete with motion—they oscillate in perpetual motion—but still, he frames the atom as, above all, matter. The atomic hypothesis, which distilled the essence of the puzzle world-view we've entertained since before civilization, continues to pump that worldview throughout our understandings of nature.

We must remember, however, an exceedingly important point. All along, because there was another choice, I have called the assumption that nature is a whole made of parts by the name "the Puzzle Hypothesis"—and have continued to do so even after establishing that it became, once mathematics formalized it, what I call our Original Theory of Everything. The same applies here. The atomic hypothesis was and still is a hypothesis. The guess it launches—that the *nature* of nature is that it is built up, from the bottom, by particles—is still a guess. There was and is still another choice of oneness, of an inherently continuous "one," that, I am suggesting in this book, accounts for all of nature as we experience it.

IN SCIENCE, matter anchors our explanations—because it both initially created and then continued to substantiate our idea that nature is designed as a puzzle. Having seen no option to explain nature as an inherently continuous one, we have scouted out what we perceive to be the parts of nature and begun our explanation of what nature *is* with matter. We find ideas unobjectionable when matter is the base, and motion happens to it or because of it. This is the very way we frame ideas. In contrast, we cannot understand an idea that suggests matter happens because of motion. Motion cannot make matter in any way we can understand. Where is the material? And where is the invariant? From Thales and Democritus to Planck and Einstein, and to Darwin, Watson and Crick, and researchers of today, many scientists find plenty to say when looking at matter. They propose profound and far-reaching theories of how nature works, how it moves and changes, based on particulate structure—based on the divisible, discrete pieces of the puzzle.

Nevertheless, motion is an integral part of the puzzle in its own right. It has stabilities and instabilities that beg for understanding. The world of motion does, like matter, appear understandable: Nature is replete with regularities, from the rising and setting sun to the predictable roll of a stone down, and never up, a hill. Motion can also become unsettlingly chaotic, however. Now we will discuss the search for that which might be responsible for the desired quality of stability—an invariant—in motion.

CHAPTER 4

Motion— How Does It Work?

FROM THE START OF CIVILIZATION, it was matter that appeared to harbor the stability of the universe. But no material thing, however solid it seemed, was totally immune from motion. Matter seemed to be the stuff of nature, easy to break into parts, but motion—though not dispensable—played a role less easily gauged.

Motion seemed to pervade the material universe in many guises. There was the motion of objects through space and time—arrows flying and rocks sliding down a hill—as well as the vital motion of animals and people. Even things staying in place experienced subtle motion, in the form of change: A tree decaying on the forest floor did not move per se, but parts of it moved, over time, so that it eventually was different. There was little in the world that could escape that fate.

The ready movement of objects through space and time, together with change in material objects, ignited our enormous need to know about motion. What makes things happen? How does change occur? Essentially, how does the world work?

Though it seems to ask about motion, this question is inescapably about *matter* in motion. Our most basic question about motion is, and always has been, tethered to matter. We think first of the world or one of its material parts and only *then* question how it moves. We do not, perhaps cannot, ask about motion without speaking of the motion of a particular object or set of objects.

The structure of the question "how does it work?" thus provides another ream of evidence of the force of our sensory experience of edges and boundaries. Our sense of matter assures us that nature is designed as a puzzle and that motion abides by the same design. We cannot and do not conceive of motion as continuous with other motion; we automatically divide it in accord with how we divide objects from one another.

In fact, because of the Puzzle Hypothesis, the thought of motion being continuous with other motion—regardless of the material objects that are moving—is difficult, if not outlandish, for us to conceive. This mental bias has been a critical contributor to our history of understanding motion.

The prevailing view of nature determined that, again as with matter, the proper way to treat motion was reductionism. "We might take reductionism to mean the breaking-down of things in to smaller and smaller parts, so that if you understand how the small parts work, that will in principle tell you how the big thing works," writes physicist and mathematician Sir Roger Penrose in an essay about how we treat matter and motion in *Nature's Imagination: The Frontiers of Scientific Vision*. "The idea of reductionism is that the behavior of the big things we study is governed by the behavior of the individual units of which they are composed. That is one strand of reductionism."

Penrose continues, "Another strand of reductionism may be related to this by asking if knowledge of the present behavior of those small units will allow us to predict the behavior of the big things in the future."[1] That is powerful stuff. Knowing the "behavior of things"—the nature of their motion and change—would allow us to predict and even control the future. The drive to rein in motion has thus always paired together with our desire to master matter, and our strategy has been to go at it full force with our puzzle-based concepts and techniques.

It took many generations to get our thinking to the point where motion became a puzzle piece comparable to matter, such that science as we know it could commence. The teachings of Aristotle (384–322 BCE), which were meant to clarify the nature of the universe, supported an idea of motion that fit in with a puzzle, but on which development then stalled.

Aristotle relied on invariants, one of the core implications of a puzzle-based universe, to give all of nature stability. He taught that the substance of matter itself is stable and unchanging: It is invariant. And he further reasoned that it is the natural state of matter to be—and stay—at rest. This account of nature's supreme stability entailed that matter moved only if some other thing, some force, acted upon it. A chariot, for example, would move only when a horse was pulling it, but in the absence of that force, the matter returned to rest. Motion, in this opinion, was not something to be considered on its own. It was something matter underwent. It also offered decisive invariance, in that moving matter would always return to rest.

And so, while various thinkers considered and reconsidered the ultimate structure of matter, work on motion quietly came to a halt.

Aristotle's idea about the nature of motion was so intuitively appealing that it persisted for some 2,000 years. People are drawn to ideas of invariance in every age, in every realm; stability, as part of a puzzle universe, is that essential. In fact, when Aristotle's geocentric (earth-centered) model of the universe was overthrown centuries later by Renaissance astronomer Nicolaus Copernicus—who said that the earth rotates around the sun—the sun simply swapped place-of-prominence with the earth. The sun became what the earth had been: the stable, invariant material center of the solar system and the universe. We clung to a central anchor, of invariant matter at rest.

The next great advance in understanding nature as a puzzle waited until someone did the same thing to motion as the early Greeks did to matter, which was, again, to treat it as a puzzle and take it apart. That someone was Italian polymath Galileo Galilei (1564–1642).

Modern science, with all its intent to practice unprejudiced thinking, is said by many to have begun with Galileo. Galileo gets this venerable credit for what is starting to become a familiar reason. He went beyond the ideas of his day and, as did all revolutionaries before him, took the puzzle-derived principles used with matter—boundaries, locality, a whole made of parts, and numbers—and applied them to a new area: motion.

Galileo's success with motion confronts us, once again, with the imposing reign of the Puzzle Hypothesis. Galileo superbly executed and, in doing so, solidified for the future the classic scientific steps of observation, hypothesis, experiment, and theory along with the application of mathematics, but his scientific approach was itself not new. It was simply based on the implicit Original Theory of Everything—that everything in nature is patterned as a puzzle—and everyone had believed *that* for time immemorial.

I am now going to explain how Galileo's most startling contribution was far more inventive. In a triumph for the Puzzle Hypothesis, he wrought incredible change on our view of what motion *is*. Galileo's concept of motion became a game-changing contribution to history that would shape every scientific field of the future.

The concept Galileo eventually developed of motion took a specific form to suit the Aristotelian idea it replaced. Remember that at that time, the everyday, common way of thinking relegated motion to something that happened to matter; the invariant quality of motion was that matter always returned to rest. That viewpoint was primarily about matter. Galileo somehow began thinking of motion as a parallel fixture in nature, with its own parts embedded within—parts that might be invariant in lieu of Aristotle's invariant, rest.

That is to say, Galileo focused in on motion itself to see what exists in *motion* that is fundamentally invariant. He regarded motion as a sum of parts, and he hoped to find within it a core invariant. In that, Galileo was the first to suggest subjecting motion itself to the process of reductionism. Going further than separating motion from matter, as has always been done when trying to understand nature as a puzzle, Galileo took a play from the book that allows separating matter into parts (i.e., separating matter from other matter) and separated motion from motion.

This was a riveting maneuver of extraordinary significance. Like the ancient Greeks who had appraised all matter from pebbles and trees to oceans in search of an uncuttable material invariant, Galileo was surveying all kinds of motion to determine *its* internal invariant.

Galileo took up matter in motion, one object at a time—a pendulum in motion, a rock in motion, a ball in motion, and more. Separating matter from motion was par for the course, for while matter was indeed present—it was objects that moved, after all—the question of "how does it work?" ostensibly detached from the question of "what is it?" was his goal. This angle was the invisible background before which he set himself to his main and groundbreaking task.

It required resourceful thinking to find an invariant within motion. "One of the most obvious traits about the world, which makes a general description of motion apparently impossible, is that everything moves differently,"[2] explains physicist Lawrence M. Krauss in his book *Fear of Physics*. There was, in accordance with different objects, all sorts of motion—fast and slow, vibrating, and random, a tree in the wind, water waves, a rock falling as opposed to a feather falling—many kinds, and none of them were endlessly steady and even. They all fluctuated. As with matter, motion appeared to occur on many different scales. It seemed complicated. But the unstated understanding of the Puzzle Hypothesis dictated that some simple invariant must be hidden within, an invariant that could clarify motion in the same way the atomic hypothesis appeared to clarify the material world. What could that invariant be?

To this puzzle-based question, Galileo introduced to the world a revolutionary idea that was, on the one hand, a mere advance in what the puzzle view already implicitly suggested and, on the other hand, an incredibly far-reaching innovation. In a word, Galileo said that the essence of motion is *linear*: the natural state of matter is to be forever in uniform (i.e., invariant) linear motion, a surprising and direct reversal of Aristotle's concept of matter naturally returning to rest.

It goes without saying that Galileo's conception of *linearity* predated any modern sense of the word in physics or mathematics. To Galileo, the linearity of motion was a genuine, idealized straight line.

Galileo's proposal was that the true nature of motion, the invariant reality just beyond our experience, is that any and all motion fundamentally moves at a uniform velocity in an idealized straight line forever. It is only when influenced by some outside force that motion changes. And that was that: It required no further explanation. Linear motion is as simple as motion can get. It is invariant. It cannot be further simplified, just as the atom was perceived by the Greeks to be the simplest form of matter.

As opposed to the ubiquitous waves and fluctuations that permeate the world, weaving all sorts of change throughout and across scales of nature, the idea of a line is indeed simple—straight and invariant, and divisible point by point—when fit to the Puzzle Hypothesis. Linear motion, as a concept, indeed marked a high point in advancing the applicability of the Puzzle Hypothesis.

Its significance cannot be overstated. Motion permeates nature; Galileo suggested a line within it that we accept to the present day.

However, though modern science is well versed in, and in fact built upon, the idea of uniform, invariant linear motion, there is no practical instance in which it is ever observed in the natural world. Galileo thought it up himself. It took inventiveness on Galileo's part to craft the idea—to come up with linearity as an invariant and then to conceptualize how it might combine with other possible parts of nature to add up to what we actually observe and experience. That in itself was no small feat. Indeed, the story we are tracing, how the Puzzle Hypothesis delivered us to the modern scientific worldview, was able to take a giant leap forward when Galileo suggested linear motion. We give Galileo credit for "discovering" a nugget within motion as fundamental to motion as the atom is to matter.

But at the same time, we are not troubled that his "discovery" never exists in our experiences. Galileo exercised an unprecedented creativity in applying the Puzzle Hypothesis, and today, we wholly accept his means for understanding nature because it appeals to our intuition of how a puzzle works. How Galileo managed to straddle the world of change and the desire to unearth an invariant within it is, for this reason, a tale worth telling. Each step was so potent that Galileo's ideology continues to steer virtually all fields of scientific inquiry today.

THE STORY OF LINEAR MOTION'S installation into science began when Galileo was a young medical student studying life but also interested in the motion of nonliving things. We all know that motion occurs in life and in nonlife. As I have already noted, Galileo considered motion through matter-in-motion; from that perspective, nonlife seemed to present simpler cases.

It is said that when he was 17 years old, Galileo observed a chandelier in a cathedral swinging in the wind. Galileo noticed that the larger swings swept back and forth faster than the smaller swings. Without a wristwatch—there were not even reliable clocks at that time—he used his pulse as his measure.

The heartbeat, of course, normally fluctuates, speeding up and slowing down, so Galileo sat very still to rid himself of those normal variations and educed a steady, regular pulse.[3] He sat as fixed and personally invariant as the swinging chandelier was fixed to the spot where it was attached to the ceiling: each unmoving and only experiencing waves while locked in place.

Immediately we are struck with the premise behind Galileo's act. At this young age, his aim was already to seek invariants. Purposely sitting still created a temporary regularity in his pulse from which he could skim off an invariant line. In a world designed as a puzzle, in which invariants are the key to understanding (a world in which Galileo firmly believed), the line held a special significance. Lines shine because they carry puzzle-compatible properties: They are simple, are divisible, are isolatable, and, perhaps most importantly, do not vary, which grants them utter predictability.

Galileo mentally placed the motion of the chandelier's swings upon this imaginary, invariant line, as if the line were divided into units with the chandelier's motion progressing along it. This is how he used his temporarily steady pulse to time the swings, and he saw that all the swings, large or small, were equally regular. What is of note in this step, beyond the incipient linearity he coaxed out of his pulse, is that Galileo had unceremoniously separated the frequency of a wave from its amplitude.

Here, in a natural wave, Galileo was for the first time measuring a frequency—how frequently it came to pass—and he saw the frequency actually in contrast to its height, or intensity, known as its amplitude. He regarded them as separate parts. That he adopted this viewpoint fairly shouts out Galileo's faith in, and reliance upon, the Puzzle Hypothesis, for, of course, neither frequency nor amplitude ever occurs without the other. Though frequency and amplitude are obviously inseparable, Galileo's reductionist thinking, as permitted by the Puzzle Hypothesis, prompted him to treat them as independent of each other.* The frequency of the chandelier's swing alone was Galileo's concern. It was of great interest to him that it remained constant despite the potentially distracting changes in the swing's amplitude.

In this single act, in construing a wave as moving along a straight line, Galileo made a portentous start into comprehending all motion in terms of straight lines. He was meanwhile setting roots for a way of thinking about waves that would eventually flourish in varied fields of science. The very concept that a wave can progress along a line, such that its frequency can be measured in terms of a wave*length*—a classic image, one most schoolchildren today are familiar with—had its beginning.

The burgeoning linear perspective offered immediate practical potential. Galileo developed something called the pulsilogium, a pendulum used to measure the relative rate of pulse beats for people who were ill. It was essentially the first heart rate monitor. Crafting this means to measure even pulsating, variable, human waves on a line significantly extended Galileo's cutting-edge slant of plotting the pendulum's waves on a line. With it, Galileo became the forerunner of Willem Einthoven, the creator of the now-familiar electrocardiogram. This will later be relevant to our discussion.

---

* Because of Galileo, we continue to treat frequency and amplitude separately today.

Another practical application was the clock. Galileo advanced the ideas that would evolve into the pendulum clock: that time is measurable if one tracks regular repetition, particularly the repetition of an oscillating pendulum's swings.

The pendulum thus helped shape the modern concept of time. Thanks to its unswerving regularity, every swing of the pendulum seemed to mark an equal unit—one of many equal units that, strung together, created an infinite line of time. The idea that time progresses *as* a line suddenly made all the sense in the world. Galileo's idea to chart waves along lines thus created a resounding point of view. What began as a human pulse–based means of measuring a pendulum became a perspective. It supported the idea of isochronism—where *iso* means equal and *chronos* is time—the idea that time progresses in evenly spaced units.

Isochronism is so entrenched in the modern mind-set that any alternative seems awkwardly incorrect, but time used to be conceptualized in a number of different ways. One long-standing conception was time as a circle, in which changes in nature repeat—a fair conceptualization, inasmuch as it is anchored onto the regularities of nature that give a framework to the passage of time. Our word *cycle* comes from *circle*. From the concept of a circle of time, one can segue into the Galilean view by envisioning the bottom portion of a circle progressing forward, instead of doubling back, beneath a bisecting line—the line that Galileo generated for measuring time. The result is an oscillating cycle, along a linear axis, as we visualize it today.

As linear as we imagine time to be, we have always depended on oscillators, including the microwave frequencies in today's cesium clocks, to provide the cycles through which we accurately "keep" time. But we settled on the idea of time moving along a divisible, invariant straight line for good reason. From a puzzle-based perspective, units on a line are a superb stand-in for regular, repetitive motion. Unlike the unruly motion of the natural world, or even of a regular oscillator, the line of time can be neatly divided and labeled by numbers on each identical unit. Why retain the curve of a repeating frequency when you can simplify it to a line? We then can use mathematics to chart changes along that straight line of time.

Though today it is generally considered difficult, if not impossible, to truly define time, there is no doubt that Galileo promoted a well-accepted idea: For every unit of motion, we believe there truly exists an independent unit of time in which it has occurred. Time *is*, to us, a frequency, moving in a straight line, as measured by a clock.

Using his pulse to measure chandelier swings had been a prodigious act. Galileo had laid the groundwork for the mechanical clockwork universe, with its intimate dependence on units of time, which Sir Isaac Newton and his contemporaries would further develop. All motion seemed to progress through linear units of time.

The future of those developments would hinge on Galileo's next steps, however, in which he took straightened-out waves even further: True to the old ways of the Puzzle Hypothesis, Galileo experimented to explore the fundamental nature of motion itself. And he hammered out a concept of motion so intuitively appealing that we retain it unquestioningly today.

As early as 1589, Galileo was conducting experiments to clarify motion; he is said to have dropped light and heavy stones from the Leaning Tower of Pisa to see which would land first. (He certainly, though less famously, rolled balls down a wooden slope laced with lute strings and tracked their increasing speed as they rolled and struck a progression of musical notes.) Galileo was observing acceleration (i.e., trying to determine how much an object's speed of motion increased as measured along a linear timeline). Both objects landed at the same time. This meant they had the same rate of acceleration.

Galileo thus initiated the modern experiment. Whether in the fabled Leaning Tower of Pisa version or in his documented trials with a ball on a wooden slope, Galileo used observation, hypothesis, and experiment to isolate and demonstrate an idea—that is, used the techniques we have always used on the presumed puzzle of nature in order to isolate and confirm the invariant properties of a single part.

In this case, the idea he proposed was that there is a downward pull on matter, later called the force of gravity by Newton (*gravis* means heavy). The downward pull was cast as a linear force that somehow participates in what we experience. Other influences also seemed to be at work; a feather fell more slowly than a rock, and the existence of air resistance was identified as the reason.

Galileo's experiment, with its scientific design and resulting theory of separate linear forces, has reverberated through our scientific history and lore—so much so that on the last Apollo 15 walk on the moon, Commander David Scott repeated Galileo's test. Scott dropped a hammer and a feather on the airless lunar surface to demonstrate that the two objects land at the same time in the absence of air resistance. Galileo had intellectually separated air resistance from the linear pull of gravity in his conceptualization of motion and used experiment to validate his theories about these hypothetical parts of nature; hundreds of years later, an American astronaut dramatically demonstrated and amplified our sense of their separateness. As he seemingly confirmed Galileo's assertion that the true motion of gravity can be idealized and treated as a straight, downward-accelerating line, Scott also seemed to prove that our devotion to Galileo and his storied experiment are well earned.

The idea of latent forces presented a new avenue of understanding to Galileo. Forces seemed to stand in their own right, for by causing change in motion, they seemed obviously separate from motion. This fledgling idea allowed Galileo to take a scalpel to motion itself. He earnestly began to dissect it into parts with what is known as his inclined-plane experiment.* Though details of this experiment are not themselves necessary to our discussion, I will recount them because they pinpoint the moment in which Galileo fully stepped into the world of linear thinking.

The experiment was designed for Galileo to observe and measure a ball rolling down and up an angular U-shape: down an inclined plane (a flat surface on an angle), then onto a horizontal surface, then, further, up a second inclined plane in the other direction. He noted that the ball would reach the same height on the second plane as the height from which it had started rolling down on the first. This was always true, no matter the angle of the second plane.

Galileo reasoned that if the second plane were lowered such that it was on a very shallow angle, the ball would travel much further to reach the same height as from which it had dropped—but it would indeed go that lengthy distance to reach that height. He then theorized that if that plane were lowered all the way, so that the ball were rolling horizontally, the ball should go on in a straight line forever. Experience shows it does not. Galileo therefore proposed that interfering forces, such as friction and the pull down to earth, are what prevent us from actually observing that reality.

In other words, Galileo reasoned, if one were to remove what he considered to be the independent pulls of friction and gravity, the motion of the ball would continue in a straight line forever. And that, said Galileo, is the fundamental, invariant nature of motion. His watershed idea had come into being. One can separate actual motion into idealized parts; from this perspective, one can say that motion is uniform, in a straight line, and will go on forever unless disturbed by separate, external linear forces.

Galileo wasn't finished, though. Now he moved on to real world matter in motion to see how linear motion might account for our everyday experience. With his nascent idea guiding him, Galileo focused on the motion of a cannonball. A cannonball flying through the air was as familiar to Galileo as a baseball flying through the air is to us; moreover, understanding its path would give valuable insight to the Italian military. Galileo began by observing that a cannonball shot from a cannon actually travels in a parabolic curve, not in two straight lines, an upside-down V, as Aristotle had said it does. (Surprising though it seems, Aristotle's view was commonly accepted wisdom at that time.)

Here was a single curve, a single arc shaped by a material projectile—a marked contrast to the many curves he had charted along a line to represent the frequency

* Again we must note that Galileo incorporated an assumption from the Puzzle Hypothesis into these experiments. Though Galileo is said to have "ignored" matter to focus on motion, one can just as easily say that he ignored that he could not actually separate matter and motion. Nor did Galileo ever sustain the possibility that motion itself might be somehow whole or indivisible. He held fast to his faith that edges and boundaries are applicable to motion and that we may dissect it just as we do matter. His a priori conceptualization of motion was that motion shatters into parts that correspond to the objects that move.

of a pendulum's swings. What did Galileo do with that curve? While his awareness of the wave was closer to the reality we experience than was Aristotle's straight vertex, Galileo again invoked his idealized straight line. Into the cannonball's curve, he intuited the same idealized linearity he had felt he experienced as he sat fixed to the floor of the cathedral and that the chandelier had seemed to experience as it held fixed to the ceiling. This was his idealized line. He combined that line with a blend of two elements: the linear pull down of gravity that he had extracted in his experiments at Pisa and the invariant linear motion he had induced from the inclined-plane experiment.

Galileo concluded that if one were to abstract away what he had conceptually separated out during the inclined-plane experiment—the seemingly irrelevant influences of the earth pulling down vertically, air resistance, and friction—motion of a projectile would continue in a straight line forever. It was the first case in which natural motion was described this way: where a curve itself was actually said to *be* motion in a uniform horizontal straight line, somehow combined with external disturbances.

If one were rooting for the Puzzle Hypothesis, it would be time to break out the champagne. Galileo had hit on the idea of ideas. Linear motion represented a true invariance within motion. It seemingly handed humanity the means to understanding all motion through reductionism.

The idea has become so well accepted that it constitutes the basis of plain scientific knowledge. We understand that reality is the way Galileo proposed it is: that projectile motion and acceleration are separate velocities, sometimes referred to as the Law of Independence of Velocities.[4] To say that an object would move in a straight line forever but for gravity pulling it down to earth—two independent vectors, one in a vertical straight line, the other horizontal, somehow adding up to create the wave we actually observe—was, from the perspective of the Puzzle Hypothesis, an ingenious way to understand the motion we experience as a sum of parts.

In this we see that Galileo's influence went further than straightening out motion, though that itself was quite a contribution. With these experiments, Galileo subtly asserted a radical proposition about how to understand nature. The motion he discussed as reality was not motion as we experience it, but rather an idealized version of it. It portrayed the downward pull of gravity and other forces of nature as if they were truly separable from horizontal uniform motion—as if they were truly independent of the straight-line uniform motion he asserted for the object itself— though they never are, in any way, that we ever, ever experience.

Galileo deemed that departure from experience to be of little importance. It was completely outside the scope of consideration that abstracting out invariants might do injustice to nature. He never addressed that there may exist an inseverable connection between the vertical and horizontal velocities insofar as there existed (and continues to exist) the possibility that nature might be an inherently continuous, ultimately nondivisible "only one." (This possibility still exists, as we will see ahead, because it is certainly possible that one could make successful and useful predictions while treating nature in a way that is ultimately not accurate.) Scientists today echo his judgment: "Modern physics becomes possible. And none of this would have been arrived at if Galileo had not thrown out the unnecessary details in order to recognize that what really mattered was velocity, and whether or not it was constant,"[5] writes Krauss in *Fear of Physics*.

Deviation from the actuality of nature was not only acceptable but also demanded by the situation, according to Galileo. He held in disdain the "many others who divert the discussion from its main intent and fasten upon some statement of mine which lacks a hair's breadth of the truth and, under this hair, hide the fault of another which is as big as a ship's cable."[6] That is to say, a small lie is an acceptable indulgence if it means greater mastery of the puzzle. To Galileo, it did not matter if the model of reality was slightly mismatched to nature as we experience it, as long as it advanced our grasp and command of the parts of nature. And that was because if nature is a puzzle, the goal is not to try to understand all of nature as an all-at-once phenomenon where no detail can be ignored, but to try to pull simplicity out of a complex design, with the implicit expectation that the pieces will eventually fit together to explain the totality of nature.

Galileo thus reversed our reliance on experience and traded it for a different starting point—for an idealized invariant line from which one can build scientific understanding. Modern science follows the route Galileo took. We accept the existence of idealized invariants and use them, in various reconstructions, to account for what we actually experience. For just as it was for Galileo, having the puzzle shape up before our eyes is a powerful experience. It makes acceptable our deviation from our experience of nature.

If it seems obvious today that it is unnecessary to be 100 percent true to nature as we know it in our foundational explanations, it is wholly because science has accepted the assumption that invariants can and should be abstracted out of complexity, piece by piece: the assumption first made by Galileo when he reduced curved motion into the two abstract lines of inertia and gravity. Such is the power of Galileo's legacy.

Galileo set the standard for good science altogether. Highly attuned to the demands of finding answers within the puzzle, he impressed generations to come with how he used the Puzzle Hypothesis's techniques and extended its premises. He tapped the effectiveness of observation, hypothesis, experiment, and theory—all puzzle-based approaches to survival—and married it to the partnership of parts and numbers. He put his blessing on the abstract technique of applying mathematics to a linearity that we foist onto nature even when—and in fact because—nature as we experience it is not quite so. This multifaceted blend appealed to our intuition of how we must treat nature.

It would be accurate to say that by studying nature with this form of reductionism, we found what we were seeking. Galileo's celebrated insights into motion were a repeat of our history. Their champion was another participant in the long tradition of finding ways to place edges and boundaries around "parts" of nature. Finding himself at a dead end, Galileo put his mind to placing edges and boundaries around idealized parts and unifying them with mathematics. Galileo's rigor evolved from an ultimate commitment to the Puzzle Hypothesis and a true appreciation of its nuances. Through his advances in how to apply it, he purified the Puzzle Hypothesis and elevated it to be a standard of truth.

The other choice of how to look at nature, namely that nature itself might be an "only one" that is inherently continuous and whose motion cannot ultimately be divided into parts, found no safe haven in the progressive approaches of Galileo. Galileo looked to nature without preconception to see what nature is—*except* for the preconception that nature is a puzzle—and his new angle of linear uniform motion empowered generations of future scientists to further dissect nature in his style. In trying to solve the puzzle, Galileo deepened its hold remarkably.

I cannot say for certain, but it seems that Galileo never did what I am doing: spelling out how an a priori hypothesis infiltrated his work. To Galileo, the puzzle was a dead-set reality. Everything discussed so far has been to help unearth and identify the roots of Galileo's intuition. If it has seemed murky at times, remember: Galileo was not following a clear or overt plan of action. Here, I am taking the time to unravel the tangle of ideas that run through his thoughts, experiments, and discoveries, and to show how it all came from his acceptance of the Puzzle Hypothesis.

It is not without irony that Galileo is remembered as an objective thinker when his work did so much to strengthen a precondition to understanding nature and block out the alternative. Albert Einstein, who was himself committed to looking at nature with as little bias as possible, writes in the foreword to Galileo's *Dialogue Concerning the Two Chief World Systems* that

> *the leitmotif which I recognize in Galileo's work is the passionate fight against any kind of dogma based on authority. Only experience and careful reflection are accepted by him as criteria of truth. Nowadays it is hard for us to grasp how sinister and revolutionary such an attitude appeared at Galileo's time, when merely to doubt the truth of opinions which had no basis but authority was considered a capital crime and punished accordingly. Actually we are by no means so far removed from such a situation even today as many of us would like to flatter our-*

selves; but in theory, at least, the principle of unbiased thought has won out, and most people are willing to pay lip service to this principle.[7]

And so, to uphold our allegiance to "the principle of unbiased thought," we must recognize that Galileo's contributions hinged on an a priori acceptance of the Puzzle Hypothesis. The "experience and careful reflection" he used as "criteria of truth" and on which he built his experiments always—always—divided motion into parts according to objects that move and then divided each instance of that motion into abstract parts. If motion *cannot* fully be multilayer divided, as he assumed it can, then Galileo imposed an assumption on every experiment he performed. And on its basis, Galileo proceeded to cement the belief that we must treat nature as a discontinuous puzzle in every scientific field.

Straight-line motion assumed a new, critical position because of Galileo. If all motion occurs through straight lines, then straight lines are, necessarily, how change proceeds. This meant straight lines serve as nature's pattern of communication. The acceptance of their existence catalyzed a shift of the Puzzle Hypothesis from a rudimentary puzzle into a more mature form: a puzzle whose parts interact through straight-line motion.

It would be with the Galilean twist on the Puzzle Hypothesis that the Western world would develop modern civilization and technology, in a continuation of what we had always been doing—seeing nature as a whole made of parts. Galileo had innovated a nearly miraculous new way, a linear way, to understand nature as a sum of parts interacting causally in straight lines. This method would be picked up by generations of scientists to come.

Most fundamentally, in his assertion that linear perpetual motion is a natural, invariant state that combines with another kind of linear motion (forces) to create the motion we see, Galileo installed linear motion as something on par with the atom. Like the concept of a fundamental, uncuttable particle, *atomos*, which has stayed with us even as scientists venture deeper into the subatomic world, linear motion was fundamental and uncuttable: the *atomos* of all motion and change. It was an ultimate invariant—a fundamental piece of the puzzle. It has been my purpose here to introduce this recognition: that Galileo's "success," the successful introduction of this new piece to the supposed puzzle of nature, profoundly affected our understanding of what nature is and how to understand it. In the same way that the atom had become the common denominator of all matter, linear motion became the common denominator of all motion and change.

Science moved forward on that premise, that idealized straight lines are the backbone of natural change.

GALILEO HIMSELF did not live to see his straight-line motion concept disseminate into everyday thinking. But disseminate it would, infiltrating and proffering its structure to every field of science in the generations that followed. It started a near chain reaction as creative thinkers found new ways to explain natural phenomena as expressions of linear motion.

Truly, there was almost nothing else one would do. Almost all motion in nature was as yet undefined. The field was wide open for people to determine how that prime linear motion combined with other parts of nature to build the different kinds of motion we actually experience. Following the introduction of one final idea about linearity, innovation after innovation brought uncharted aspects of nature under the dominion of the line.

That final ingredient was the straight line of space.

Linear motion required a suitable venue through which it could occur. The concept of linear space was developed by a contemporary of Galileo—French philosopher and scientist René Descartes. Descartes based his concept on one developed by ancient Greek mathematician Euclid: the axiomatic idea of a straight line connecting two points.

Descartes applied lines to points in space. He showed that straight lines that intersect at 90-degree angles, à la Euclid, could geometrically label space, just as they did for motion. The lines stood in formation in the now-classic three-dimensional grid of length, width, and height.

This grid was very practical for the mathematically minded investigator of nature. It allowed people, for the first time, to do something we do all the time today, on and in everything from maps to geometry: to reduce a material object to a point, locate the object on that grid with numerical coordinates, and, when appropriate, track its motion on that grid linearly. Here was a representation of matter in space that suited the Puzzle Hypothesis in both its divisibleness and its linearity. It allowed for the application of numbers. Galileo's linear motion now had a linear stage on which to proceed through linear time—one on which the motion could be tracked mathematically.

One can practically reel off the subsequent historical progression of the line's march through the natural world, through the work of Newton, Faraday, Maxwell, Fourier, Carnot, Thomson and Clausius, and Einthoven.

Newton, born in 1642, about a year after Galileo died, was Galileo's first natural successor. The medieval mind-set, in which nature was lawless but for the word of divine authority, still dominated Newton's world, but Newton was inspired by the invariance of straight lines. He determined that there must be a consistent (i.e., lawful) way to describe all motion, as linear, throughout the universe. Galileo had contemplated only earthly motion. Newton took it further. He conceptualized a force as that which moves things, in straight lines, through straight lines in space—even beyond earth—and, instant by instant, through straight-line time.

All these lines allowed Newton to apply mathematics, step by step, to explain motion as a sum of linear forces, as we will further discuss in the next chapter. Some of Newton's most historic contributions—the law of gravity, the first three laws of motion, and the concept of the clockwork universe—were all ways to describe motion, with math, as a combination of lines. The work so extended and deepened the Puzzle Hypothesis that, like Galileo's contributions, they seemed revolutionary.

Yet as revolutionary as these ideas were, we can see that they were, at heart, also just new ways to formulate motion as traveling along Galileo's ideal straight lines. They marked an advance in the puzzle to be sure, but they were also what naturally followed. The unheralded predictions they now made possible, however, such as determining the speed and path of objects as they move, ensured that the formulations seemed astounding.

And even beyond the predictions they now permitted was perhaps their biggest impact—how they reshaped people's thinking about change.

Every object in the world seemed subject to Newton's laws of motion. That meant that each object surely moved in a fundamental straight line due to some other straight-line force. And it would not move otherwise. That meant that straight-line forces, whether a push or a pull, were responsible for all change.

The mechanism was what we now call local, linear cause and effect—and it burst on the scene with incredible power. The effect was grand. The question had always been about how "it" works, and Newton had an answer: It moves and interacts in straight lines. Deterministic cause and effect between two bodies seemed to drive each micro and macro change in the puzzle of nature. In fact, an interaction—an *interaction*, meaning the action between two objects—was freshly conceptualized as a straight-line cause-and-effect change, over time, that transpires between two bodies.

Guided by the mighty schema of motion progressing on straight lines, people imagined a mechanistic universe that proceeded like clockwork (inasmuch as the precise progression of a clock was a paragon of deterministic—i.e., linear cause and effect—mechanisms). Of course, the idea of a clockwork universe incorporated and was built upon the ancient Puzzle Hypothesis. But it clarified the nature of the puzzle: that it is made of invariant, localized parts that move via invariant, linear motion. French mathematical physicist and astronomer Pierre-Simon, marquis de Laplace, conjectured that in a clockwork universe, a being who knew the position of every bit of matter and every force that influenced linear motion could (if he possessed the ability to so calculate) predict the future and account for all the past. Newton's equations were so precise that they were time reversible, meaning that if one were to undo each decisive cause in an interaction, one would theoretically return to its initial circumstances.

The fresh feature of mechanistic change rejuvenated the Puzzle Hypothesis's ability to explain nature. History may say that our worldview was changed with this development, but it would be more precise to say that our worldview became a further, deeper articulation of how a puzzle works. When Newton expanded the lines of motion and causality to reach beyond the earth alone, he served humanity the hope that we can understand the entire universe. It seemed that lines permeate the terrestrial and celestial realms, unifying them as one giant, understandable arena of linear interactions.

This familiar picture, taught in today's schools before advanced courses refine it with Einstein's relativity, may seem obvious, which stands in testimony to how completely we accepted this framework for understanding our experiences. Even when further developments went beyond the model of the clockwork universe, local causality would remain the underpinning of all scientific investigation.

Newton's formulas had in fact painstakingly defined for the scientific world what appeared to be the only acceptable means through which change could be understood to occur. Changes were understood to happen through mathematically trackable cause and effect. And therefore, due to the linear transmission through time of local causes, change—the effects—could never be simultaneous at a distance or across scales.

Thus, while deterministic cause and effect dazzled everyone with its effectiveness, and seductively promised to explain all things yet to come, it put a harsh spotlight on an old alternative: action at a distance.

Action at a distance is the term for when two (or more) things, separated in space, change in unison without some form of direct contact between them to cause the change. There is no accounting for the simultaneous process of action at a distance in a part-by-part way. It does not sit well, therefore, with the understanding that nature is a puzzle. While action at a distance had, for generations, quietly lurked as a transcendental, supernatural phenomenon, now thinkers trying to explain all of nature through direct, local cause and effect realized that it had to be eliminated.

Newton understood the two concepts to be incompatible opposites. No one ever dreamed about the possibility that action at a distance and direct causality might somehow coexist in orchestrating daily changes in nature, as they can when nature is "one." It was one or the other: an either-or choice. Therefore, action at a distance became—and continues to be—the pariah of the local, linear causal universe.

Yet the fact remained that Newton was left without a cause when calculating gravity as operating along a straight line—what was causing the pull? This was a problem. We accepted straight-line motion and independent forces, but in doing so, created a mystery of how it all connects. The effect of gravity was clear, but the absence of some physical impetus to cause an apple to fall from a tree—or even more so, to connect the pull of the moon to the earth, for there is no direct contact—could not be explained by the model of mechanistic change. It appeared that objects were affecting one another without direct contact. To use Newton's own word, this was "inconceivable."

"I do not feign hypothesis"[8] for how gravity's effect comes to pass, he wrote in the essay "General Scholium" as an appendix to the second edition of his *Philosophiæ Naturalis Principia Mathematica*. But even without that hypothesis, he was certain of one thing: The idea of gravity acting from body to body without some sort of mediator was "so great an Absurdity that I believe no Man who has in philosophical Matters a competent Faculty of thinking can ever fall into it."[9] It was incomprehensible that a person with "a competent faculty" could believe in action at a distance. Absolute simultaneity appeared outright impossible. It does not jibe with the concept of a puzzle, and therefore no person in his right mind could believe that nature itself could operate that way. Newton, like Galileo, seems to have been unaware of the Puzzle Hypothesis. He would not have known that the mystery of gravity was a created one, set up by his crafting a linear, isolated formula to measure gravity's effects.

As disturbing as the specter of action at a distance might be, the mathematical law of gravity nevertheless gained acceptance. The math worked. And application is powerfully convincing, as it has always been with the Puzzle Hypothesis. Indeed, even without a mechanism to account for gravity, the practical applications of local, linear cause and effect made the choice clear. Action at a distance was booted out of the arena.

It took well over a hundred years, but eventually, in 1831, a model was developed by English scientist Michael Faraday to address Newton's problem of how a force works without action at a distance.

It should come as no surprise that Faraday too used the ingrained Puzzle Hypothesis, our Original Theory of Everything, to develop his model. Faraday proposed using "electromagnetic fields" in space: "Lines of force," in Faraday's words, could spread out to create the field, and then, so-called local concentrations of the lines of force were responsible for directing the forces (i.e., through cause and effect).

No one can visualize a field, but it is accepted as a mathematical description of forces in space. It is a way to explain how forces—including electromagnetism and gravity—work in space over both short and great distances, through conceptually straight lines, without direct contact between objects. Fields are understood to account for the pull between the moon and earth or an apple and the ground. We still use this model today to keep action at a distance out of our explanations.

Simultaneity had become a pariah. But what had happened to the natural waves of nature in which we had always lived? Rhythms permeate nature on and across every scale, without boundary, progressing simultaneously. Where did they fit into the clockwork universe?

Waves were regarded, literally and figuratively, as immaterial. Lacking substance, they were the opposite of matter; they seemed to be a form, a pattern that moves up and down. And as waves, they were the opposite of lines, as well. Their failure to click neatly into the puzzle cornered waves into a deferential role. Scientists defined them as a "disturbance in a medium"—a disturbance in what was considered to be the "real" world of matter and motion in space and time.

The spirit of Galileo, flowing through Descartes, Newton, and Faraday, directed people to think about waves as truly moving and interacting along lines. We treated them the same way we treated matter in motion. They were present in nature, to be sure, but they were in no way extraordinary. There was no undue concern about what they were as long as they could be measured and calculated.

Galileo's ideas, in fact, prompted the sense that mathematically working with waves—measuring them and calculating them—was not only akin to understanding but actually the climactic event that satisfied what it means to understand.

It was a gratifying approach. To understand waves in this capacity, people developed the now-familiar ways of thinking about waves in terms of lines: ways that broke waves into parts and reassembled them along an idealized line. These included wave*lengths* that measured waves from crest to crest, averaged-out frequencies of waves over linear time, *heights* of amplitudes, and calculating positions for electric and magnetic waves in terms of straight lines and right angles. (Measuring and tracking each wave on a line, including assigning a length and a height, only makes sense when the wave has been pinned on that abstract line.) Over time, all these ways of casting waves in terms of lines facilitated "innovations" that were but new ways of seeing and calculating natural wave phenomena as lines.

In essence, the act of measuring and calculating became identified as understanding. And we wanted to understand.

In 1861, Scottish scientist James Clerk Maxwell famously derived four laws of electromagnetism. These laws were akin to Newton's laws of matter in motion, but this time they applied mathematical linearity to waves. Maxwell charted waves on a linear trajectory—that is, he construed their path as invariant lines—and then measured them and crafted equations to describe their behavior. Maxwell discovered that waves of electromagnetism moved close to the speed of light and proposed that light was a kind of electromagnetic wave.

The lines of Galileo were in clear play as Maxwell employed them to brace his electromagnetic waves on their travels through space. Moreover, Maxwell recognized a reciprocal relationship between electrical and magnetic fields. In the tradition of his predecessors, Maxwell plotted the electric field and the magnetic field as oscillating at right angles to each other, asserting that both fields moved together as waves of electromagnetism along the nexus of a single straight line. With Maxwell's equations, people could now predict the movement of waves along a line, and that was tacitly held to be as good as understanding.

In the early 19th century, Joseph Fourier commandeered the straightening of waves and combined it powerfully with the treatment of waves as a sum of parts. He proposed that any wave—any kind, including sound waves, light waves, and water waves, no matter how complicated and fluctuating—can be understood, that is, calculated, by adding sine waves together on a line. The Fourier method of wave analysis involved taking the wave and placing it along a line and then deriving a number of individual waves whose amplitudes, through interference, either add up to or subtract from one another to form the amplitude of the original wave; at the same time, the frequencies add and subtract to form the frequency of the original wave, too. Any wave could be treated this way by linear superposition. Mathematically adding and subtracting amplitudes and frequencies piece by piece—as if the vertical amplitudes and horizontal frequencies are independent of one another—marked the final induction of natural waves into the world of lines. Real changes in waves could be grasped as a sum of theoretical linear waves on a line, with their curvatures left behind. Even a wave was treated as a whole made of parts. Fourier had taken Galileo's discontinuous, linear treatment of waves to an ultimate conclusion.

THE AFTERMATH OF NEWTON'S WORK took us far from the realm of natural experience—and deeply into abstract rules, which appeared to govern the seemingly linear changes of nature. In daily life, local, linear cause and effect had an absolute hold on people's understanding of how change transpires. It seemed to be the one and only way that nature works.

Its powerful directives shaped the theories that flowered during the enormous changes of the Industrial Revolution: changes that brought us to the brink of the modern worldview. We have been listing the changes sparked by Galileo's work, and we find in the Industrial Revolution a pivotal development whose resultant philosophy still rings true today.

The all-important question of that time—the question that shuttled us to many of our present understandings of nature—was how the steam engine worked. The sooty world of the steam engine represented motion in perhaps its grittiest but most valuable form: It promised labor beyond what people could perform by physical effort alone. (Modern steam turbines remain the source of most electrical power today.) Yet the steam engine had been developed not by foresight but through trial and error. No one had come up with a predictive theory or rule to explain why it did what it did.

The steam engine complicated things because Newtonian mechanics, the science of the day, somehow did not work to explain how the gas—the steam—managed to actually power the engine. Newtonian equations sketched a universe of perfect determination, where interactions can equally happen in either direction; like the chiming cogs of a miniature music box, the world need only be wound in the other direction for even time to reverse, due to the precision of what came to pass. That meant we ought to observe every interaction happening in both one direction and the other. But certain interactions, the steam engine being a prime example, did not conform to those rules. Even a schoolchild could tell you that a hot object, including the steam engine, gets cold (change happening in one direction) but never goes in the other direction to heat up on its own. If dying embers returned to flame, if tepid cups of tea grew warm, if shattered glass spontaneously reformed—sometimes, somewhere, people would observe it. These events would show Newtonian laws working in both directions, as calculations showed they should. But such things never happen. Reversible Newtonian equations notwithstanding, these events did not ever reverse. Something was happening in the engine that seemed to represent a natural truth we all experience, which differed from the clockwork calculations of mechanics.

What was a scientist to do? The obvious move, à la the Puzzle Hypothesis, was to corral the phenomenon and break it into parts.

The steam engine seemed an obvious starting point. The engine—and not living organisms, in which heat and motion also obviously exist—was, plainly, the simplest place to look. It seemed an ideal case in which to observe how, in the puzzle of nature, a local, linear transfer of heat relates to the motion we call work.

French military engineer Sadi Carnot led the way. Though his primary concern was not matter but motion, heat, and work, Carnot began with material wholes and parts—with matter in motion—just as Galileo had. He isolated the steam engine as a closed system. Then he broke it into two boxes (i.e., two distinct parts) based on what he observed about the motion within the engine. All working steam engines, Carnot saw, required a temperature difference between regions of hot and cold to do work: The design incorporated a separation of what came to be called a hot source (the burning fuel, gas, or coal, or eventually oil and atomic fuel) and a cold sink (such as the smokestack, which emits waste into the cooler surrounding environment). Using the fundamental tenets of the Puzzle Hypothesis, Carnot boxed off the hot source and cold sink as two totally separate parts of the engine.

There was a problem, though. Carnot, like Maxwell and others who worked on this problem, understood that the gas in the steam engine—the steam itself—was likely made of billions of atoms, though atoms had not been proven by experiment to exist at that time. If indeed the gas was made of atoms, the actors in the steam engine problem had progressed from one body, the steam engine, to two bodies, the hot source and the cold sink, to n-bodies: innumerable atoms. And n-bodies lay outside the realm of Newtonian mechanics, so it would not be possible to calculate how those atoms might move.

Carnot was not deterred. Because a hot source and cold sink had to be separated for a steam engine to work, Carnot envisioned that there was a flow of heat transpiring in the engine. He proposed that the heat, whatever it was, did not move freely in any direction; rather, it went in a straight line from hot to cold only. And from that unidirectional direct flow of heat in the engine, we harnessed some power that was converted partially to work, and what could not be harnessed went to waste in the cold sink. "The production of motive power is therefore due in steam engines . . . to its transportation from a warm body to a cold body," wrote Carnot in his 1824 book, *Reflections on the Motive Power of Fire*.[10]

The history of science takes little if any note that Carnot assumed that heat flowed from hot to cold in a direct line. There seemed no other option—how else would it travel? Curves or waves, for example, were never an option. Just as Carnot was, we are all so well taught by Galileo that the assumption of fundamental straight-line motion is ingrained. Following it warranted no discussion. But here we note: If heat "somehow" moves from hot to cold, the motion was and is automatically imagined as a straight-arrow line.

Nevertheless, I want to point out a reality never discussed, which is that the steam engine's "straight-line" flow from hot to cold cannot commence without first heating the hot source. You must stoke fire to build the heat that only then runs through the steam engine. You must first go from cold to hot. Intellectually, however, even though it was in no way dispensable, that rise of heat preceded the starting point to which engineers turned their attention. Carnot and the scientists who built on his work omitted this upward side of the process. They divorced the upswing of heat from their conceptual straight line down from hot to cold, much like ignoring the initial ride up a roller coaster and looking only at the plummeting drop that ends it. From there they developed their theories.*

It is no wonder Carnot took this liberty without fanfare. He simply followed Galileo and deviated a "hair's breadth" from reality. The scientific world admires Galileo's gall, as is evident in the language mathematician Steven Strogatz uses to explain Galileo's achievements in his book *Sync*: "Galileo would not have discovered that a body in motion tends to stay in motion (the law of inertia) if he had been content to describe what really happens (friction causes things to stop). By disregarding

---

* Carnot got implicit permission to do this from his instinctive reliance on Galileo. Galileo had wed us to the belief that matter fundamentally moves in uniform straight-line motion and that it comes to curve only under the influence of other straight-line forces, such as gravity and friction. Galileo viewed the parabola traveled by his projectile cannonball as two fundamentally straight lines, inertial motion and linear gravity, and considered the one line of inertial motion to be the true motion of the cannonball; Carnot likewise separated the two sides of the natural curve of heating and cooling into lines and worked only with the one line of cooling as the true flow of heat. The warmup of the engine, the upswing in which a coal fire generated the tremendous heat which only then would dissipate into cold, was shunted out of his thermodynamic arena, just as Galileo stripped gravity from the motion of the cannonball itself.

the inessential, he discovered the most fundamental law of mechanics."[11]* As per Galileo, Carnot "disregarded the inessential." Building the hot source was considered inessential to running the steam engine, and so the heat that runs through the engine was taken as a given—just as Galileo treated gravity and friction as if they were of no consequence and required no explanation.

We must note that though the plummeting heat of the steam engine was the sole focus of Carnot and other Industrial era engineers, the history of science is not without discussion of that upswing of heat—only, it is treated completely separately.

In the year 1900, on the tail end of the development of thermodynamics, Max Planck explained blackbody radiation: the thermodynamic patterns of electromagnetic radiation that bodies emit as they are heated *up*. As I noted earlier, with that explanation, Planck unleashed the startling new world of quantum physics.

It may come as no surprise that here, again, Galileo's idealized lines shaped how new ideas were conceptualized. The heating-up process, which resulted in a blackbody absorbing and emitting electromagnetic waves of radiation, was also conceptualized as moving in a straight line: from a cold state to a hot one. We might even, with a wry smile, consider it to exhibit a cold source and a hot sink.

Never do physicists speak of this inherent relationship—the relationship between the heating up of a blackbody, including the internal walls of the steam engine, and the flow of heat from hot to cold, epitomized by the steam engine itself. Experientially, these acts of heating up and cooling down, which science treats separately, occur together; they are no different from many other phenomena, such as waking and sleeping, exercise and recovery, and others covered in the first chapter, that occur together in our natural experience. It is a fact that you cannot heat up without cooling down, and you cannot cool down without first heating up. Yet science demands we separate each phenomenon for isolated study.

Recognizing that it is necessary to build up heat before that heat can flow from hot to cold in the steam engine turns an astonishing spotlight on how people have chosen to view nature. Here were accomplished, brilliant scientists—who relied completely on the Puzzle Hypothesis's tenet that everything in nature is discrete. Planck started a new scientific field when he heated up blackbody radiators from cold to hot, while Carnot started a new field by isolating heat as flowing from hot to cold, yet the inherent connection between quantum physics and thermodynamics—or that there is indeed any connection at all—has been overlooked by history.

---

* Galileo's determination that one may "disregard the inessential" revamped the puzzle-derived idea of boundaries: It permitted scientists to imagine idealized (not 100 percent experientially real) phenomena and to dissect them through abstraction since reality did not permit completely separating out parts. That is, it was okay to speculate about hidden, nonisolatable parts and treat them in theory and by mathematics as if they were separate. If the equations worked, then that "part" was said to have been discovered. (It was by this reasoning that Galileo "discovered" the idealized line he said underlies all motion.)

And it is not just the history of science that ignores the steps taken by innovators. The methods and ideology of science itself inherently negate the possibility of considering an unbreakable connection between the two fields. Two distinct sciences emerged because we applied Galileo's straight lines to motion.

What I am calling the upswing of heat, the indispensable warming up phase, found a blind eye not only in Carnot but also in others who developed thermodynamics. The originators of the laws of thermodynamics, William Thomson (later known as Lord Kelvin) and Rudolph Clausius, shared the implicit conviction that motion occurs in straight lines. This conviction left them no choice but to conclude that the linear change described by Carnot—the collective going from hot to cold, as observed in closed systems like the steam engine—was the way of the world. It seemed indisputable. They noted, moreover, that like a tag-along sibling one would rather be rid of, waste accompanied the flow of heat from hot to cold under all circumstances. As work was done, in fact, the majority of the heat went to waste and could not be recovered.

By the mid-1800s, Lord Kelvin expanded on Carnot's straight-line flow of heat to include this unavoidable waste. He suggested that such is the way of all of nature. He and Clausius formally identified this flow from hot to cold and waste as the second law of thermodynamics. Though their new science was called thermodynamics, where *thermo* is heat and *dynamics* is motion, there was only one kind of motion they entertained. It seemed an indisputable law that the motion of heat is a straight line, through which it dissipates to cold.

History calls this time period the Industrial Revolution, but now we see its apparent trailblazers were hardly revolutionary. They simply applied and reapplied the old tenets of the Puzzle Hypothesis. The thermodynamic science that emerged from their efforts was likewise nothing new; it was the inescapable result of the a priori hypothesis that nature is designed as a whole made of parts. Ideas that had already matured—the ancient atomic hypothesis, the idea that motion fundamentally moves in straight lines, and the idea that we can apply mathematics, the so-called language of the universe, to try to figure motion out—led directly to it.

Thermodynamics is part of the modern worldview, in which the entire universe is viewed and treated as a great heat engine. As a scientific discipline, it perpetuated and extended the fundamental acceptance that everything on the whole goes in straight lines. Whether from hot to cold or cold to hot—according to thermodynamic theory, cold to hot can happen in local pockets, but overall it is indeed hot to cold—the universe was understood to progress through straight-line motion.

Moreover, the straight line of time submitted to a modification with the advent of thermodynamics. Thermodynamic interactions are not reversible, as Newton had proposed, but travel in one direction only from which they cannot be undone: a one-way "arrow of time." We must note that whether Newton's model or the thermodynamic model had prevailed, time would still have been conceptualized as a line. As it was, however, the thermodynamic model became the model of the universe.

For scientists, the second law is considered immutable. It has become so ingrained in our thinking that astronomer, physicist, and mathematician Sir Arthur Eddington said, "if your theory is found to be against the second law of thermodynamics I can give you no hope; there is nothing for it but to collapse in deepest humiliation."[12] It expresses our utter commitment to Galileo's straight-line motion as the truth about nature that we plainly embrace its final incarnation, a pronouncement that irreversible linear dissipation is the eventual fate of all order in the universe. Closed systems containing straight lines will lead us there.

As the legacy of Galileo-based Newtonian mechanics gave way to thermodynamics, quantum mechanics, and relativity, our ultimate concept of motion has remained linear. The forces of the world—gravity, electromagnetism, and the strong and weak forces in the quantum—are understood to be and treated as linear, just as they were in Faraday and Maxwell's day. Fields are still envisioned as "lines somehow curved," to use Galileo's expression, in which the description is of straight-line motion that will stay linear forever unless acted upon by some other, linear, force. Descartes's use of lines on space persists in the puzzling realm of the quantum—despite the fact that "the idea of the path has to be abandoned [and that] a particle cannot even be talked about as having existence as it jumps from one state to the next," as science writer John P. Briggs and physicist F. David Peat explain in their book *The Looking Glass Universe*. "Nevertheless, embedded in quantum mathematics are [Cartesian] coordinates with their notion of continuity, paths, and infinite divisibility."[13] In the same way, the lines of Cartesian space have survived the advent of relativity theory: "Although the space-time order of relativity is no longer absolute, its Cartesian description has been retained," Briggs and Peat further note. We hold on to invariant lines as the way to understand the universe.

PERHAPS THE LAST STRONGHOLD of the wave in nature was in the living human being. Rhythms are a powerful, variable force in human life.

Galileo, who went to medical school, did his best to linearize his pulse when looking at the chandelier in the cathedral. Einthoven formally took care of these types of waves by developing the electrocardiogram in 1903, and he won the 1924 Nobel Prize for his efforts. Einthoven graphed the electrical impulses of the heart as they correspond to the motion of the heart muscle itself.* The waves are traced on graph paper, along linear axes. This linear graphing renders the waves and rhythms of the human organism fit for mathematical analysis. The same is done in the electroencephalogram, the EEG, as well as the electromyography, the EMG, whereby brain waves and muscle waves are plotted in linear fashion, respectively.

The most fundamental understandings of biological change, in fact, are today based on straight-line, local causal interactions. American mathematician and philosopher Norbert Wiener's cybernetics, which arose in the 1940s, popularized a model of feedforward and feedback cybernetic loops: Changes are construed as a step-by-step sequence of up versus down signals. The same idea applies in genes and cancer: There are "on" genes, called the oncogenes, and "off" genes, or tumor-suppressor genes, together creating a linear on-off model of how cancer works. Similarly, signal transduction pathways model biochemical change as a cell converting a signal or stimulus from one form to another, as a lock and key, in a linear fashion. And homeostasis and balance are linear ideals that the body strives toward.

Interestingly, the "on-off" model of cybernetics bears a strong resemblance to Aristotle's idea that a projectile flies straight up and then straight down. Both cases draw on the idea that there is only, actually, straight-line motion, as opposed to the curve from which Galileo elicited a straight line.

The central dogma of molecular biology, which states that information flows on one direction only, from DNA to proteins, likewise champions the notion that there is a linear flow that is irreversible. The so-called building blocks of life and their interactions are a one-way ticket, just like the second law of thermodynamics. When "DNA makes RNA makes proteins," as the popular description goes, it is a one-way flow of information. Taking pieces apart and putting them together may be the way to learn, understand, and tap into the biology of nature, but the way the interactions transpire in life is considered to be one-way only. Even today, the field of epigenetics (which is the understanding that genes are influenced by local, environmental molecular phenomena) is still perceived similar to everything from cybernetics back to Aristotle: straight-line motion going up and down.

Local, linear cause and effect is, as Galileo first intimated, the modus operandi of nature to us. Action at a distance and top-down organization have been eliminated

---

* Consistent with our historic approach to understanding nature through the lens of the Puzzle Hypothesis, Einthoven's innovative tracking of the heart's motion only followed thorough analysis of the material of the heart. The history is: In 1628, William Harvey had conceptualized the travels of the blood as a circle (circulation) and described the anatomy and physiology of the heart. (Here is another illustration of our tradition of prioritizing matter over motion: Rather than begin with visualizing the motion, Harvey described a finite, physical loop along which motion could travel.) Knowledge proceeded to grow through literal dissection: careful study of the physiology, anatomy, and pathology of the heart and blood vessels. Medical observations developed awareness of blood pressure, heart murmurs, and respiration. The structure of the heart muscle, including congenital heart defects, earned further detailed study. But still the electrical activity in the heart muscle, and irregular heart rhythms, was yet to be isolated. Einthoven developed the electrocardiogram in 1903 to measure it.

from the model of life. For science and even in our general thinking, it seems so obviously correct today that it is hard to imagine that there could ever have been another choice—or that it could be right.

The ultimate theory about the biological evolution of life, Darwin's theory of natural selection, also incorporates and relies upon a one-way, irreversible chain of cause and effect. In full concert with the view that life is built from building blocks, the motion of these blocks is also always seen as bottom up: changing by, and only by, local, linear cause and effect. This resulted in the understanding of "descent with modification," as it is known: a linear tree of life arising from a buildup of incremental linear changes, with no top-down organization.

This perspective negated the unwelcome specter of action at a distance—unwelcome because without a puzzle-based means to validate it, it would have hinted at intelligent design. Certainly there was and is no acceptance of an élan vital, or life force. There is "no pilot," to use the phrase of British biologist and science writer John Maddox.[14] The parts are understood to interact locally and causally, and that is the only way change comes about. The vibrant rhythms of nature have been shuffled aside in favor of seeing them as the product of linear causal interactions.

THE PUZZLE HYPOTHESIS has wrestled the variant wave into submission, *in every aspect of nature, in every realm of our experience.* Science construes all waves and motion, no matter how variable in the natural world, only and always as moving along straight lines.

The enormity of the impact of Galileo's simple idea of motion continuing along a straight line forever unless disturbed cannot be overstated. It granted impeccable credibility to the idea that nature is a whole made of parts—not just matter, but motion, space, and time, as well. Motion was as divisible as matter, and it too had a prized internal invariant—uniform linear motion—as fundamental as the atom is to matter. It could be abstracted into lines just as matter could be abstracted into numbers. It promoted the idea of three-dimensional linear space and, through the invention of the pendulum clock, linear time. Uniform linear motion gave us the dazzling ability to analyze events using local cause and effect, which became our cherished explanation of how everything works, and along with it, the mighty power of prediction.

Invariant linear motion made it appear that it is no accident that the Puzzle Hypothesis is our underlying Original Theory of Everything. Nature seemed to agree.

CHAPTER 5

Laws—What Holds Everything Together, and Why Do Things Fall Apart?

IF THIS BOOK WERE A HISTORY of the ascendance of the Puzzle Hypothesis, we would now crown our progress with the scientific law. After Galileo and Newton, an unprecedented era commenced. Thinker after thinker found new ways to describe nature mathematically. These descriptions, the laws, continue to reign as supreme explanations of nature.

But I am not telling such a history; I am in the business of bringing to light how the Puzzle Hypothesis pervaded and determined the course of our progress. Scientific laws were no exception.

Though scientific laws hold an august role in our thinking, their rise was merely the next step in the progressive way we have understood nature when beginning with the Puzzle Hypothesis. Like the Galilean innovations that fostered them, laws are, on the one hand, a tremendously eye-opening tool for dealing with nature as a puzzle and, on the other hand, a mere extension of the puzzle-based techniques we have been using for all time.

To fathom the hold laws have on us, we must plainly understand what laws are. We must also see why they necessarily followed Galileo's suggestion that there are abstract lines in nature.

Laws, like atoms and lines, are all about invariance. The term *law*, from the Old Norse *lagu*, means something laid down or fixed. The purpose of a law is to spell out subtle consistencies—what seem to be invariants—in natural change. We reserve the word *law* for when an event establishes a discernible, seemingly invariant pattern. For example, when matter in motion, through space and time, seems to always behave the same way under a fixed set of conditions, we call that a law.

Laws may command royal respect nowadays, but the fact remains that they owe the fabric of their existence to the everyday regularities of nature. The opening sentence of ecologist Robert H. MacArthur's book *Geographical Ecology* bluntly states this truth: "To do science is to search for repeated patterns, not simply to accumulate facts."[1]

AS I HAVE EMPHASIZED from the beginning, ordered patterns and regularities have always been a critical aspect of nature. Our most basic sense of order has rested upon that treasured observation: that all patterns of motion around us, in general, progress in an orderly way. In the first chapter, I showed how we have lived within, and depended upon, manifold ordered regularities as long as we have been on earth. The regular rise and set of the sun, the rise and fall of the tides, the reliable turn of the seasons, patterns of plant growth and animal migration, and the regular cycles of one's own body are all regular, rhythmic patterns of motion. Without them, chaos would fog our lives.

Patterns of motion were too valuable to leave alone if we could express them as invariants. That was what laws were meant to be: invariant expressions of stable patterns, so we know what to expect and, hopefully, predict. The repetitiveness of patterns assured us that we were justified in hunting for something in nature that organizes it and grants it stability.* Ultimately, laws attempt to answer the great question "what holds things together, and why do things fall apart?" and to do so by describing nature's regularities as invariants.

The marriage of regularities and laws has been long understood in science. At the heart of every law is a regularity that people derived from the intricate dance of what we take to be the parts of nature.

Richard Feynman explains in *The Character of Physical Law*, "There is . . . a rhythm and a pattern between the phenomena of nature which is not apparent to the eye, but only to the eye of analysis; and it is these rhythms and patterns which we call Physical Laws."[2] In the words of K. C. Cole from her book *The Universe and the Teacup*, "The fact that patterns repeat allows us to formulate laws of nature—really, recipes encoded in equations that describe relationships that repeat over and over again. . . . The equation is a shorthand for a relationship with an enduring quality."[3] In her book, Cole quotes American physicist Frank Oppenheimer's raw assessment

---

* One could even say that linear motion owes its "discovery," in part, to our certainty that natural rhythms will repeat. Even Galileo would not have construed a line from his pulse had he been unsure whether his pulse would carry rhythmically forward.

of the predictive power of laws: Prediction "is dependent only on the assumption that observed patterns will be repeated."[4]

As people tried to articulate the orderliness of nature, the Puzzle Hypothesis quietly dictated the acceptable parameters by which one could do so.

First, order meant not just any order, but, specifically, the orderly motion of matter through space and time. Order, in other words, was the order of a puzzle, subject to the local causality that such a design entails. This assumption, as it had often done in the past, completely discounted the possibility of the type of inherently continuous order discussed earlier.

Next, the Puzzle Hypothesis determined that we may treat the presumed parts of nature as abstractions whenever we try to describe such patterns of order.

Abstractions offered a way to discuss pure relationships between parts of nature without attachment to the parts themselves. Abstraction, from the Latin *abstractiō,* meaning separation, detaches us from tangible experience. We have discussed that one of the earliest abstractions was using numbers to describe units of matter; numbers allowed us to abstract out, and idealize, the puzzle-assured individuality of each part of nature. This way, one apple and one apple could be understood as one plus one, focusing on the dematerialized, idealized *one*ness of each apple and leaving the apples themselves behind. This was, in retrospect, a first step toward laws.

Historically, our ease with abstractions took off when Galileo produced his work. He taught that an idealized straight line is embedded in every instance of motion. Though it was not present in natural experience, meaning it was abstracted out of experience, the line did seem a correct way to depict motion—it let a person think about nature as a sum of parts and could be described with numbers to good effect, as well.

Abstractions escalated as the Galilean concept of linear time shed its origin, which was the waving pendulum, and stepped forth as a conceptualized freestanding line along which instances of motion occur. They strengthened further with Descartes's proposal that motion proceeds in straight-line space.

The scientific law was something of a crescendo that followed this mounting trend of abstractions from nature. With matter reduced to a point* and motion reduced to a line traveling through space and time, we had a ready language with which to describe the patterns of order we saw in the puzzle of nature. Things were happening, and many things happened dependably and consistently: Was there a way to speak of that regularity that wasn't tethered to real objects? How could we

---

* Mathematics began with numbers, as discussed in Chapter 2. It moves beyond plain counting to reunite numbers in relationships. Even the simple equation $1+1=2$ represents a consistent relationship that can be found in nature. It is that consistency which makes it and other, more elaborate, equations so attractive to the mathematician and scientist. Mathematics has no equal as a clutter-free language for describing patterns in parts of nature.

leave the specific bits of matter and particularities of each instance of motion behind and speak only of invariances?

It took only one more element to create a new way to craft such statements. That element was measurement. Measurement, so common in modern life that it seems as natural as nature itself, is a specific sort of act: one made possible by the Puzzle Hypothesis and the abstractions that follow it.

Every act of measurement, as we all know, requires applying numbers to space and time. That is how we can count units. In accordance with the abstractions we drew from thinking of nature as wholes and parts, we measure an item moving in space, instant by instant, in its unit-by-unit path. The revelation was that if there was a consistent pattern to its motion—say, repeated measurements of a steady acceleration or of a specific change in velocity from some force—we could describe that pattern of motion mathematically.

That's what a law is. The purest scientific law is a statement of something that invariably happens—an invariant—expressed as a measurement of changes that describe how an entity moves, in a predictable way, through space and time. It is a mathematical expression of that invariant, dependable pattern of change. It articulates a regularity in nature.* And we adore these laws because their pinpoint declarations of natural regularities allow us to make predictions.

To science, measurement of invariants is indistinguishable from the process of assessing nature itself. "Measurement, it's probably fair to say, is the cornerstone of knowledge. It allows us to compare things with other things and to quantify relationships," states Cole in *The Universe and the Teacup*.[5]

If measurement is the cornerstone, then the mathematical relationships between what is measured are the edifice. The act of repeatedly measuring that which we believe to be invariant gives us the material with which we build laws.† Measurement allows us to expose and match up what anthropologist and social scientist Gregory Bateson called "patterns which connect" through their shared relationships regarding how things move in nature.

Every equation, from the most basic laws taught in physics 101 to the intricate equations used by physicists, relies on this framework of consistent (i.e., invariant) measurements.

---

* I use the word *law* to represent all laws, theories, models, and so on. As physicist James S. Trefil says, "The laws of nature are not always called 'laws.' . . . A law is just as likely to be known as a 'theory,' a 'rule,' a 'model,' or a 'principle,' or, reflecting the fact that laws are often stated in the language of mathematics, as a 'relation' or an 'equation.' I suspect that whether or not something is called a law has more to do with quirks of history than with the logic of the sciences." (Trefil, *The Nature of Science: An A–Z Guide to the Laws and Principles Governing Our Universe* [Boston: Houghton Mifflin, 2003], xxix.) All laws, theories, and so on are identical in their attempt to state invariants in nature.

† It is not a single measurement from which laws are created. Measurements dominate when they are compared and contrasted. When a number of measurements are compared, the accumulated data possesses the power to verify that a pattern is regular. Putting consistent measurements together is the way to demonstrate a pattern (i.e., to state a law).

Once measurement was made possible by the historic combination of numbers and linear dimensions, it unleashed the explosive potential of laws to predict motion. It armed scientists with penetrating field glasses through which they could scout the universe for fixed patterns of linear motion. Thanks to the unending string of number-units offered by measurement, the line surpassed the rhythm as the archetype of regularity. Lines seemed to indicate a deeper order to natural change. If natural change could be construed as a network of lines—and it surely was—that meant nature followed hidden patterns that could be found by measurement and uniquely stated by mathematics. Mathematics has the ability to make "the invisible visible," to use mathematician Keith Devlin's phrase—but only in such a universe.[6]

HISTORICALLY, IT WAS GALILEO, Descartes, and Newton who launched mathematics as the language in which laws are articulated. They pioneered the application of numbers to linear spatial dimensions and to linear time, such that matter in motion could be measured in their terms.

Nature seemed beautifully organized—highly ordered—and these men showed we could devise mathematical descriptions of that order through experiments on it. By repeatedly trying different activities and measuring the outcomes, they demonstrated that we can sift out certain regularities and state them as mathematical laws. To Galileo, mathematics was the language of God: "Philosophy [i.e., physics] is written in this grand book, the universe, which stands continually open to our gaze. But the book cannot be understood unless one first learns to comprehend the language and read the letters in which it is composed. It is written in the language of mathematics."[7]

This viewpoint has persisted, and even flourished. In his essay "What Are the Laws of Nature," physicist and science writer Paul Davies quotes 13th-century English philosopher Roger Bacon as saying: "Mathematics is the door and the key to the sciences. . . . For the things of this world cannot be made known without a knowledge of mathematics." Davies himself continues, "From the inverse law of gravitation to the abstract gauge groups of modern unified field theories, mathematics is the language that most succinctly encapsulates the workings of nature."[8] Lawrence M. Krauss espouses a similar view in *Fear of Physics*, "Language, a human invention, is a mirror for the soul. . . . Mathematics, on the other hand, is the language of nature and so provides a mirror for the physical world. . . . Like it or not, numbers are a central part of physics. Everything we do, including the way we think about the physical world, is affected by the way we think about numbers."[9] Albert Einstein wrote to a friend in 1916: "Never before in my life have I troubled myself over anything so much, and I have gained enormous respect for mathematics, whose more subtle parts I considered until now, in my ignorance, as pure luxury!"[10] Years later, in 1933, Einstein summed up his view at the Herbert Spencer Lecture at Oxford University, "Our experience hitherto justifies us in believing that nature is the realization of the simplest conceivable mathematical ideas. I am convinced that we can discover by means of purely mathematical constructions the concepts and the laws connecting them with each other."[11] The list could go on. Certain regularities in nature have permitted the application of mathematics; mathematics, in turn, promises to give voice to ever more deeply hidden regularities.

As delightful as mathematics may be, nature does not set the table with a smorgasbord of straight lines. Mathematics portended a feast of invariants, but to make it possible, a fair amount of prep work was required: We had to figure out a way to convert all real instances of curved motion into linear mathematical statements of straight line motion. Newton and German polymath Gottfried Wilhelm Leibnitz each made this possible with their independent inventions of the calculus. Even calculus exemplifies the omnipresent role of the Puzzle Hypothesis in the development of modern science.

The calculus converted the unruly curves of natural motion into abstract, measurable lines. The variability of the curve, due to its changing nature, cannot be easily measured and calculated; it cannot even be construed as a series of points because points are static and curves change. The calculus mathematically reduced the curve to what we could manage. (In a differential equation, a curve was divided into what were called infinitesimals—theoretical, infinitely small points. The creators of calculus then made straight-line tangents to those points and worked with those lines instead of the actual curve. Areas under a curve were construed as a series of minute rectangles.) Once it was clothed in points and lines, the "curve" could be treated as any line would be—mathematically—to create equations to describe it. Newton and Leibnitz, in other words, grasped Galileo's concept of the idealized straight line and firmly applied it, point by point, to the instance of curved motion. Galileo's parabola received its proper linear treatment in the pointlike divisions and abstracted lines of calculus.

Through calculus, matter and motion—or, rather, their abstract doppelgangers, numbers and linear vectors—were unified to create equations. Those equations spelled out patterns of order that repeat and were therefore predictable.

For calculus, the field was wide open. Curves abounded in nature, just waiting for mathematical treatment. The method could be (and was) applied everywhere from the stars, moon, and sun down to an apple falling on a person's head here on the earth. No curve could defy the linearity that calculus pressed on it. No action or interactions between one and two bodies could evade being described, step by measurable step, with calculus. Though once imperfect, nonlinear, and unpredictable, the motion of nature was broken into parts, linearized, and measured. It was set up to be read as equations, describing what consistently happens through local, linear cause and effect. At last, the variable motion of the universe could be broken up and made to relinquish its hidden linear patterns. Causality reigned supreme.

Newton thus ushered in the law as we know it. His laws of motion and law of gravity described some of the consistent ways motion happens: for example, that a falling object always accelerates 32 feet per second with each passing second. Note the absence of which particular item might be falling—the matter was abstracted away—and that the regularity was stated as a measurement of motion along linear space through linear time. Newton had created the first abstract mathematical model that described hidden patterns of connectedness between regularities in nature. It was a landmark event.

Newton's laws were a fabulously simple way to measure and predict change over distance and time. Perhaps their most important quality was that they worked. Their predictive power was invaluable. With this power came control. It was an exciting prospect. As with many developments in the long line of practical puzzle-based ideas, the fact that they worked offered enormous support to the sense that laws revealed truths about nature. These laws birthed the concept of the mechanical clockwork universe.

To Newton and his contemporaries, the emergence of the law and its predictive power gave shape to their understanding of how God relates to the universe. "The early stages of this science retains strong theological roots, developing the view that the world could be understood by rational inquiry and must therefore be the manifestation of a rational deity. Thus the scientists of the sixteenth and seventeenth centuries proclaimed that God invented mathematical laws of nature to make His creation run smoothly," explains Davies in "What Are the Laws of Nature." "In the centuries that have followed, Newton's laws have been replaced by others. Notwithstanding the changing fashion regarding their precise form, the belief that mathematical laws of some sort underpin the operation of the physical world is now a central tenet of scientific faith."[12] Laws that described the regularities of nature were the language of God, for God presided over nature: God governed the laws that governed nature.

Even today, in a most basic way, nothing has changed since Galileo's time. With or without a god behind them, laws are understood to govern nature, and the language of laws is mathematics.

Indeed, mathematical laws remain our modern attempt to answer the final questions we have had about nature: What holds everything together, and why do things fall apart? They have, from the beginning, boldly revealed hidden, mathematically describable, apparently invariant patterns of matter in motion. They may not have illuminated the reason behind the patterns, but they did imply a causal determinism.

This is why, even as religion and science parted ways, laws bolstered our sense that we understand the stability behind nature. The way they satisfy our longing to know stability—through mathematical formulations of the way changes consistently transpire—was (and is) so dependable that for many scientists they seem to be the reason the changes happen.

Our goal in understanding nature, after all, has always been a two-helmed quest, for personal illumination but also for prediction and control. Once these patterns could be described mathematically, the prospect of prediction tore far ahead. A theme common to many developments in our history reemerged: The usefulness of scientific development offered immense satisfaction for our sense that we understand. Practical mastery seemed to equal knowledge.

In our journey of understanding, we thus came to a place where the stability we sought to comprehend—the underlying reasons for order and chaos—was being mapped out, as much as possible, in the language of mathematics.

And that is where we stopped.

This subtle shift was a significant one. The fact is that today, when we find the mathematical description of a pattern, we regard that description—that equation— to be the reason for the change. In other words, we mistakenly believe that a *description* of change is the same thing as an *understanding* of change. Without ceremony, the appearance of laws marked a segue in our worldview of what it means to understand nature through the Puzzle Hypothesis. Prediction surpassed its old partner, illumination, and quietly became the raison d'être of investigation. It quietly took on the role of the sole bearer of understanding.

And it *is* satisfying to know patterns. In a universe that is designed as a puzzle, things that change erratically seem random and confusing. A pattern, however, suggests some sense of understanding. Patterns in the order of nature, and also patterns lurking within disorder (we even hope to understand chaos on some level through patterns buried within it), ground us with the masterful feeling of knowledge. Finding a regularity, however finely its thread may be woven through nature, satisfies our sense of having discovered a piece of the puzzle.

But though we regard a discovered pattern as a cause, a description it remains.

When it came to a head in quantum physics—where the predictions and results were inconsistent with our worldview—the position of the physicists was decisive. The general sentiment was that mathematics and experiment are all we need for truth, and the rest is for the philosophers.

Laws had altered our understanding of what it means to understand.

It is a curious reality that mathematical laws, which, intangible as they are, are nowhere to be found in nature itself, stand solo as the way to express truths about it. They so viscerally encapsulate stability that they even seem to surpass matter itself, to which we had originally clung for stability. Matter's edges and boundaries and invariance were the ancient basis of the Puzzle Hypothesis, but it stepped aside as numbers stepped in to represent individual units—and the law stepped up as the modernized version of our belief in edges and boundaries (i.e., closed systems) and invariants.

A law ends up being, interestingly, something that is indisputably super-added to nature as it exists. Laws govern from a "higher" place. In other words, laws do not exist in nature in any way that we experience; rather, they are a conceptual construct that helps solve the puzzle of nature. In an opinion piece for the *New York Times*, Davies writes that physics rests on a "faith" that there exists "an unexplained set of physical laws" that are "inhabiting an abstract transcendent realm of perfect mathematical relationships." This transcendent existence, argues Davies, is immutable in the eyes of the scientist: "the universe is governed by eternal laws (or meta-laws), but the laws are completely impervious to what happens in the universe."[13]

While we have come to see them as part of nature, perhaps more real than the substance of nature itself, we should not lose sight of the fact that laws are an innovative way of extending our Puzzle Hypothesis. Laws say: Nature is a puzzle, and there exist intangible, super-natural/transcendental entities* called laws that piece together the other parts of nature (matter, motion, space and time, and order and chaos) to make the puzzle be the way we see it. Laws are a fundamentally different statement of the physical reality of what *is*.

When you look back at our history of trying to understand nature, it may be surprising that we understand it based on something that cannot be found in the puzzle itself but is, rather, transcendental. But you can understand how it came to be this way, by following through the logic of the Puzzle Hypothesis. That hypothesis did more than compel us to break apart nature to look at it piece by piece. When invariance was nowhere to be found, it became acceptable to deviate a "hair's breadth"† from nature as we experience it so that we could read further invariance into fluctuating things (i.e., so that we could isolate further invariant parts of nature). That is how Galileo came to read lines into natural change and how everyone who followed culled out repeating patterns, mathematized them, and ended up with invariant laws. The practical predictive success of those steps emboldened us to speak of nature *thoroughly* in terms of the invariants we cleave to and to convert all our natural experience to the terminology of laws.

Coming from the other direction, our confidence in mathematics is not diminished by the fact that it is an idealization of the natural world on which it is based and also imperfectly matched to it. We accept that mathematics depends upon a departure from nature as we experience it in favor of idealized numbers, straight-line motion, and idealized dimensions. In his book *Equations of Eternity*, David Darling

---

* By *super-natural*, I mean, of course, applied over, or superadded to, the natural.

† Remember, Galileo had minimized and justified his deviation from nature as we experience it by calling it merely "a statement of mine that lacks a hair's breadth of truth"—a small sacrifice to get at what he believed was the greater truth: of solving the puzzle by breaking motion into invariant pieces. (Galileo Galilei, *Dialogues Concerning Two New Sciences*, trans. Henry Crew and Alfonso de Salvio [New York: Dover, 1914]).

readily explains, "And in the roots of mathematics, too, we conveniently overlook this fluxlike essence of our surroundings. Because in abstracting the notion of 'oneness,' as applied to anything, we are making assumptions: first, that the thing can be treated apart from its surroundings; and second, that there is a quality about it that is permanent and consistent."[14] Devlin, writing in *Life by the Numbers*, concurs:

> *The mathematician views the world in terms of perfectly straight lines, perfect circles, geometrically precise triangles, squares, and rectangles, smooth, flat planes, instants in time, and the like. In the real world there are no perfect circles, no perfectly straight lines or smooth, flat, planes, and, to all intents and purposes, no instants in time. If you don't believe me, take a look at a glass plate (a "smooth, flat plane") through a microscope. What looks to the naked eye like a flat plane will reveal itself to be far from smooth and flat. The mathematician's circles, lines, planes, and so forth are all fictions, figments of the imagination. They are idealizations of the "almost circles," the "fairly straight lines," the "smooth-and-flat-to-the-naked-eye planes" et cetera, that we find in the real world.*[15]

It was our search for invariants that gave us license to treat nature as a combination of perfect lines, circles, and borders. We could not manipulate the units of mathematics if we did not operate this way. Even something as apparently simple as a straight line of space—a bare-bones dimension—is an idealization. All this is justified by our original choice that nature is designed as a puzzle, a whole made of discontinuous parts. Our faith in mathematics extends our "faith that nature," as Einstein said, "takes the character of a well formulated puzzle."

Nevertheless, our commitment to idealized mathematics as the way to solve the puzzle seems invulnerable, even, and perhaps especially, when placed in contrast to its alternative, a "oneness" of nature.

If nature is *only* one, then oneness for individual items is a practical perception on our part that does not match deeper reality. It would not matter that that is the concept on which the entire scientific system is based. It would mean that abstracting out idealized lines on which to measure these items is, again, practical but not ultimately true to nature.

The implication would be that we gain a valuable measure of control by mathematically predicting patterns in our world. But that that intellectual path will never lead us to fathom nature as the one that it really is.

This possibility, whose viability was crippled by its impracticality, especially relative to the empowering Puzzle Hypothesis, was long forgotten. And so we became certain that the practicability of our laws attests to their ultimate fit to nature.

Our sense of understanding the mechanical universe blossomed as laws to describe it proliferated. Many mathematical formulas were developed to capture the motion of objects through space, gravity, electromagnetism, and more. Such formulas brought us to our modern sense of scientific mastery.

IN CLOSING THIS CHAPTER, I must mention three specific examples. These overarching laws covered all the bases of the Puzzle Hypothesis and shape our current picture of the universe. They are the first law of thermodynamics, Darwin's evolution, and Einstein's special relativity.

The first law of thermodynamics is a sweeping statement about the universe— one that seeks to explain a particular aspect of the puzzle. That aspect is the relationship of the whole and its parts.

The first law states that the energy of the universe is conserved, meaning that no matter what events may transpire, the energy of the universe is constant. This law is an updated, mathematized way of saying the universe is a whole made of parts. In fact, it shines as an example of how readily we use numbers to label those parts. It's only a tiny step to go from saying the design of nature is a whole made of parts to saying the whole equals the sum of the parts. That we segue so easily into this expression shows how we so easily slipped into mathematics to discuss nature. Science has done it for centuries.

In thermodynamics, the whole quite literally equals the sum of the parts. Thermodynamics wholesale borrowed the idea of a closed puzzle with absolute edges and boundaries; you might say it took the idea and ran with it. All the laws of thermodynamics innately depend upon science's ideal closed system.

An idealized closed system is a complete puzzle-made-of-parts, totally separate from the rest of the universe. In each closed system, matter, energy, heat, and motion are construed as discrete parts and are expressed as interchangeable quantities of energy. Picture a puzzle within a set frame: Pieces may move and change into one another, but as everything changes, the total remains the same.

That picture is the law of conservation of energy (i.e., the first law) in a nutshell. It speaks of a closed system in which the parts will always equal the whole, no matter what transformations they undergo. They interact through cause and effect, and all the parts always add up.

In fact, if we ever doubted that the borders between pieces were absolute, the first law of thermodynamics stabilized them for us. It was extraordinarily practical and broad-sweeping. It stepped forward as one of the most practical theories in all of science by treating heat and energy (the "ability to do work") as pieces of the puzzle, assuming their motion was local and linear and transpired through causal interactions, and by always adding them up to the whole.

Charles Darwin likewise borrowed the idea of a closed puzzle but applied it to the realm of living creatures. That is to say, just as thermodynamics did, Darwin's theory took the idea of a puzzle with pieces, in which there is no influx of influence that might cause change. Every change was seen as a fixed set of pieces playing out in a closed system.

With this framework in place, Darwin brought on a theory akin to Newton's mechanical universe. Just as the clockwork worldview indicated that every change in nature results from an interaction between material parts, natural selection indicated that changes in living creatures also emerge from interactions between parts. In his book *Entropy*, economist Jeremy Rifkin describes the parallel.

The mechanical world paradigm experienced its greatest triumph in the aftermath of Charles Darwin's publication of On The Origin of Species *in 1859. Darwin's theory of biological evolution was every bit as impressive as the scientific discoveries of Newton in physics. It could well have pushed the mechanical world view off center stage and claimed hegemony for itself as a completely new organizing principle for society. It never happened. Instead, Darwin's theories became an appendage to the Newtonian world regime . . . that further legitimatized the mechanical worldview.*[16]

Nevertheless, Darwin's commitment to, and application of, the ideals of the Puzzle Hypothesis ensured its lasting influence.

Einstein's law of special relativity was founded on a complementary aspect of the Puzzle Hypothesis: the invariance of laws themselves. Einstein derived relativity by beginning with an assumption, that all laws are the same on all different scales of uniform motion. (This, of course, drew from Galileo's concept of straight-line uniform motion, which also drew from the Puzzle Hypothesis.) Einstein's commitment to laws' invariance prompted him to determine that if laws do not change, it must be motion that does change (i.e., is relative). It was that determination that oriented him to calculate his famous theories. His commitment to the invariance of the law, which is a pillar of relativity theory, was one of Einstein's attempts to make a statement that would help us understand all of nature.

THERE IS AN APPEALING invitation in the way Einstein describes laws. If all laws have an invariance to them, then there is some commonality to them all. They apply anywhere in the world; they are universal.

Our deeper goal, at last, would seem attainable. We hope to unify the laws. We hope to find their underlying invariant: to make a theory of everything. For laws themselves, as you can see, have ended up being as much a puzzle to be pieced together as wholes and parts were when we first tackled the mysteries of nature.

Today, one of our greatest living physicists is Stephen Hawking. Hawking has said, as is true, that all laws are partial: "It turns out to be very difficult to devise a theory to describe the universe all in one go. Instead, we break the problem up into bits and invent a number of partial theories. Each of these partial theories describes and predicts a certain limited class of observations, neglecting the effects of other quantities or representing them by simple sets of numbers."[17] Think about it: Laws themselves are now the pieces of a puzzle.

Science is not without hope of finding a grand, sweeping explanation of the universe. We seek patterns with the hope of discovering some essential kernel: something invisible but imperturbable that extends stability to the universe. Mathematically described, regular patterns do seem to feature that kind of hidden stability. The dream is that one day it will be possible to unify the laws to understand the big picture of nature, which seems designed as a puzzle made of those pieces we keep finding.

Scientific inquiry has found parts of what it is looking for, but not because that is the only option—rather, because it is the only sort of option that that mode of inquiry has been willing and able to find.

PART 3

Where Do We Stand Today?

Prelude: Do We Have a Theory of Everything?

From the perspective of material progress, science has been extraordinarily effective. We can predict and control many natural phenomena. Science has made lifesaving discoveries such as antibiotics and has sent people to the moon. We have everything from automobiles to computers, from effortless lights to devices that allow us to speak to people around the world without so much as a connecting wire. In many respects, technology's usefulness seems to validate science's Puzzle Hypothesis even further.

But the Puzzle Hypothesis was supposed to do more than that. It was supposed to reveal to us the nature of nature.

Has it?

The answer, in short, is no. We still do not know what reality *is*. There is no unified Theory of Everything based on the Puzzle Hypothesis.

It was Albert Einstein who set in motion the modern hunt for a Theory of Everything. Einstein famously spent the last 30 years of his life looking for something—a law, a theory—that would unify the parts. Matter, motion, space, and time, in the form of laws of physics, seemed somehow ripe for being put together as a single, mathematical equation. Leon Lederman described this desired fruit—something simple and clear—as a formula you could put on a T-shirt.

Today, in place of a single Theory of Everything, we have a number of different opinions about where we stand with regard to such a theory. Some say it is right around the corner. At the other end of the spectrum, scientists such as Nobel laureate physicist Robert B. Laughlin reject even the possibility of ever weaving together a coherent image of nature from its details.

What goes without saying is that there is no known blueprint that accounts for what nature is, how it works, and its relative degrees of order and disorder. So we are still at square one in that respect.

An impenetrable irony infiltrates even our grandest dreams of a unified theory. When scientists speak of a single Theory of Everything, they refer to a mathematical expression: a law. The mathematics are a tenable description of nature because we are certain nature is designed as a puzzle, and so the patterns found in the actual, physical puzzle of nature can be abstracted out and reduced mathematically. Nature itself, the physical reality we live in, is something other than the mathematical treatment *of* it. Thus persists the inescapable irony that even if scientists were to find the unifying theory for which they search, it would still stand infinitely separate from the reality of the nature we experience. They cannot be put together. The ultimate unifying theory, science style, will necessarily fail to unify the rules of nature with nature itself.

While this state of affairs is so familiar it may seem difficult to understand that it could be otherwise, the fact remains, as I suggested before, that nature could be viewed in a different manner, not as a puzzle made of parts but as an inherently continuous "one" phenomenon. In his book *The Dreams of Reason*, physicist Heinz R. Pagels says, "The transcendental and the natural viewpoints are two complementary representatives of a unitary reality, the full expression of which is still elusive. . . . Part of the answer, I believe, will turn out to be that we are asking the wrong questions, making a false distinction between the transcendent and the natural world."[1] That dichotomy to which Pagels refers is the dead-on result of adopting the Puzzle Hypothesis.

A unitary understanding of nature—the "oneness" to which I allude—would mark so profound a shift that Pagels concludes: "But to see that *that* is the answer would be quite an accomplishment. And when that day comes, it will transform our civilization."[2]

IF WE DO NOT have a final law, where, at least, do we stand with answering the four basic questions about nature? What is it? How does it work? What holds everything together? Why do things fall apart?

Good answers to these questions would indicate that science is on the right track to identifying the nature of nature, which is what the Puzzle Hypothesis would lead us to expect. In the following three chapters, I will discuss the extent to which these expectations have been met.

This exercise will allow us to evaluate whether science and the scientific method substantiate, in the most basic way, the reason we began using them in the first place. Do the findings we've gotten by using the scientific method justify its claim to being the best and only way to study and understand nature itself?

This delicate question orients us toward the original good reasons we adopted the Puzzle Hypothesis and then science. Science is a tool we began using to search for truth and understanding—according to what made sense to us, according to how nature appeared to be. It was our own reasoning which compelled us to embrace, cradle, and nurture it so it could in turn serve us.

Human reason told us that science is the right way to look for truth, and I propose that, also according to reason, the criteria science sets up for seeking truth cannot be used to validate that science is the proper way to look for it. You can't do a scientific experiment to test whether such an experiment is a good fit with nature. It would put the cart before the horse to use science to determine whether the Puzzle Hypothesis is true to nature, because the Puzzle Hypothesis is the premise behind science. We will have to go with our thinking—which was what led us to trust science in the first place.

We will use reason to evaluate whether science is still the most reasonable, or only, option for understanding the reality that nature is.

CHAPTER 6

Where Do We Stand Today with Regard to Matter?

IN THIS CHAPTER, I will dive right into an assessment of where we stand today in our knowledge of what matter is by pointing out a rather remarkable fact: In a variety of ways, science has come to view matter as being dematerialized and not discrete particles at all. In other words, scientists no longer understand and treat particles as the physical stuff they once were considered to be.

The dematerialization of matter emerges against the backdrop of a peculiar phenomenon that I now bring to your attention, which is that the atomic hypothesis has never been proven. Atoms were theoretically validated by Einstein in 1905 and later experimentally confirmed by French physicist Jean Baptiste Perrin, but it was only through circular reasoning that science used them to confirm the atomic hypothesis.

Science had for years worked with nature as if it was built of tiny parts but was actually awaiting information that would validate this approach. Perrin won the 1926 Nobel Prize in physics for this "work on the discontinuous structure of matter," as his citation reads.[1] Yet the presenter of his Nobel Prize congratulated him on a different aspect: work whose aim was to "put a definite end to the long struggle regarding the real existence of molecules."[2] When the existence of atoms was scientifically confirmed, it automatically was taken to confirm the idea that nature is built *of* them.

The atom Einstein and Perrin had found was not an *atomos*, an uncuttable particle. It was, however, close enough to what science already believed that it was shepherded in as a confirmation of the system that predicted it.

No one has considered the possibility that atoms exist in some form yet nature is not built of them—that they do exist but the nature of nature might be other than a puzzle. I reiterate this possibility in the spirit of retaining the other choice of understanding nature: It is possible that atoms exist, but not as *atomos*, and that nature is not built of them—because nature is not designed as a whole built of parts.

ONE WAY IN WHICH matter has become dematerialized in recent thinking is that physicists today treat and work with particles as abstract mathematical quantities of energy. This is well illustrated in the famous equation $e=mc^2$, an equation that relates mass and energy.* It means that "mass is energy and energy has mass," as Paul Davies puts it in his book *Cosmic Jackpot*.[3] Though it may astound those not familiar with particle physics (or, equally interestingly, attract little attention from those well versed in it), the mass is dematerialized—truly not material in an everyday way— and is treated instead as a mathematical abstraction; it is easily swapped out in equations for a sum of energy divided by the speed of light squared. As Heinz R. Pagels states in his book *The Cosmic Code*, "The visible world is neither matter nor spirit but the invisible organization of energy."[4]

I think that in a general way, people may have grown comfortable with the idea that matter is energy, so let me emphasize here that the popular term *energy* is nothing more than a mathematical model. In no way is it palpable like matter is. The scientific world knows and accepts this, as Paul Davies and Julian R. Brown explain in the book *The Ghost in the Atom*:

> *Many of the purely abstract, mathematical concepts employed become so familiar that they assume a spurious air of reality in their own right . . . energy for example. Energy is a purely abstract quantity, introduced into physics as a useful model with which we can short-cut complex calculations. You cannot see or touch energy, yet the word is now so much part of daily conversation that people think of energy as a tangible entity with an existence of its own. In reality, energy is merely part of a set of mathematical relationships that connect together observations of mechanical processes in a simple way.*[5]

In the most blunt terms, matter, which had been understood to be the solid stuff of nature, has been reduced and dematerialized into sets of useful mathematical equations.

Science thought up the Puzzle Hypothesis eons ago because our senses told us matter had boundaries, and now, amazingly, science retains the boundaries even

---

* The letters stand for: energy (e) is equivalent to mass (m) times the speed of light squared (c^2).

without the matter. The concatenation of the Puzzle Hypothesis with the atomic hypothesis has progressed so far, and has such a grip on us, that as long as we can calculate changes effectively, we do not care if the matter itself no longer has the mass that started us thinking that way.

And yet, even if this fact does not bother physicists, I am suggesting that we should at least recognize that it negates the reason people, including, of course, scientists, began treating nature as a whole made of boundaried parts in the first place—because it made sense.

The dematerialization of matter doesn't stop there. It also stands center stage in one of the most shocking observations of particle physics, as revealed by the famous double-slit experiment. Under certain experimental conditions, particles appear to behave—indeed *do* behave—as if they are waves. The particle is a wave; the wave is a particle. This phenomenon, known as wave-particle duality, effectively dematerializes matter. Waves are not material in any way known to science, and yet, matter is, at times, waves nevertheless. It is one of the reasons quantum phenomena are sometimes described with the phrase "quantum weirdness."

The weirdness arises, plainly, because particles and waves are not compatible in a way that would help us easily grasp their relationship. How can something that was shot from a particle gun as a particle end up on a detector in a way that indicates it has traveled as a wave? How does the outcome change depending on whether or not someone is watching? For without getting into the experimental findings, which are not necessary to get into for this discussion, we know that this occurs.

Some physicists broach this duality by saying the particle has a dual nature, which changes depending on whether or not one observes or measures it. Niels Bohr called this coexistence of particle and wave characteristics complementarity, a term that restates the paradox rather than resolving it. No matter what one calls it, the fact remains that two seemingly incompatible phenomena ("waveness" and "particleness") present as one, and no one understands how this can be. Yet so it is. Matter and waves have seemed different since the beginning of time yet somehow join together at the bottom of it all.

Less obvious but also contributing to quantum weirdness is the fact that matter has always been understood to be discrete while waves have been understood to spread out. Fencing in waves as discrete entities originally put scientists in an uncomfortable bind: It is mathematically feasible to do so and the equations work, but no wave in anyone's experience is discrete in that way. This is why Planck called the quantum an "act of desperation" when he introduced it to explain the waves of blackbody electromagnetic radiation. Never before had energy waves been confined to discrete packets.

It remains difficult to conceptualize what a discrete wave even means. Nevertheless, because the equations work, scientists accept that the wave packet is the alter ego of every subatomic particle.* It is simply another bizarre but tolerated aspect of today's idea of "what matter is."

A few years after Planck showed energy waves can be emitted as quanta, Einstein went further and showed that light not only is emitted but actually exists as either a particle or a wave. Louis de Broglie then demonstrated that all material particles, such as electrons and protons, and even atoms themselves, can be understood as waves, as well. Energy, light, matter, waves—the interrelationships were staggering. Planck's equation e=hv meant waves at the quantum level are equated with energy, just as Einstein's e=mc^2 would later equate matter with energy.† All this opened a Pandora's box. Mathematical equivalence between light, matter, and waves of energy meant each type could switch into another. This entailed, among other things, that quantum particle waves are dematerialized energy, as well.

Yet the underlying relationship between waves and energy also has yet to be explained by science.

Strange or not, quantum physics does exhibit the clear mark of originating from the Puzzle Hypothesis in that it is 100 percent consistent with other disciplines across the board. Discussing atomic physics is just like discussing cell biology or even, as we will soon see, singularities in space like black holes or the big bang. Everything in science is treated in the same way: as an entity with boundaries.

It is amazing that the dematerialization of matter has not undone the Puzzle Hypothesis in the most elemental of ways. Science still *treats* everything as discontinuous. Regardless of what anyone thinks they "actually" are, all natural phenomena are treated as boundaried systems.

"What is it?" It's hard to say. We keep the boundaries we originally imposed, but matter itself appears to be immaterial.

So perhaps, one might reason, we may have lost our sense of the stuff of matter but at least we have our ultimate particle. However, this is not the case. Physicists have unearthed instead a superabundance of subatomic particles. That is what we have found in lieu of a single, ultimate particle. "What is it?" has no single answer. The subatomic world has been dubbed a particle zoo.

The particle zoo is a unique exhibition, far from what anyone dreaming of an *atomos* might have imagined. To begin, it houses a variety of particles that may or may not have mass. And the physicists developing the current general theory of particle physics, the Standard Model, look not only at material particles but also at forces in the environment as being discrete. Forces are understood as particles, too, adding to the menagerie.

That is to say, the invisible forces that are recognized as the four forces of nature,

---

* The wave packet was also called a wavicle to indicate that it is boundaried like a particle.

† This equation states that energy (e) equals Planck's constant (h) times the wave frequency (v); it is a mathematical way to convert a measurement of a wave into energy.

mediating all attraction and repulsion—gravity, electromagnetism, the strong force, and the weak force—are understood to operate in the form of independent particles. These force particles, called bosons, stand on equal footing with actual particles of mass (such as fermions, which are divided into quarks and leptons and make up matter, and other particles, as well). Yet these particles generally do not have mass. What renders them particles is the way scientists describe and work with them: with discrete boundaries. Scientists also work with "virtual particles," which are said to come in and out of existence in force fields.

These numerous, often immaterial, particles have been organized by physicists into a sort of order. The Standard Model is neither complete nor final, however. Not all species of the particle zoo fit neatly, and there are still particles that cannot be found. The graviton, for example, which is supposedly responsible for the force of gravity, eludes detection.

This is not meant to sound pessimistic; it is simply what scientists have found when probing nature in search of discrete invariants. In his book *The God Particle*, Leon Lederman says, "In the past few decades in particle physics, we have been in a period of such curious intellectual stress that the parable of the Tower of Babel seems appropriate. Particle physicists have been using their giant accelerators to dissect the parts and processes of the universe." Lederman continues with the assessment that "the issue is whether physicists will be confounded by this puzzle or whether, in contrast to the unhappy Babylonians, we will continue to build the tower and, as Einstein put it, 'know the mind of God.'"[6] Scientists persist despite the unsettling superabundance of particles and even though the particles themselves do not provide the answers they'd hoped for.

Neither does the recently discovered Higgs boson resolve the model's problems. Years before it was discovered, Lederman had called the Higgs boson the God Particle. It was hoped that, just like a god, this particle would be the source of mass—and therefore matter, substance, and stability. Great fanfare surrounded its discovery, which was considered definitive support for the Standard Model. But the Higgs boson does not explain what makes the massless photon discrete; it does not explain what makes a particle a wave or a wave a particle; it does not account for what scientists call missing mass or dark matter; it does not deliver the graviton; and it does not account for the existence of the Standard Model as it is.

A final example of matter's dematerialization in the world of science is the black hole. Black holes are, according to theory, formed when a star collapses on itself. The gravitational pull of this condensed mass is so strong that nothing, not even light, can escape. Of course, no one has ever actually seen a black hole, just other effects that would indicate something is there.

Scientists describe black holes mathematically—not as material chunks of blackened star but in the abstract, as a point with no dimensions, known as a singularity.

(A similar singularity is considered to have existed at the origin of the universe, known as the beginning of the big bang.) That material stars dematerialize into singular mathematical points again raises no alarm bells. Rather, this feature, strange as it is, is readily accepted because it has the character of every other finding of the scientific method. It fits with mathematical constructs we use as a result of our Puzzle Hypothesis.

The bizarre reality is that our concept of matter has set us up with a style of answer that in the end leaves us without matter.

And, even more, it also leaves us wanting for answers. In fields across the board, it has generated mysteries it itself cannot solve.

We cannot always find matter where our calculations tell us we should, for example. The mass that we see as "galaxies" does not exert enough gravity to account for the stability of those galaxies. They are spinning so fast that they should, according to all known laws of physics, fly apart.

To compensate for this startling fact, scientists say there must be missing mass, also known as dark matter. This invisible material would assert the necessary gravity to hold the galaxies in place. So great is the discrepancy that scientists estimate that 23 percent of mass is missing, and what we see and measure is only 4 percent of what exists. The remaining 73 percent is thought to be dark energy, the supposed cause of the universe's spreading out; it is also undiscovered. Thus 96 percent of the universe is thought to be undetected.

The absence of expected mass in the galaxies surely marks a bewildering contrast to the superabundance of subatomic particles that we were not expecting to find.

The principles of the atomic hypothesis continue to be applied in the face of the dematerialization of matter in hopes of discovering "what it is." Beyond the quantum, into depths of the world we cannot practically reach, the strings of string theory are a proposal for what might account for matter's existence. The theory posits the existence of loop vibrations of strings in multiple dimensions that "somehow" make particles. Far beyond the reaches of scientific testing, string theory pioneers different ways to extend the same model. Its concepts derive from mathematical structures that are meant to account for real occurrences, but the point here is the mind-set that allows for its proposals—and the outcome of doing so. In this theory, matter is even further dematerialized into discrete waves.

WE EXPERIENCE MATTER on the macro scale in our daily lives, of course, and our desire to understand its nature twines intimately with our desire to live and thrive. To understand "what it is" with regard to the human body, we have used the same model and the same techniques—and have come up with many of the same outcomes.

Medical strategy overtly adopts the atomic worldview. Researchers target discrete marks under the assumption that controlling or killing wayward molecules and genes will eradicate disease.* People do this because everything is understood to be built of parts—so it makes sense to presume that those localized parts are responsible for problems that plague us. The model supports itself by pointing not only to the atomic hypothesis but also to the design of the computer universe, in which knowledge is built from morsels of information (i.e., what physicist John Archibald Wheeler called "it from bit").

The realm of health and disease also offers a large-scale counterpart to the particle zoo. The gene is parallel to the particle in our understanding of living organisms, medicine, and biology. Scientists believe many different genes are responsible for different diseases, just as many particles are responsible for matter. Thousands of genes have been identified and associated with disease. We have encountered what you might call a disease-gene zoo, just as we have encountered a particle zoo, as we follow our map for a mechanistic explanation of biology and life.

THIS, THEN, is the current state of our answer to the "what is it?" question about matter. The simple atomic principle of Democritus, which would have elegantly dispensed with the mysteries of matter, has instead led us to dig up an unruly number of evermore minute particles on a road riddled with deep questions. We work with a model bereft of stalwart blocks of *atomos* matter. No solid unit remains when our analysis is done. Matter in science's current model dematerializes into waves and energy and combines with equally immaterial force particles and virtual particles to "build," as it were, the universe.

It is generally accepted that the current flavor of our essential understanding is not something at all familiar. We have ended up with what has been called the uncommon sense of and unnatural nature of science. The book by developmental biologist Lewis Wolpert, titled *The Unnatural Nature of Science*, addresses the fact that science has been turned into a different type of thinking than what is natural to most people. Wolpert celebrates science's piece-by-piece approach and subsequent astonishing findings. In this discussion, I suggest that the unnatural quality of science actually contradicts the original reason we adopted the premise that underlies science itself. We had thought about nature as a whole made of parts because it seemed natural, and now our findings about the parts do not appear to be natural.

We no longer know what is real. Niels Bohr's assistant Aage Petersen wrote that Bohr would answer questions about the underlying quantum world by saying:

---

* Once again, this model may be so familiar that an alternative is hard to imagine. One easy possibility to imagine, even within the framework of the Puzzle Hypothesis, is that the dynamic between parts—rather than a specific part itself—is the source of problems.

"There is no quantum world. There is only an abstract quantum physical description. It is wrong to think that the task of physics is to find out how nature *is*. Physics concerns what we can say *about* nature."[7] That is quite an about-face from why we started speaking the language of science in the first place: as the method of finding and piecing together parts of the whole, with precisely the intent to show "how nature is." It is hard to say which is more bewildering: that we have backed into a position where we abandon the natural material world we experience in favor of a mathematical reality, or that scientists have had to accept this unnaturalness for want of a better choice.

In 1954, in his Nobel Lecture, physicist Max Born addressed the very root of the problem.

> *To come now to the last point: can we call something with which the concepts of position and motion cannot be associated in the usual way, a thing, or a particle? And if not, what is the reality which our theory has been invented to describe?*
>
> *The answer to this is no longer physics, but philosophy. . . . I will only say that I am emphatically for the retention of the particle idea. Naturally, it is necessary to redefine what is meant. For this, well-developed concepts are available which appear in mathematics under the name of invariants in transformations.*
>
> *The latest research on nuclei and elementary particles has led us, however, to limits beyond which this system of concepts itself does not appear to suffice. The lesson to be learned . . . [is] that somewhere in our doctrine is hidden a concept, unjustified by experience, which we must eliminate to open up the road.*[8, 9]

Until such time, however, we are stranded with the mathematical methods of understanding the particle.

The simple hope of understanding what matter is—the "what is it?" of our most rudimentary questions—has thus run into a dead end. The physical reality of matter has ceased to be as significant as mathematical expressions of energy are. Science can calculate and predict, but it cannot answer the original, simple question. What is it?

Science does not know.

CHAPTER 7

Where Do We Stand Today with Regard to Motion?

RECALL THAT GALILEO'S PROPOSAL was that the true nature of motion, the reality just beyond our experience, is that any and all motion fundamentally moves at a uniform velocity in an idealized straight line forever, and only changes when influenced by some outside force. Science moved forward on that premise, that straight lines are the backbone of natural change. Linear motion seemed as basic to motion as the atom is to matter.

In the previous chapter, I asked whether science's observations of nature confirmed the expectation that it is constructed from recognizable building blocks (particles). You saw that that expectation went unmet. In this chapter I will ask a similar question of motion: To what degree have straight lines been observed to be the backbone of natural change? Do we now know "how it works"?

This question is critical because studies of motion incorporate a bias: Motion is as important for daily life as matter is, yet we do not think of it in its own right. Namely, we always think of motion in a particular way: as "matter in motion." There's always a "thing" that moves. People never talk about motion alone, or motion in relation to other motion, without reference to matter.* Motion is, moreover, treated as if it has matter's characteristics and in segments that relate to matter.

That we fundamentally accept dividing motion the way we divide matter is well illustrated by Zeno's paradoxes, the famous questions first stated by Zeno of Elea in ancient Greece. A popular version of the paradoxes—which are all about motion— asks how one ever gets from point A to point B. To travel from A to B, you will travel

---

* For example, if a person would discuss a beautiful pass in American football, even when describing the perfect arc and spin, it would be *of* the ball.

halfway from A to B first. But to travel halfway, you must first travel a quarter of the way, and to travel a quarter of the way you must travel an eighth first, and so on, such that you could not take the smallest step without dividing it first, and would never move at all.* Of course we readily set these conceptual paradoxes aside in favor of experience; we all know motion does happen. Yet the problems tickle our sense of ambiguity. I want to draw your attention not to a solution† but to our fundamental regard for them: On first introduction, Zeno's paradoxes impart a profound sense of surprise, wonder, and mystery.

This sense arises because we implicitly accept that motion takes the design of a puzzle in the same way matter does. When it does not conform, we are surprised. Said another way: Zeno's paradoxes perplex us because we take for granted that we can treat motion the way we treat matter. Matter seemed to be the stuff of the universe in a way motion could never match, so we projected onto motion (and the rest of the universe) the qualities we perceived in matter. The idea that motion might exist as a phenomenon in its own right, with distinctive qualities different from those our senses tell us are the qualities of matter, has not been considered.

Science cannot consider it, in fact, because of its founding hypothesis—this is the bias I mentioned on the opposite page. The precondition of discontinuity will always prevent us from looking at motion in relation to other simultaneous motion. Every idea we have about motion relies on discontinuity. And because discontinuity requires invariance in order to characterize each part of nature, it prompted Galileo to "discover" straight-line motion.

Science has not questioned whether Galileo, in his formulation of the nature of motion, was homing in on a true understanding of it. Prediction now fills the slot that used to be reserved for a genuine sense of reality. It was an either/or choice at the time the scientific revolution was beginning: Stick with reality as we experience it, or deviate that "hair's breadth" from it and know the future. There was no reason at the time to think the ability to predict motion would ever come to clash with understanding motion itself. Newton proposed his first law of motion, indicating that mathematics is the language of nature, and science as we know it took off.

As time passed and science progressed, people began to observe curves—the antithesis of lines—in their research. Yet it did not dissuade them from treating motion as if it happens on abstract, mathematical lines. And at the same time, something odd began to happen. Just as matter "dematerialized" into energy as we tried

---

* Another incarnation of Zeno's paradox regards the motion of an arrow in flight: At any given sliver of a moment, the arrow must be in a certain location in space, which makes it the same as being at rest in that spot, yet the same is true of the next instant, and the next. If at any given instant, an object is at rest, the existence of motion is impossible.

† Mathematicians deal with these sorts of problems by treating them as sums of infinite series; that approach, however, does not eliminate the conceptual paradox. It addresses how to calculate the problem mathematically, whereas Zeno meant to demonstrate that the one and the many are irreconcilably different. That is to say, for the mathematician, being able to calculate is a form of understanding, as is often the case with modern science. Many philosophers of science feel the paradoxes continue to present valid metaphysical problems.

to understand "what it is," motion has done what you might inelegantly call "de-motion-alized" into energy as we tried to understand "how it works." Just as dematerialization means that matter itself is no longer matter as we would recognize it, demotionalization means motion is no longer motion as we would recognize it.

Fundamental motion is no longer treated as objects moving through space and time. The toss of a softball and flight of a hawk have given way to energy at the bottom line: A physicist would say there is potential energy, kinetic energy, and so forth, each a quantity that changes freely into the others through the phrases of math.

It is a strange reality that all change is understood by science to be an exchange of energy. Energy is defined as the capacity to do work. It is a mathematical phenomenon, based on lines, which is calculated—through direct linear transformations—to represent the changes we experience, thus shedding no light on our original question about "how it works."

Though we treat energy as that something that causes motion or change in a closed system, energy is not some "thing," or any "thing" at all. Even as it forms the cornerstone of our explanations of motion, we do not, as I discussed in the previous chapter on matter, know what energy is. We have embraced the model so deeply that it is commonly said that energy drives the universe. Yet what that energy is—a definition or even description of its nature—no one can say.

Energy presents what is known as an operational definition. Instead of revealing the underlying nature of the part of the puzzle called motion, it simply describes tests through which we may measure units of change. We know what happens from (our idea of) energy, in the form of a mathematical equation. We know the measurements in which it does it. But we never know what it is. What energy actually describes is hidden.

And the fact that we now rely on operational definitions, peculiar as it is, is not restricted to motion and matter. Other parts of nature have fallen to the same fate. We operationally define the space and time through which matter moves.

In his paper on special relativity, Albert Einstein spells out exactly what he means by time and space. Time, he carefully explains, is an expression of how we measure it. He devotes paragraphs to state exactly how one properly measures time—because time is defined as what one measures with a clock. Space is similarly treated, with its definition being what one measures with a measuring rod.[*1] I would like to point out that Einstein takes care with his language specifically because no one knows what time or space *is*. All we can do is measure them: measure them along the straight lines that seem to uphold the clock's and the measuring rod's progressions. Our operational definitions are based on straight lines but

---

* Galileo led us to envision a timeline of evenly spaced intervals, for he was the first to construe a line within regular waves of nature—a swinging pendulum and his pulse—as I mentioned in Chapter 4. The straight lines of space were Descartes's contribution.

correspond to nothing we can identify. The Puzzle Hypothesis told us space and time are the necessary partners to matter and motion, but we no longer know what they are either.

Where, then, are we today with understanding matter in motion—with knowing how "it" "works"? We do not know what the matter that moves is; it has dematerialized into energy. We do not know what motion is; it has demotionalized into energy. And we do not know what the space and time through which motion occurs are; we define them by their linear measurements. Somehow, matter and motion—two parts of nature that had seemed worlds apart—both tap into, or emerge from, this phenomenon we've discovered called energy. Energy may hold all the answers—but energy is of an unknown nature, defined only by the mathematical, linear measurements through which we work with it.

If we could understand what makes energy be what it is, we might find sweeping answers to our questions. Such an understanding of energy would have to come from a different framework, however. Science's approach committed us to finding the answers we have gotten.

The way science treats energy has contributed to the current state of affairs in another way, too. Scientists embrace energy as the cause of motion: It is defined as the capacity to do work. But it has gone unnoted that we have allowed ourselves to understand energy only within a closed system. That choice restricted our potential to understand.

That is to say, of all the motion in the universe, our paradigm for motion has been the steam engine of the Industrial Revolution. We set out to get a handle on the work of that closed system rather than the natural motion that abounds in the universe. It was no coincidence that the word *energy* comes from the Greek *en ergon,* to cause work: Our focus on motion was on the work we could extract from that machine.* Mechanical work, and never the lively realities of a running child or a beating heart, was our choice example for understanding all motion of the world. Energy as a concept came from that closed system, of human creation.

It is, as ever, possible that there is a way to understand how work and motion are generated that is not on the basis of a closed system. It may be that a continuity accounts for motion—not something to be found in a so-called open system, which is a closed system whose boundaries allow interaction with the environment, but rather, some form of motion that is truly not closed or divided. This possibility is

---

* Our perspective has even created an odd glitch in the operational definition we accepted for energy. We call energy a capacity to do work. Yet everyone agrees that no transformation of energy is 100 percent toward work; energy is overwhelmingly directed toward waste. (This is the second law of thermodynamics.) We sought that which creates motion and change, but the closed system in which we studied it left us a majority that cannot cause motion or change.

We could just as easily, and perhaps more properly, define energy as that which has the potential to go toward work or waste (though that more precise definition would not be any more enlightening). The weighted definition we use simply confirms the outsize role of the steam engine in our thinking.

not entertained, however. If motion somehow emanates from a reality that is not and cannot be confined to a closed system, our models will neither detect nor explain it.

The demotionalization of motion into energy puts an awkward emphasis on how we have come to understand nature. We are sure nature is a puzzle and continue to explain how the puzzle works, yet we find ourselves unable to grant the same level of recognition and definition to its parts—matter, motion, space, and time. Some scientists try to brush this away by saying we do not need to know "what is" as long as our equations work. But they fail to recognize that they do indeed rely on "what is": They believe and work with the idea that nature is a puzzle. We are settled with the idea that nature is a whole made of parts, but even if we can calculate many mathematical relationships between those parts, not having a real identity for them affronts that way of understanding.

ONE MIGHT HOPE that the motion of fundamental particles of the quantum would shed light on the nature of motion throughout the universe. The most elemental motion ought to instruct us on the big picture.

Yet quantum motion does not answer "how it works." It corresponds to nothing in our experience.

On the most basic level, quantum motion is discussed in a way where we can calculate it without even being able to identify it. In this way, it is like the operational definitions we use for energy, space, and time. It is treated through mathematics alone.

This less-than-ideal situation arose because quantum motion is wholly unlike the motion we originally studied on the macro scale. People had, from the time of Democritus until the beginning of quantum physics, innocently envisioned atoms as stalwart pellets. Their motion was not supposed to differ from the motion of large-scale objects. If we shot them from a gun, we thought they would move like tiny bullets whizzing through the air.

But the famous double-slit experiment I mentioned in the previous chapter, and its many variations, conducted with particles such as the electron, show unflinchingly that quantum motion cannot be reconciled with our most basic ideas of natural motion. Particles shot out of a particle gun and through a barrier with two small slits arrive at their destination in an interference pattern—a pattern made only by waves—thanks to wave-particle duality. (The detector reads them as particles when they land.)*

And more, and if possible more astonishingly, the pattern happens even if just

---

* If electrons were the thoroughly solid particles we'd imagined, the pattern would feature two collections of spreading circles like bullets would make, centered near the holes they'd passed through. The rippled pattern that appears instead demonstrates that somehow the electrons have the nature of waves; waves that cross one another, like ripples on a pond, canceling and reinforcing each other as they cross, creating this type of striated pattern.

one electron is shot at a time. With a long delay between each firing, one would think that each particle would simply travel as a bullet would. Yet the interference pattern accrues over time. Each electron seems to "know" where it can and cannot land to create the final, collective pattern, the result of interacting waves. This means that waves somehow interfere with each other over time, even when electrons travel alone.*

Adding to this bizarre situation is that physicists cannot observe what is happening because observation (i.e., measurement) changes what happens. When the particle is measured, the interference pattern on the detector vanishes. All particles land as if they'd traveled as bullets. Only when unobserved does it switch—however it does so—to a wave.

The evolution of the quantum concept of motion was progressive. Early on, Niels Bohr had explained the hydrogen emission spectrum with the daring idea that electrons can only occupy certain energy levels. This meant they must "jump" from one location to the next, rather than travel in between. Then, physicists refined Bohr's model and dared to eliminate the orbits he had suggested while retaining the jumps the electrons must make. German theoretical physicist Werner Heisenberg cemented the chasm between familiar motion and quantum motion with his famous uncertainty principle, through which he demonstrated that it is impossible to measure both the momentum and location of a particle at the same time. (According to this principle, electron behavior is inherently undetermined and becomes fixed only when we observe or measure it; the act of observation or measurement ruins whatever wave characteristics the particle had had.)† Those and similar developments are how we came to the conclusion that electrons move in strange ways: They jump between states instead of traveling smoothly, have no orbits, and prevent us from knowing where they are if we know their momentum, and vice versa.

Suffice it to say, this description in no way resembles our experience of a classical object moving through space and time. It is accepted, nevertheless, because the math works and experiments come out as predicted.

Scientists are not without a way to talk about quantum motion. They can predict what a large group of electrons will do, just never the behavior of only one in the group. That is to say, how each individual electron participates in the big scheme can be stated as a probability: as a mathematical wave of probability, known as a wave function. (We will return to the mathematics of probability below.)

---

* After enough electrons have landed on the detector, the detector shows the striated pattern characteristic of wave interference. So that eliminates the possibility, whatever it would have meant, that electrons interfere with one another; apparently one electron at a time can bring about the final interference pattern, suggesting it somehow interferes with itself or is bound to the travels of the other electrons despite the gap in time.

† Moreover, physicists agree that the uncertainty principle is not due to a deficiency in our mathematical knowledge. More math or better technology will not solve how scientists can know both the location and the momentum of a particle. According to modern physics, it is actually impossible to know both.

Physicists work with particles as having probable paths and momentums, instead of working with them as if they have definite paths of motion. And—somewhat for lack of a better model—many think that that probability wave interferes with itself to create the wave interference pattern. They believe the act of measurement forces the particle to commit to a path, thus collapsing the wave function of its probable behavior and destroying its interference pattern.

If this sounds confusing, don't be dismayed—you're probably right on target. Richard Feynman, who was certainly in a position most likely to understand it, said in *The Character of Physical Law*, "I think I can safely say that nobody understands quantum mechanics."[2] And Heisenberg quotes Bohr as saying, "Those who are not shocked when they first come across quantum theory cannot possibly have understood it."* [3]

All one can say with certainty is that it is accepted in physics that quantum matter and motion somehow relate in a way we had never suspected and that we do not understand in any way other than through calculation.

Physicists are clear about that: Quantum particles do not exist or move in the way we have accustomed ourselves to identifying matter in motion in nature. Amazing though it is, the particles of quantum physics exhibit no set motion or matter in the way we have used those terms throughout our history to refer to a boundaried object moving through space and through time. The way we have a priori understood the universe to exist ceases to apply at the bottommost layers of nature. Somewhat astoundingly, "what is it?" and "how does it work?" not only don't have answers—they lose intelligibility as questions in the quantum context.

HERE'S ANOTHER MYSTERY of motion in the quantum realm: the phenomenon of action at a distance, which I introduced in an earlier chapter when discussing gravity. Instead of being chased away by linear dogma, action at a distance has settled comfortably into documented science.† In other words, a phenomenon that *should not occur* according to science's current model of nature has, in fact, *been observed to occur*. No one knows "how this works" either.

You may recall that Newton fretted about gravity as he formulated it; his law of universal gravitation made it seem that action at a distance is common throughout nature. Yet even with no alternative mechanism for gravity, Newton rejected the possibility of action at a distance. Its nonlocality clashes with the local cause-and-

---

* Renowned physicists' attempts to give meaning to quantum findings may sound to the layperson to be more like science-fiction speculation than physicists talking. Some proposals—mainstream proposals—are that there is no reality, that reality is created when you observe it, and that there are multiple universes. Others simply rely on the mathematics as reality, stating that the probability function—clearly a mathematical construct we use to predict what an electron might do—interferes with itself as an electron passes through the double slit.

† Action at a distance occurs when two or more entities, separated by space, act together, simultaneously, without an intervening message or common cause.

effect mechanisms that Newton was certain drive nature.* The either/or nature of the dilemma—either local cause and effect or nonlocal action at distance—left him with a point-blank choice. He sided with locality.

Centuries later, Einstein finally seemed to abolish what he called the "spooky" action at a distance of gravity with his proposal of curved spacetime. However, he inadvertently created new action-at-a-distance problems in another area.

Because he disliked several aspects of quantum theory, Einstein wanted to negate the claim that it was complete. Action at a distance was his leverage. Together with colleagues Boris Podolsky and Nathan Rosen, he proposed a thought problem that showed that action at a distance was possible within the quantum model.† It is known as the Einstein-Podolsky-Rosen (EPR) paradox. Einstein's intent was to use the patent absurdity of action at a distance, as shown in this paradox, to strike a blow to quantum theory.

Yet physicist John Stewart Bell showed that instead of revealing quantum theory to be wrong, Einstein's outlandish situation could indeed happen. Subsequent tests have actually proven that such nonlocality—action at a distance—exists.

Action at a distance exists even though it is not compatible with our idea of how things come to pass, piece by piece, in a puzzle. It is motion that transpires by some means other than local cause and effect. Yet cause and effect is the only way we have understood motion from our perspective. We have no idea what such motion is all about.

Quantum findings fail to vindicate—and if anything, call into question—our method of understanding nature as a whole made of parts. Science has therefore come to accept that experiment alone is the dictator of truth and reality. "Experiment is the sole judge of scientific 'truth,'" said Feynman, and he put *truth* in quotes because there is no truth outside experiment, as far as modern science is concerned.[4] This is the mainstream view. Whether or not it makes sense, we must accept the outcome of experiment. We are instructed by quantum physics' greatest practitioners that we should not even try to understand further—that, as Bohr reportedly said, "it is wrong to think that the task of physics is to find out how nature is."

I would like to suggest that it is paradoxical that we entrust our core understanding of nature to the findings of experiment even when it defies our sense of reality. It was our natural sense that experiment *matches* the reality of nature that led us to trust it in the first place.

---

* If nature is built up from the bottom in the design of a puzzle—and science has ever trusted that it is—then individual invariant pieces must interact locally; things affecting one another without signals carried over space and time would eliminate cause and effect, leaving us with an entangled mess.

† The EPR thought problem states that according to quantum physics, if one would devise a quantum system whose parts are separated from one another in distant space, the effect would be action at a distance: that when you measure one part of a quantum system, which causes certain effects on the system, the effect would, by quantum rules, in the same instant appear in the other, distant part. There would be no time lapse in which a signal could travel.

ONE OF THE MOST popular arguments in favor of quantum mechanics is that the math works. This fact brings us to another round of assessing where we are today—to get a good look at how we rely on mathematics to understand motion.

Despite the confusing findings on the quantum level, we have not changed course—and have not had to—because mathematics spans the gap between the Puzzle Hypothesis and the wayward findings it has led us to. Mathematics is a language of discreteness: It speaks of relationships and connections between separate things, where the "things" are certain abstractions. Abstractions enable math to address the strange quantum world with the familiar model of a puzzle. It enables us to put edges and boundaries around amorphous phenomena—to treat entities in the quantum world as if they have edges and boundaries, even when natural experience does not indicate that a boundaried object is there.

If the math works, and it does, then that is taken as an understanding of nature. When it comes down to it, scientists choose to stick with math (and thereby the Puzzle Hypothesis) over natural experience, even if the commitment presents a tough, if not impossible, intellectual challenge. Max Planck reaffirmed this choice, to great effect, over a century ago.

You may recall that Planck had significant personal hesitations about applying mathematics to motion when there was no discrete entity for the math to refer to. He called it an "act of desperation." Yet even though he did not like it, and even though it referred to nothing people can imagine or experience, Planck went ahead and bundled the waves emitted by a blackbody into discrete mathematical packets and measured them as linear frequencies. The equations worked, and so the puzzle model was upheld.

Wave motion in the quantum was suddenly forced to depart from what we think of when we think of waves. Until that point, waves had always been a forward flow of unbroken, undulating motion. Scientists do not know how there is a disconnect between waves moving forward along a linear axis on the quantum level. Physicists proceeded with Planck's approach anyway. It helped them describe and predict nature as a whole made of parts, albeit mathematical ones.

AND NOW WE TURN to the mathematics itself—what we use to address "how it works."

Newton's laws were the first to describe regularities in nature mathematically. At the time, they gave hope that we would eventually understand all of nature's motion through similar equations. We seemed on the brink of spelling out "how it works" through mathematically describable, local, linear cause and effect. But further work on motion overtly showed us, many times over, that we do not live in a clockwork universe.

Local and linear causality did empower us remarkably, of course. One can only imagine Newton's wonderment if he knew his laws of motion allowed us to send human beings to the moon whose gravity he'd pondered. But Newton would likely have been equally, if not more, dismayed by the limitations to what we can calculate. Even as James Clerk Maxwell, Joseph Fourier, and other 19th-century scientists successfully formulated Newtonian-style equations, they found they could not make linear equations for anything with more than two bodies interacting at a time. Beyond that was complexity.

Nevertheless, the laws achieved something further, which has had lasting impact: They convinced us we ought to be able to predict everything with mathematics. Mathematics fleshed out connections between parts, even if the parts themselves were receding from our understanding. As long as people could predict mathematically, we felt our understanding improved. The failures of classical mechanics in the face of complexity simply left us feeling we must find a different mathematical solution to the question of "how it works."

And here's the clincher. We do have a different mathematical solution today. While linearity was indeed the regularity scientists preferred to use for predicting motion, when it could not be found, we continued searching for some other pattern that would allow us to predict—and found a curve, namely the bell curve of statistics.

The bell curve describes what appears to be a secret order in nature, an order that characterizes large numbers of anything and everything. Things often happen in this pattern instead of being random or evenly spread out. For instance, if you were to measure and plot the height of every individual in a town (meaning, not one or two, but a large number of people), it would create a curve. At the top of the bell-shaped curve, you would find an "average" height that is most common, with fewer people shorter or taller than average falling on the left and right slopes of the bell. Bell curves are extraordinarily common in nature, describing everything from human life expectancy to global temperatures to the average return on stock market investments. The phenomenon tells us something about the nature of nature. It is a regularity in nature, a consistent pattern, that we know with certitude will happen. Instead of a straight line, which the current model of nature says will guide us in understanding, things happen in accordance with this simple but elegant curve, helping us anticipate the future.

The bell curve is found everywhere yet is not a particularly good fit with the way science has understood nature as a puzzle. Its broad-sweeping curviness little resembles the local, mathematically describable interactions through which Newton's laws indicated we might tame nature. It shows that something happens with the way things move—something that infuses a pattern, of a curve, in the way things work. We can reconcile this something with our approach to a certain extent, yet we can never truly get it to achieve harmony with our other strategies.

There are barely words to convey what an unusually remarkable phenomenon the bell curve represents. It transcends any specific place and time. What we draw on paper does not do it justice: The drawing represents real change, in a pattern, that happens no matter the objects, location, or events that transpire. So long as they are similar enough to compare, and so long as there are enough instances of them, a bell curve pattern emerges. It is the definitive pattern of group motion.

Before statistics came to describe nature, when all that existed was the Newtonian perspective, large numbers had been indescribable and chaotic. Polymath Sir Francis Galton, in his book *Natural Inheritance*, published in 1889, explained the scientists' delight in the order the bell-shaped curve revealed.

I know of scarcely anything so apt to impress the imagination as the wonderful form of cosmic order expressed by the "Law of Frequency of Error" [i.e., the bell-shaped curve]. . . . It reigns with serenity and in complete self-effacement amidst the wildest confusion. The huger the mob, and the greater the apparent anarchy, the more perfect is its sway. It is the supreme law of Unreason. Whenever a large sample of chaotic elements are taken in hand and marshalled in the order of their magnitude, an unsuspected and most beautiful form of regularity proves to have been latent all along. The tops of the marshalled row form a flowing curve of invariable proportions; and each element, as it is sorted into place, finds, as it were, a pre-ordained niche, accurately adapted to fit it.[5]

I want to drive home the importance of this approach to understanding "how it works." The beautiful bell-curve pattern is as impressive today as it was in Galton's time. Intuition would suggest that "the huger the mob," as Galton said, the greater the chaos, but in fact, greater numbers sharpen the bell curve. The more players involved, the more the facts fill it in. Twenty thousand coin flips or dice rolls will reveal a more robust bell curve than if only a hundred are recorded. The more instances of motion and change we observe, the more pronounced a wave we see. It works a remarkable order, as if a gentle hand guides the moving members of what seems to be turmoil into an overarching form. It flowingly, eloquently, and forcefully reveals that, even if it is a curve, there is an unmistakable pattern to what large numbers of things do: a curved pattern of motion.

Because it could be calculated, statistics became a new mathematical way to describe how things work. It was of no great consequence that the source of the curve was unknown. Even if it could not show what a Newtonian equation would show—what happens to two things that interact—this new kind of prediction did foretell what will happen for all members of the group, as a whole. There was no denying it accurately described a way things happen, a way motion unfolds. The bell curve immediately stirred the breeze of hope that we will yet make sense of how nature works, even in the absence of deterministic, linear laws.

By offering an alternative explanation for the machine work that powered the Industrial Revolution, even without the mastery offered by Newtonian equations, statistics eventually earned the widespread acceptance it could not earn through mathematical precision. The acceptance came about because people working with the steam engine in the 1800s could not successfully apply Newtonian calculations to the engine's behavior. It was beyond anyone's ability to chart and calculate the paths and collisions of billions and billions of gas molecules (which were as yet unproven but presumed to compose the gas in the engine). Maxwell therefore suggested a possible alternative way to account for the engine's behavior: through statistics.

Austrian physicist Ludwig Boltzmann developed Maxwell's idea. Boltzmann accepted the fact that "how it works" in the natural world is represented by the bell curve. Then he fused that curve with Sadi Carnot's linear idea that the gas, as a whole, within the closed system of the steam engine, goes from hot to cold and never spontaneously reverses. Last, he made a statistical argument. Statistically, gas molecules have many, many ways to end up in the arrangement we perceive as cold (relatively disorganized) and very few for hot (relatively organized). Boltzmann reasoned that that is why the engine runs from hot to cold, because that is the astronomically more likely outcome. He used statistics to "explain" the behavior of the gas.

We can see that this was a curious turn on how to apply the Puzzle Hypothesis and its offshoots. A curve appeared where we were looking for a line, and what did we do? We simply drafted it to serve the linear perspective. When linear calculations did not work and then a wave appeared, people incorporated the wave into the linear approach instead of suspecting that that curve might call the primacy of straight lines into question.

In the realm of mathematics, the defiant qualities of the statistical curve are impressive. Clearly, statistical descriptions of what happens in a drop of water or with molecules in a steam engine differ vastly from the neat determinism of direct cause and effect. As a way to explain nature, statistics represents a tremendous turnaround from deterministic laws. It is undeniably valuable anyway because it represents nature as we experience it when the usual mathematics fails. N-body problems and the irreversible processes of thermodynamics offer glimpses of their secrets not through linear equations but through statistics. Its calculations describe behavior of a collective—and collective behavior does not bow to Newton's calculus. People came to realize that calculus cannot be the only mathematics.

Thus, beyond its impact on prediction, the bell curve made a mark on mathematics itself.

The dawn of statistics marked a move away from the absolute reign of calculus, in fact. To this day, the two branches of mathematics, calculus and statistics, simply coexist. They do not harmonize in a complementary way. Mathematician Ian Stewart explains in his book *Does God Play Dice?*

By the end of the 19th century science has acquired two very different paradigms for mathematical modeling. The first, and older, was high-precision analysis by way of differential equations. . . . The second, a brash young upstart, was statistical analysis of averaged quantities. . . . There was virtually no contact, at a mathematical level, between the two. . . . As the 20th century unrolled, statistical methodology took its place alongside deterministic modeling as an equal partner. . . . The two paradigms were equal partners, equally accepted in the scientific world, equally useful, equally important, equally mathematical. Equal. But different. Totally irreconcilably different.[6]

The statistical wave forced us to sharply depart from our original style of using straight lines. But its power is so great, and so reliable, that it stepped in, with authority, to illuminate otherwise unsolvable problems. It was never reconciled with determinism; it just reveals patterns that determinism cannot. Its reach is no less impressive. "How things work" as a collective is properly described by a wave.

AND NOW COMES A STING. Long ago, science bought Galileo's vision that straight lines would explain all. But now, inasmuch as science also values curve-based statistics, it ought to straddle a commitment to both the line and the curve. That's what one would think, anyway. Yet science has firmly kept to Galileo's vision even as statistics' curve was accepted.

As successful as statistics was and is, it honors science's commitment to the line. It always presents the bell curve along a linear, mathematically calculable Cartesian grid. The curve rests on axes of space, time, and length. Scientists and mathematicians separate out frequency and amplitude as independent, linearly measurable qualities. They also combine multiple curves by superposition, adding and subtracting them linearly to determine constructive and destructive interference. This approach forever rooted the bell curve to a line.

It is an incredible truth. Science insists that nature follows lines but cannot calculate how it does so with more than two bodies interacting at a time. Then when we do accept a curve that appears—and that suddenly shows us a pattern of motion and change not limited to only two players—we enslave it to the line.

Science in fact used the statistical wave to validate math as the language of nature.* We string it along a line, disregarding, if indeed we ever recognized, that its curve arose from a line-free nature. Lines do not naturally exist in nature—curves do—but we mount the statistical curve on a line, analyze it, and give credit to mathematics.

The bell curve remains stranded in this position. It connects to nothing. As a linear, abstract representation of what happens with large numbers in individual instances, it simply, repeatedly, appears on different scales and hierarchical levels. Indeed, though totally untouchable by the prized lines of Newtonian mechanics, the majority of natural outcomes unresistingly submit to the curve of statistics. Direct linear comprehension gave way to curves out of the blue; we nevertheless continued and continue to treat them with lines. Lines seemed more reasonable. The curve was simply, unignorably, if majestically, *there.*

The mystery of that wave has stood the test of time. It is one instance—a powerful instance, and the first instance I am describing here—of the persistent emergence of curves even as we treat nature with the lines science believes are fundamental. Physicist Hans Christian von Baeyer beautifully describes the bell curve's unfathomed power in his book *Maxwell's Demon.*

The way the bell curve emerges is nothing short of magical. . . . The general features of the bell curve make intuitive sense, but why it should have precisely the shape it does . . . and not some other shape . . . remains a mystery. What power guides the pennies [of a penny toss], whose individual motions are completely random, to fall in such a predictable way? What intelligence orders them? . . . How does this exquisite order emerge from the chaos of pennies tumbling pell-mell out of a cup?

The literature on the calculus of probabilities is vast, but it is powerless to dispel the magic. Since the eighteenth century mathematicians have derived increasingly general propositions, collectively known as "central limit theorems," to explain the astonishing universality of the bell curve, but its reappearance, over and over again, in the most diverse circumstances, still excites our wonder.[7]

In the natural universe, the bell-shaped curve seems to hold hidden reins, secretly guiding normal phenomena to occur in their places within its arching wave, over and over and over again.

THE CURVE OF STATISTICS is in fact but one example of a far-reaching but scarcely recognized outcome that has resulted from our study of motion. Our Original Theory

---

* In Einstein's Brownian motion experiment, he used statistics to validate the atom, which in turn was used to validate that nature is built of atoms.

of Everything told us to view and treat motion as fundamental straight lines, but curves—all sorts of curves, waves, and fluctuations—pop out and robustly resist simplification into lines.

This happens on all hierarchical levels of the universe. The unremitting presence of curves and waves is the grand finale of our story of where we are today with understanding motion.

I have been surveying the history of science, tracing how science has searched for something fundamental in nature, and come, at last, to waves and curves. This mystery stands quietly behind all the recognized mysteries of science. It is an unnoted but astonishing truth that the Galilean presumption of lines as fundamental to nature has not warded off and cannot ward off the emergence of waves and curves throughout nature even as we study it on the basis of lines. At every extreme, as scientists try to home in on absolute aspects of nature, they find waves and curves. If science were a musical piece, lines have long been the dominant refrain, but if we listen carefully, we will hear waves and curves presenting the opening notes of a different theme.

Nature is rich with waves and curves. We have already touched on some of the scientific findings that suggest their enormous scope: Quantum physicists have determined that particles are waves (i.e., they exhibit wave-particle duality). Einstein showed that light is particles and waves, and de Broglie showed that all matter at the quantum level is waves. The equation $e=mc^2$ even invokes waves traveling at the speed of light to equate matter with energy. Single electrons, even those that go through the double-slit experiment with a significant time lag between each one, somehow land, over time, in a wave interference pattern—seemingly joined in a statistical wave pattern that unites the large number of particles' behavior. In the same vein, one cannot pin down "particles" to a specific location with a specific momentum; the best scientists can do—and ever hope to do, in accord with their theories—is to predict the likelihood of where a particle will appear, based on a purely mathematical probability wave. Einstein, moreover, used a statistical wave to validate the very existence of atoms. Planck charted his blackbody radiation frequencies, which started quantum physics, as moving on a straight line, but they plot as a curve when one graphs them collectively. It has become known as the Planck curve, though no one understands its origin.

And more: Deep within the quantum world, zero-vacuum fluctuations infiltrate scientists' best attempts to create a vacuum. The best machinery cannot get rid of fluctuations and curves, and theory confirms that those flutters cannot be eliminated. In fact, these fluctuations seem to permeate the entire so-called vacuum of the universe. Together with that is the fact that even the most perfectly controlled laboratory on earth with near-zero fluctuation is, in any case, hurtling through space on our planet; the hurtling is not a straight shot but waves within waves, as the earth rotates the sun that rotates in the galaxy, which flies through space. At the other extreme from zero-vacuum fluctuations, cosmic microwave background radiation reaches to the edges of the macro universe as we have observed it.

In the phenomenon called quantum tunneling, subatomic particles manage to cross barriers though their energies are inadequate to do so, and they are said to travel on the undulations of quantum-mechanical waves of energy. In superconductivity, the electrons—which are waves—are coupled, synchronized, as rhythmic patterns of waves. String theory, an extreme attempt to discern what might make up quantum particles, rests on the idea that the strings vibrate (i.e., curve and loop) to somehow make particles. And curves appear in other unexpected places. Force fields are curved. Cybernetics, which models systems as regulated through straight lines of feedforward and feedback loops, entails curves and cycles. Power scaling laws, which span scales to show similar processes on different scales, are curved. The sun and the planets are known to be pulsating. Our planet even emits a constant, oscillatory vibration, called the earth's "hum" (if it were pitched higher, so we could hear it, it would be quite loud).

On the other end of the spectrum from the quantum, gravity as explained by Einstein, considered one of the great discoveries of the past century, showcases exceptional curves. To explain gravity, Einstein showed that space and time are a continuum, which he dubbed spacetime, which is itself curved. That is to say, much as a bowling ball on a trampoline will distort the trampoline such that a Ping-Pong ball will roll inward, so too the massive sun distorts spacetime such that the earth appears to be pulled in, by gravity. Einstein relied on direct cause and effect in his explanation of gravity instead of the statistical curve he used to explain Brownian motion, but once again, and here in the most basic way—in the actual fabric of spacetime—the straight lines preferred by science rebounded into curves.

This is what we find at the extremes of the universe as we try to reduce nature to its elemental parts: Everything is rhythmic, everything is cycling. From zero-vacuum fluctuations to microwave background radiation, from quantum particles to curved spacetime, from the big bang to the forces it entailed, scientists hunt for ultimate puzzle parts—things discrete and linear—and come up, instead, with waves.

The outcome of curves everywhere is so startling it is almost bewildering. Modern science aches to tame curves into lines. Galileo launched science as we know it by drawing idealized straight lines—within his pulse to measure a frequency for the swinging chandelier, within the curved motion of the flying cannonball, and as a projection for the ball on an inclined plane. He rejected the curve as a candidate for the fundamental nature of motion. And yet here we are, finding the opposite of what Galileo promised on every level of the universe. Burgeoning curves confront us everywhere we turn even as we try to box them into lines. Though full of allegiance to

Galileo's fundamental straight-line motion, and with full intent to stay on his ideal-ized path, the scientific community somehow cannot avoid swerving into a plethora of waves throughout nature. People look at curves as something to get rid of, and yet whenever we try to get rid of them, even more emerge.

The waves I have mentioned here are but a few outstanding examples of the irony that drapes the modern portrait of motion. As thoroughly as science has tried to prune variability out of reality in favor of straight lines, curves appear, back in action. Despite our best attempts to linearize motion, curves and waves persistently reemerge. In the end, we find ourselves able to define neither motion nor the space and time through which motion occurs, and our most rudimentary understand-ing—that motion proceeds along a straight line—gives way as those lines curl into waves.

If one would say—correctly so—that the definitive answer to the basic question "what is it?" is "we do not know," then one would have to say the same for motion. How does it work? We do not definitively know. But we do know that waves are everywhere.

It would appear that there is a language of nature whose primacy has not been addressed: waves.

CHAPTER 8

Are the Laws Bringing Us Closer to a Theory of Everything?

AS SCIENCE'S WAY of understanding nature shifted to abstract description, through mathematics, so did our ultimate goal of understanding become abstract. Science abandoned its dependence on nature itself and became a search for the law. Even though experiments draw from and corroborate theories through nature, discoveries are considered to be deepest and most "real" when they are of a law. Laws transcend the reality of nature. (The fact that laws don't exist in nature at all is of no consequence.) Invariant mathematical laws are science's favorite way to understand nature.

As abstract statements of regularities in nature, laws should be compatible with one another. Nature itself works seamlessly; laws should mirror its unified elegance. That is the reasoning behind the search for a single law.

Therefore, though we still have no unitary understanding of nature, and though we do not know what matter is or what motion is, one would hope that the laws we do have are bringing us closer to an eventual unification. They should show patterns in nature that, though disjointed, are compatible enough that they may yet add up to a final law.

Unfortunately, the opposite trend unfolded. Not only is science not unifying nature with its progressive discoveries, but we are actually going in the other direction.

Physics consists of many laws that are disparate: laws of quantum, gravity, thermodynamics, relativity, complexity theory, and more. In *The Character of Physical Law*, Richard Feynman put it this way: "What turns out to be true is that the more we investigate, the more laws we find, and the deeper we penetrate nature, the more this disease persists."[1] He later muses, "What of the future of this adventure? What will happen ultimately? We are going along guessing the laws; how many laws are we going to have to guess? I do not know."[2]

The wedge dividing us from our goal goes deeper than that. On top of the progressive disease of laws that disunites science, a split characterizes the laws and other natural phenomena. Nature seems to be plagued by opposites. Rather than understand "what holds things together," we are frequently stymied by dichotomies and opposition—in essence, the bane of unification.

This discouraging outcome of our pursuit to understand nature as a whole-made-of-parts extends to every corner of modern science, so let's examine some of the most outstanding rifts.

MANY SCIENTISTS REGARD QUANTUM and gravity as the most pressing pair of opposites confronting physics today. The micro realm and macro realm skirt any attempts we make to unify them. Unlike, for example, the laws of electricity and magnetism (which James Clerk Maxwell's equations reunified, as per the scientific ideal), quantum physics and Einstein's relativistic explanation of gravity find no common mathematical ground. Einstein, who searched for such a unification for decades, was but the first of many physicists who tried to no avail.

The failure to mathematically unify the two realms exposes a schism deeper than one of calculation alone. The difficulty in harmonizing the calculations is this: Whereas differential equations are used to predict relativity's cause-and-effect interactions, probability is the best scientists can do in predicting quantum motion. Calculus and statistics do not mesh. But the mismatch in the means of calculation exposes the deeper rift in the philosophy behind the two realms of study.

General relativity—the part of relativity theory that deals with gravitation—relies wholly on traditional cause and effect. Einstein unified space and time into spacetime and curved it such that objects fall toward one another. That approach allowed him to rely on straightforward, local cause-and-effect interactions between bodies, including planets and stars, to explain what we perceive as a gravitational effect. Quantum mechanics, on the other hand, does not and cannot describe straightforward interactions to predict quantum behavior. The particles themselves seem simply not to move in the same way planets do.

The immense gap between these disciplines was described by the playwright Tom Stoppard: "There is a straight ladder from the atom to the grain of sand, and the only real mystery in physics is the missing rung. Below it, particle physics; above it, classical physics; but in between, metaphysics."[3]

Several pairs of opposites appear when one tries to tie together macro- and micro-world phenomena. One is the character of particles and waves. It goes without saying that in the macro world, particles and waves are distinct phenomena. The micro-world phenomenon of wave-particle duality, however, indicates some sort of fluid union between the two. Similarly, we cannot reconcile the kinds of motion that exist in the two realms. Motion in our everyday experience appears to be continuous (though we do break it up to work with it), which intuitively seems normal to us; in the quantum arena, our calculations show that particles somehow jump from one state to the next without traveling the road between. The quantum world also exhibits nonlocal action at a distance, which contrasts starkly with local causality as we understand it in the macro world. Einstein, in fact, purposefully got rid of action at a distance as the mechanism for gravity when he curved spacetime, replacing instantaneous attraction with a model in which things fall toward each other causally, yet nonlocality plays an active role in the quantum world. Last, the two sides present the dichotomy of certainty versus uncertainty. In the classical macro realm, we measure position and momentum simultaneously without any problem, whereas in the micro realm, the uncertainty principle declares we can never be sure of both at once.

These are just a few of the crucial opposites that contribute to quantum physics being called "weird" and illogical. The macro world of physics, in contrast, seems "real" and logical.

Yet, I must mention, even gravity has its strangeness. Scientists have found that stars are born near black holes (which they locate through mathematical projection, not direct detection). The gravity of black holes is thought to be so strong that it bends spacetime such that the dying star that is forming the black hole collapses on itself and forms a singularity (i.e., a point particle with no mass). Fledgling stars and black holes have been called the "odd couple." This is yet another unresolved contradiction: that stars are born in the periphery of black holes, around which, according to our calculations, things ought to be disappearing.

Even beyond the quantum and gravity, we face many more pairs of opposites in our ultimate scientific findings. Most noticeable is thermodynamics and complexity theory. This set of contradictions rings uniquely with our personal experience because it addresses the age-old question of the origin of order in the natural universe. Why do we exist in a boisterous and wondrous diversity of living things when, at the same time, there is good reason to expect everything in the universe to degenerate into disorder and randomness? Each side of this dichotomy suggests outcomes that are intimately meaningful to humanity, obviously, but are irreconcilable with each other. Because of its centrality to human existence, I will explore this dichotomy at some length here, with complexity science as the source of order on one side and thermodynamic theory as the tendency to disorder on the other.

Thermodynamic theory reflects the idea that the whole equals the sum of the parts and that change comes to pass through local, linear interactions of those parts. The first law codifies this model.* The second law outlines our understanding of what happens, in a closed system, during those interactions.† It says that, thanks to the loss of heat in every exchange of energy, every interaction bumps up the total unusable waste energy of the universe. Physicist and mathematician Rudolf Clausius and physicist and physician Hermann von Helmholtz, summarizing the opinion of Lord Kelvin, phrased it as the entire universe is headed for eventual "heat death."

This thermodynamic prediction, which points to a final, universal state of disorganized chaos, derives from our observations of the steam engine. Because some energy is understood to be lost in every interaction, the portrait of the universe as painted by the second law is a dismal one. It plunged 19th-century scientists into an upsetting return to the chaotic view of nature that Newton's mechanics had so recently expunged. There was no beautiful clockwork order to the universe; there was, instead, a one-way ticket to chaos. If the universe is driven by energy, and some energy goes to waste in every interaction, there can only be eventual dispersion into utter formlessness: from work to waste, from hot to cold, from order to chaos, from concentration to dissipation—to Lord Kelvin's "heat death" of the universe. It is a sweepingly pessimistic view of the universe. It casts the universe as the ultimate machine, which will eventually run down to waste.‡

In the face of the second law's dire predictions, stability stands as a crisp opposite. Stability means things insistently hold together. And beyond basic stability, complexity affronts the second law's predictions even more. Complexity, the scientific perspective would say, is when what emerges from a system is greater than the sum of the parts (i.e., when we cannot account for what we observe by our understanding of parts and their interactions). It is also called emergent complexity. Science affirms the model that the output of a system is proportional to the input, yet at the same time, it recognizes that complexity bursts forth with more richness than is accounted for by its structural organization. In his book *Emergence: From Chaos to Order*, complexity theorist John Henry Holland describes emergence as "much coming from little."[4]

This complexity is described, by science, through what is known as complexity

---

* The first law states that the energy of the universe is conserved, meaning that no matter what events may transpire, the energy of the universe is constant.

† I remind you that in nature, there are no truly closed systems, any more than there are true straight lines. The second law states that the universe will tend toward disorder (increase in entropy) as time passes. The second law stipulates a closed system even though none exist in nature.

‡ Even for a law of nature, the second law was a particularly sweeping pronouncement about the universe at large. I remind you that the steam engine was subject to what scientists have always done—treating nature as a whole made of parts. The steam engine was just another closed system of the Puzzle Hypothesis, subject to the same process of being conceptually isolated from its surrounding environment. Thus those scientists who took that system as their starting point made a nondescript move in a scientific sense. However, meanwhile, unknowingly, they resoundingly rebirthed our weltanschauung. They achieved great practical success in working with the steam engine as a closed system, as a whole made of parts, and thereby renewed the puzzle worldview with vigor. The second law unfortunately came along with this approach.

theory. But the theory is not what one would recognize as a theory. It is more a statement that complexity exists, with attempts to reconcile it with its nemesis, the second law. From the scientific perspective, the dichotomy here is the utter unlikeliness that complexity would arise from the local, linear interactions that seem to propel the universe toward the opposite outcome—toward the monotonous heat death predicted by the second law. Everyday experience abounds with instances of complexity: the first crocuses of spring, a whirling hurricane, human life. Such complexity records the wellspring of novelty, of organized and durable phenomena in nature, which cannot be explained by the simple laws of physics. In life, in the environment, in the solar system, and in the galaxies, glorious proclamations of tremendous complexity seem to suggest some source of renewal. Yet in the face of all that, our understanding of laws fatalistically paints a universe that is headed toward a still, cold, undifferentiated monotony.

For human beings throughout the ages, the persistence of stability has been no small issue. A paragraph or two in a book cannot convey its enormity, because in terms of our lives, our experience, and everything we care about, the origin of life and the origin of complexity are of grave importance. It has been both a practical question and one of personal reassurance. We have always wanted to know how to nurture and manage this sort of complexity—the sort where the whole is greater than the sum of the parts. Any failure of such complexity threatens our very lives. Therefore, the ideological incompatibility of the second law with stability and complexity takes the conflict miles further, into the realm of pessimism versus hope. According to the laws of thermodynamics, the sort of emergent organization found on earth should not happen to begin with and will surely fall apart wherever it does somehow exist. The thermodynamic laws we have derived baldly negate the existence of our physical selves; the growth of grasses; the enormous dynamic changes through which summer breezes give way to life-sustaining rains; the tenuous atmosphere that, somehow, stably cocoons our planet as we breathe. But no one wants to see those go.

Science has always regarded this dichotomy as a serious dilemma. James Clerk Maxwell suggested Maxwell's demon, an imaginary creature who can influence thermodynamic reactions at the atomic level, as a model for something—something deeper—in nature that can generate complexity in defiance of the second law. It was a thought problem of the most profound kind. It set forth, into our model of a whole made of parts, a tentatively optimistic suggestion: that perhaps some part can cause an increase, not only a decrease, of complexity and organization.

Yet, throughout the decades, various scientists have resurrected the demon and found new ways to show that he cannot circumvent an increase of entropy. Complexity remains unharmonizable with thermodynamics. Our models actually cannot account for complexity's existence or endurance.

Moreover, all these ideas are based, like our findings about matter and motion, on energy. Specifically, the second law predicts an ultimate, uniform dispersal of energy, whereas complexity indicates increased concentrations of energy. No one knows what energy is, however. We feel, once again, a sense of edification because we can calculate changes according to a law, but we do not know anything about the energy that disperses according to that law.

This is the problem of thermodynamics versus complexity—and the way science frames it comes right from the playbook of the Puzzle Hypothesis. As far as science is concerned, nature is a puzzle and nothing shapes its complexity from the top down; no design that we might call "intelligent" orders the universe. But neither has there been detected a source of the existent order within the parts themselves.

Therefore, in his book *What Is Life?*, Erwin Schrödinger suggests that the complexity of life on earth is a form of negative entropy stimulated by the waste heat of the sun.[5] This approach casts life and all other complexity on earth as a blip of local order that feeds off solar waste energy. Physical chemist Ilya Prigogine won the 1977 Nobel Prize for his work on dissipative structures, which made a similar suggestion, that energy that is input and dispersed into chemical systems can locally and temporarily reverse the entropy increase predicted by the second law. Other like ideas, such as polymath Stuart Kauffman's spontaneous self-organization or the generally accepted idea of the availability of free energy, skirt the question by stating—without explaining—that structure and organization happen in localized pockets even while all, overall, ultimately still goes to waste.

TWO INHERENT LIMITATIONS diminish the value of these explanations (beyond their reliance on energy, whose identity evades us). First, an issue undermines all the accepted models: Every one fails to clarify how emergent complexity and stability actually hold together. Scientists and mathematicians are able to model certain versions of emergent complexity on local scales, but no mathematics backs a coherent model of increasing order. None of the proposed explanations adequately counter the statistical argument Ludwig Boltzmann used to explain thermodynamics. Instead, people accept that there are such regular emergent complexities and then proceed to pose mechanisms through which such emergence might defy thermodynamic laws.

That is to say, there is no theory as one would normally recognize a theory. A standard scientific theory of how complexity arises would be subject to the usual rigors of scientific experiment, test, and control. The many faces of what is called complexity theory—fractal geometry, self-organized criticality, chaos theory, evolutionary biology, cybernetics, synergetics, computational theory, solid-state/condensed matter physics, complex adaptive systems, game theory, artificial intelligence—all feed into one another to point to emergent complexity but do not reconcile with one another into a coherent, unitary explanation. There is no mathematical argument or mechanism proposed for what propels the parts of nature, as we see them, to blossom into the complexity we literally eat, breathe, and sleep.

In his book *The Fifth Miracle,* Paul Davies put the dilemma this way:

> *As a simple-minded physicist, when I think about life at the molecular level, the question I keep asking is: how do all these mindless atoms know what to do? The complexity of a living cell is immense, resembling a city in the degree of its elaborate activity. Each molecule has a specified function and a designated place in the overall scheme so that the correct objects are manufactured. There is much commuting going on. Molecules have to travel across the cell to meet others at the right place and the right time in order to carry out their jobs properly. This all happens without a boss to order the molecules around and steer them to their appropriate locations. No overseer supervises their activities. Molecules simply do what molecules have to do: bang around blindly, knock into each other, rebound, embrace. At the level of individual atoms, life is anarchy—blundering, purposeless chaos. Yet somehow, collectively, these unthinking atoms get it together and perform the dance of life with exquisite precision. Can science ever explain such a magnificently self-orchestrating process?*[6]

We are cornered into a singular approach: Without a single ordering principle of the universe, our attempts to explain a vibrant ecosystem or even the growth of a single tree limit us to local and causal interactions and statistical probability. Complexity doesn't follow a Galilean-style causal law or a Boltzmann-style statistical law; it emerges from a causality science cannot pin down.

Current explanations for complexity are thus entirely unlike other scientific theories or laws that might allow, through experiment, for prediction and control of nature. We can build a machine that follows Newton's laws as a sum of parts and wholes and build an insurance company that runs on probabilities, but the same cannot be said for emergence. No understanding of complexity can, for example, allow anyone to build even the most primitive living organism, let alone a creature like the myriads that naturally inhabit earth.

The second limitation of complexity theory reflects a theme of science we have repeatedly seen: the unintentional closed-mindedness of the scientific perspective. Our perspective is built into our laws, but its contributions are not recognized. It was no accident that we chose a closed system, the steam engine, as our model for how things work, to the exclusion of any other possibility. But the entire contradiction of thermodynamics versus complexity arises specifically in a closed system—for it is from a closed system that we derived thermodynamics' second law. What if we did not start with a closed system? Many, many processes exhibit a different way of being. Maybe a closed system is not the appropriate model for complexity. If our model had been a blade of grass or even the human organism instead of the steam engine, perhaps we would not have these "problems" at all.

So here we see that the dichotomy of thermodynamics versus complexity theory is itself a direct outcome of the Puzzle Hypothesis and not necessarily an inherent problem of how nature works. We could not viscerally or intellectually deal with the overwhelming panoply of nature eons ago, and so we mentally broke nature into parts, as a puzzle. And we continue to believe we should be able to break the whole into its parts to understand. Simplifying nature is the unstated goal in all our research, whether in our most primitive interactions with raw nature or in our sophisticated laboratory experiments. The name of the problem itself, complexity, shows our slant: Complexity, as a phenomenon and as a theory, goes against the grain of the most fundamental premise of how we study nature. The organized world does not bow to reductionism. It displays irreducible complexity. That is why we see it as complex.

One could even say that reductionism fails complexity at the outset. Beyond its overall failure to explain complexity, reductionism eliminates its object of study when it is put into practice. As one reduces, for example, a living organism down to its parts, and those parts down and down into molecules and atoms, one loses the emergent qualities of life that are being studied, for life is not visible on those scales.

Some scientists do look for what one would call "something else" that could be responsible for the order and complexity of nature. Louis de Broglie and American physicist David Bohm proposed the existence of what they called a guide wave or pilot wave—real waves, instead of probability waves, that somehow organize and give shape to local phenomena and interactions. Though the idea was thought to have promise, no one has come up with a way to connect such a wave to the real world. Another suggestion, this one from Paul Davies in *The Cosmic Blueprint*, is the need for what he calls an "optimistic arrow"—an arrow of time going up, in the opposite direction of the second law of thermodynamics.[7] Geneticists acknowledge that there is a local environmental influence of a biochemical nature in the function of DNA and genes; the one-way function of biology's central dogma does not account for these top-down influences, called epigenetics. To date, though, despite these ideas, no one has come up with something in nature itself that could account for the whole being greater than the sum of its parts.

BEFORE WE MOVE ON to further opposites in science, I want to introduce you to a startling, even unnerving, possibility: that the problem lies in our definition of stability. We think we know what stability looks like. We tend to think of regular, invariant patterns, such as those of a crystal, as stable. And they are. But in the complexity of life, the opposite appears true. Normal healthy behavior exhibits variation.

The famous Framingham Heart Study showed, for instance, that decreased variability in heart rate was the single common risk factor for elderly people dying.[8] Animals in captivity experience decreased variability and exhibit autistic symptoms. Ary L. Goldberger, MD, a professor of medicine at the Harvard Medical School, has drawn attention to the fact that people with regular, invariant heartbeats—with low heart rate variability—have *less* stable physiologies and are physically more vulnerable.[9] We have thought that regularity and invariance equals stability, but studies show that these people are more prone to chronic disease. Dr. Goldberger likes to hold up two graphs at a lecture—one with regular, even heartbeats and one with them jumping all over the place. He asks the audience to guess which one of the patients died shortly after the graph was taken. People are shocked to hear that the steady heart was the failing one.

I call this observation the Ary Paradox. It is a profound if not unsettling realization. It forces the hard question: Do we really understand nature? We think stability is one and the same with regularity. Yet a regular heartbeat leads to death, and a variable heart equals life. Variability is stability, in this case at least, and we have no principle to account for it.

BEYOND COMPLEXITY AND THERMODYNAMICS, even more opposites polarize the overabundance of laws we've found. Another set, which also involves complexity, is complexity versus gravity. Complexity is said to be present when the whole is greater than the sum of the parts, whereas gravity showcases the dilemma in reverse. There is not enough mass in the galaxies to exert the gravity needed to prevent their flying apart. From the scientific perspective—though no one phrases the problem in such elemental terms—in gravity, too, the whole is greater than the sum of its parts. But, the scientific solution zeroes in on the parts themselves (mass) instead of on the whole (gravity). Scientists propose that up to 96 percent of galactic mass and energy is missing. They have searched for, but not found, this missing, or dark, matter and energy.

The very idea of missing mass is a fascinating testimony to our inability to see nature as anything other than a whole made of parts. Science's trust in our Original Theory of Everything, where the whole *equals* the sum of the parts, leads to trying to make a fit no matter what: When the whole is greater than the sum of the parts, scientists' solution is to propose more parts, albeit ones no one can detect.* Rather than frame the mystery in terms of missing parts, one could as easily say that gravity

---

* That solution means the puzzle model is comfortably maintained, for the whole now equals the sum of the parts. It is a puzzle answer to a puzzle problem. Of course, coming from the scientific perspective (in which the Puzzle Hypothesis is taken for granted), the missing parts remain a huge mystery.

emerges in the same way that complexity does: It is greater than the sum of its parts.* This perspective would orient science to the possibility of understanding both phenomena through an insight into emergence.

FRACTALS IN NATURE PRESENT yet another set of complexity-based opposites that science cannot explain, making us seem all the more distant from a unified law. Fractals are a mathematical class of complex geometric shapes in which any scale can be magnified to resemble the whole.† That is to say, fractals exhibit self-similarity across scales, a phenomenon found throughout the natural world: the budding heads of cauliflower and broccoli, snowflakes, coastlines, mountain peaks, and clouds are instances of self-similarity in nature. (A cloud's puffy edge, for example, will remain true to the same pattern whether viewed from afar or magnified many times over.) In all these cases, there is no given scale from which you can define the shape of the object. From whatever scale it is viewed, in fine detail or from a distance, a fractal will roughly exhibit the same shape.

While, as we have seen, scientists try to account for emergent complexity within a given scale, with ideas like spontaneous self-organization through feedback loops and negative entropy, no one can account for the complexity of self-similarity *across* scales. In his book *How Nature Works,* Danish theoretical physicist Per Bak says: "The importance of [fractal mathematician] Mandelbrot's work parallels that of Galileo, who observed that planets orbit the sun. Just as Newton's laws are needed to explain planetary motion, a general theoretical framework is needed to explain the fractal structure of Nature. Nothing in the previously known general laws of physics hints at the emergence of fractals."[10]

But fractals are not easily resolved with the idea that nature is built up as a whole made of parts. In *The Fifth Miracle*, Davies muses:

> *Could life be like this: apparently complex but actually very simple, like a fractal, and therefore the product of a simple, lawlike process? . . . Personally I do not believe it, not least because it demands a view of nature that is incredibly contrived. To claim that there is "a code within the code," generating living creatures on demand from simple formulae, is just too far-fetched.*[11]

The overt presence of fractals across scales in nature foundationally opposes our idea that the parts of nature become organized exclusively through local cause-and-effect interactions.

---

* It is reasonable to assume that scientists prefer missing mass over the idea that gravity resembles emergent complexity because the former explanation is more compatible with a model where the whole equals the sum of its parts.

† The Koch curve, or Koch snowflake, is a popular example of a fractal. However close you zoom in or far you pull out, the shape is the same.

SYNCHRONICITY IS ANOTHER FACET of complexity, unaccounted for by current scientific knowledge, which, in its opposition to the second law of thermodynamics, steers us away from a unified law. Synchronicity is when the same things happen at the same time. When it happens without an apparent conductor, it is a scientific mystery. Superconductivity reveals that electrons, which usually bump shoulders as they crowd through a wire, somehow organize and couple at very low temperatures, speeding in synchronized pairs through the wire with no resistance. Myriad events in the living world display mesmerizing synchronicity: pacemaker cells firing in the human heart, hordes of fireflies flashing by a Malaysian river, schools of fish shifting in silvery turns, and flocks of birds swooping and turning as one. Scientists at Lawrence Berkeley National Laboratory have found that photosynthesis depends on quantum entanglement: Groups of chlorophyll molecules synchronously alternate between excitation and relaxation.[12] Ecosystems, such as a woodland forest or tropical jungle, feature countless distinctive elements, from sprouting trees to creeping fungi and worms, from mosses to birds and other creatures, which work in harmony as a unified, flourishing ecosystem. "Unlike many other phenomena, the witnessing of [synchronicity] touches people at a primal level," says Steven Strogatz in the final sentences of his book *Sync*. "Maybe we instinctively realize that if we ever find the source of spontaneous order, we will have discovered the secret of the universe."[13] Emergent complexity and organization twinkles with the promise of an ultimate understanding of nature.

Complexity, as a phenomenon, ultimately also goes against the grain of the first law of thermodynamics. The first law, as I outlined in Chapter 5, is a mathematical version of our a priori belief that nature takes the design of a whole made of parts. Also known as the law of conservation of energy, the first law states that the energy of a closed system remains constant no matter what changes it undergoes—even as grand a system as the entire universe. It is thus, to a certain style, a theory of everything: a mathematical statement of our Original Theory of Everything. But emergent complexity does not mesh with this model. Something is happening in life, for example, that cannot be mathematically modeled this way. Everything is supposed to neatly add up, but complexity breaks the mold.

LET'S LOOK AT A FINAL SET of mismatched phenomena that distance us from a unified theory: quantum physics versus thermodynamics. I illustrated in the previous chapter that historically, quantum physics and thermodynamics both derived from a single phenomenon of heated blackbodies. Each boxed in one half of the process and left out the other. Planck drew the original idea of the quantum from blackbody radiation, where a body was heated up and radiated energy, but he ignored the inevitable cooling-down process that follows it. Carnot, on the other hand, launched thermodynamics by isolating a cooling body and shut out the indispensable heating-up process that precedes it. Both sciences forced a divorce between two sides of one, singular process for study. Thermodynamics and quantum physics stand as yet another pair of opposites that science fails to unite in a single theory.

ALL THESE OPPOSITES are a troubling outcome of how we have gone about under-standing nature. We believe nature is a whole made of parts, but the parts are often irresolvable. "Niels Bohr, the Danish physicist and philosopher-king of quantum theory, once said that a great truth is a statement whose opposite is also a great truth. . . . [This] seems to me to sum up much of the history of science and philoso-phy," states science writer Dennis Overbye in an article for the *New York Times*. "Is nature discrete or continuous? Is the universe infinite or finite? Is life inevitable, or is it a lucky accident? Will we ever find company in the cosmos? . . . A final answer to any of these questions would be a landmark of human progress," he continues. "But it might be in the nature of being human that we will never answer them but have to hug them both in a kind of Hegelian surrender. And so we live in the tension between opposites."[14]

Laws were supposed to tell us why things stay the way they do and why they sometimes also lose their organization. They draw their power from the one fact alone: that regularities exist in nature. Every regularity is a pattern we try to distill. We attached numbers to the regularities, dispensed with the particulars, and abstracted out mathematical statements of them. We treated these statements as the rules of nature—as our laws.

And somehow, though it is a true recognition that regularities exist in nature and the laws are abstractions we have pulled out from them, this recognition has not been maintained and incorporated into our regard for laws. A tremendous turn-around in our perspective happened once Newton formulated his equations. His laws meant people could use laws to predict and control via cause and effect. And that seemed to reveal, in laws, some innate power. The perspective has become that the laws are the cause of our natural experience.

In fact, as matter, motion, and even laws have been reduced to statements of energy, this perspective has become all the more extreme. Where some scientists simply wonder at why the math works—Hans Christian von Baeyer wondered in his essay "Weyl's Rule," "How can pure mathematical speculation produce equations that accurately describe the world? No one knows"[15]—many scientists feel that the math is the ultimate reality. MIT cosmologist Max Tegmark, for example, proposes in his book *Our Mathematical Universe* that "our physical world is not only *described* by mathematics, but that it *is* mathematics." He grounds this belief in the puzzle model, of course, saying, "We live in a *relational reality*, in the sense that the proper-ties of the world around us stem not from properties of its ultimate building blocks, but from the relations between these building blocks."[16]

As surprising as this proposal may sound to someone not familiar with modern science, it is not unusual. Ages ago, philosophers and thinkers began by conceiving of the universe as a whole made of parts. But when this perspective was followed all the way down, our modern experiments—and the laws that describe them—led many to believe that mathematical relations between parts are what makes nature. This is yet another bizarre outcome of our laws. However, even those who do not embrace this extreme opinion cannot deny that the assertion is open for argument.

In a certain sense, it is no surprise that many respected scientists propose that such a possibility is true. Though it defies the very reason the nature-is-a-whole-made-of-parts notion was accepted in the first place—because that model seemed natural, logical, and simple—it does harmonize with the other outcomes of following the Puzzle Hypothesis.

The story of laws, in other words, has turned out to mimic the stories of matter and of motion. We took both matter and motion apart in hopes of pinning down essential parts we could unify into one. We likewise tried to separate out laws, studying one localized part of nature at a time, in hopes of securing local laws that will add up to a "one." Every endeavor shattered in our hands: We ended up with a particle zoo of immaterial point particles instead of a single *atomos* in matter; a host of waves instead of a single linear rule for motion; and a proliferative disease of laws from the macro to the micro realms. But they are all akin to one another.

And still compelling to the scientific thinker is that the laws are immaterial. We accepted their transcendental nature without question, but oddly enough, this ended up being one of the reasons people feel comfortable saying nature may be made of math: Our experience with laws resembles our experiences with other aspects of nature. Matter and motion have ended up as mathematical abstractions of energy; the premier laws of the universe, the first and second laws of thermodynamics, are mathematical statements of energy; and the first law in particular states that the whole energy of the universe always equals the sum of its parts. In this context, it is understandable that some hope that a mathematical model of nature will uncover what holds everything together (while the second law mathematically answers why things fall apart). Everything seems to mathematically relate through energy. Our ideas about energy should, at least, give us a hint.

Unfortunately, however, as we have repeatedly had to face, we do not know what energy is. We can mathematically assess a capacity to cause change but do not know what our numbers describe. Energy cannot answer what it is, how it works, why things stay together, or why they fall apart, though it does seem to relate to all these questions.

And though it is never called out for doing so, the predictive successes of mathematical statements of energy set us into a circle of reasoning similar to the circular reasoning through which the discovery of the atom was used to confirm the Puzzle Hypothesis. I explained earlier that from the time of the ancient Greeks, the atomic hypothesis prompted people to divide and reduce nature to study it as a sum of building blocks, and when the existence of the atom was confirmed, it was also taken as confirmation that nature is built of them. No one seriously considered that perhaps atoms exist but nature is not "built of" them as a whole made of parts. In the same way, the regularities of nature, when viewed through the lens of the Puzzle Hypothesis, prompted people to divide and reduce the regularities to simple patterns, which were most readily visible in the simple, nonliving realm of physics. The success of mathematics in those localized areas was taken as confirmation that all— meaning, all of nature—is a sum of those laws. No one seriously considers that perhaps linear regularities exist in those realms but that nature's greater regularities are not built of them.

THAT LATTER POSSIBILITY is worthy of notice nonetheless. For even as we work to tame all the divergent features of nature, mathematics—this prime analytic tool— has drifted from our grasp. Scientific acumen has changed. What began with the seemingly prophetic insights of differential equations moved to the probabilistic predictions of statistics and then to the inscrutability of chaos. That is to say, we have gone from direct prediction to probabilistic prediction to computer-generated approximations for what we cannot analyze and understand.

Nobel laureate physicist Eugene Wigner describes the unreasonable effectiveness of mathematics in the natural sciences in a famous paper by the same name. Wigner says "that the enormous usefulness of mathematics in the natural sciences is something bordering on the mysterious and that there is no rational explanation for it."[17] Yet the increasingly complex details of nature defy Wigner's pronouncement.

We discussed the mathematically underdeveloped aspects of complexity theory a short while ago, but now we must look at it from the perspective of assessing mathematics itself. Mathematics does not work where there is high variability, or a high degree of complexity, within or across scales. Our natural experience of nature includes, on a personal level, everything from breathing to sleeping to eating to enjoying the sunshine. All around us, we experience a teeming world of living and nonliving entities and phenomena of every assortment and variety—from gusts of wind to children laughing. Mathematics works only on a very specific corner of this world. It takes on cases with little variability, such as stones dropped from towers, rockets shot with enormous speed, and trajectories of particles. Mathematics in science is thus self-selecting: It restricts to where its practitioners think it will work. Though calculus copes with a curve, it is hopeless to address the many changing variables in an open system. Biological variability, which furnishes our planet with wondrous complexity, lies outside the realm of mathematics. The best anyone can do at this time is partial, simplified modeling with computers.

Rather than Wigner's "unreasonable effectiveness of mathematics," our data point to a different outcome: mathematics becoming progressively ineffective. We have accumulated a tremendous amount of data and knowledge, but putting it all together is very complex. Computers can organize and correlate lots of data at phenomenal speeds, but even supercomputers cannot clarify the ever-increasing quantity and complexity of data. This is a flip side of the problem of complexity. We have a never-ending story of data collection with no reunification. One could say we have encountered a data zoo, just like we found a particle zoo in matter, a host of curves in motion, and a proliferative disease of laws across scales.

Underlying the difficulty of managing the sheer quantity of data is a further problem, that the plain facts themselves are often incommensurable. That is to say, the facts we have found about nature usually do not harmonize with one another, even though we consider them all to be verified invariants. The disconnect is most conspicuous when we separate invariants from their close group of facts; trying to connect the existence of a muon with the details of cardiac arrhythmia is all but meaningless. This incompatibility of facts across scales thus represents another sort of complexity in its own right. It is further evidenced in that people specialize in specific fields rather than practice science in numerous fields. And more, within fields, individuals become involved in subspecialties. No one tackles the big picture, other than philosophers of science.

For centuries we have gone full force at nature in our mission to collect invariants, but now the only common denominator between facts about nature is *that* they are invariant. The thread of being suited for the puzzle is truly the only characteristic that runs through all our knowledge. Though we have derived many invariant descriptions of localized, invariant patterns in nature, rampant dichotomies thwart our desire to unify what we have found.

Our commitment to mathematics as the language of nature also allowed a fantastic irony to slip in as we opened the door to the modern landscape of atoms. It was statistics that, as a tool in the hand of Einstein and applied to Brownian motion, proved to the scientific world that the atoms of Democritus actually exist. People automatically took Einstein's theoretical validation that atoms exist to also validate the idea that nature is built from them. His work was also taken to confirm the belief that atoms exist in isolation, separate from space. (Einstein based his math on a model of atoms as spherical balls, each a closed system.) That idea had held sway since the ancient Greeks but was believed to be validated at last—by the mathematics that describe the bell-shaped curve.

But the reverse was true, as well. Statistics had exacting demands on what it could calculate. Specifically, it worked only on boundaried, discrete, near-identical separate objects.

Herein lies yet another graceful circle of reasoning. Atoms—as real, discrete, boundaried particles, separate from space and time, which jiggle about in perpetual motion—seemed to be validated by statistics, yet that selfsame statistics was considered suitable for explaining nature through thermodynamics only because atoms and particles were assumed to be discrete. One cannot escape the unsettling power of the Puzzle Hypothesis in this development. It set forth the idea that something called the atom exists and that it necessarily exists as a discrete unit, with the properties necessary for "building" nature as a whole made of parts. It then set forth linear mathematics—and those mathematics validated the idea of the *atomos* atom. Both came from the same hypothesis.

That they both come from the same hypothesis suggests an important but overlooked possibility: that they are consistent with one another not because they are indisputably true but because they share a common origin, a common assumption. They are woven through with a host of compatible premises that, in each, has been used to show that the other is true. The atom validated statistics and statistics validated the atom.

When one steps outside this snare of reasoning, one can begin to glimpse another possibility—that small particles exist, but differently than was assumed (i.e., *not* as discrete, boundaried units). Perhaps what Einstein saw as atoms were something real, but not tiny balls. And if they were not discrete balls, they could not rightly be used to validate that nature is built of discrete particles. In the same way, it is possible that the bell curve—being an emergent, unexpected curve—exists differently than we assume it does (i.e., not anchored on a line, as we anchor it when we graph and analyze it). If this assumption is not correct, the bell curve cannot rightly be represented in mathematical linear dimensions. And in that case, it was an error to use its mathematical treatment (statistics) to validate atoms as discrete particles.

Perhaps something else is going on.

I'D LIKE TO PULL BACK for a moment and touch on how our a priori ideas infiltrate our daily lives, for the effects of our presumptions polarize more than the facts we know of nature—they affect our experiences in the greater world, as well. Life and health rank at the top of our priorities. One would think that ideas for fostering life and health would revolve around nurturing and cultivating complexity in the system as a whole. From the perspective of nature as a whole made of parts, however, complexity isn't in the picture, and it seems to make sense to isolate the relatively small part causing the problems instead of working with the system as a whole.

Medicine therefore seeks to destroy the unwanted parts of nature, using a strategy with disease that people historically have used against each other: divide and conquer. It is commonly said that we have a war on disease, such as the war on cancer or the war on multiple sclerosis, and we experience a victory when we kill misbehaving cells and wayward molecules. We imagine health and survival could, ideally, be attained through a magic bullet. (Yes, a bullet.) We target and destroy aberrant genes. These common metaphors seem unexceptional because our perception of nature, as a whole made of parts, sets us up to see parts as independent from one another. It seems natural, if not necessary, to pit one part of nature against another in order to get what we want.

An alternative to destroying disease could be to try to cultivate health. But again, that is far more difficult from a scientific perspective because we understand our bodies as a whole made of parts, and we do not know how complexity emerges in the whole system. Epidemiologically, we can point to trends that encourage health. But when confronted with disease, the mind-set that we are built from the bottom up steers us to attack the diseased parts. The idea behind a cure is to get rid of the enemy.

Many people find it somewhat shocking to hear that we have no definitive cause or cure for any chronic disease. We have attained significant success against what we would call foreign invaders—many bacteria and some viruses—but for chronic, organic disorders, such as cancer and autoimmune disease, our approach has not led to a cause or cure.

We also find ourselves at war with the environment. Though many more people today are involved in actively caring for the environment than ever before, the reality of our interactions with it has been harsh. People often treat nature as something utterly separate from ourselves: They strip forests, use insecticides and pesticides, and genetically modify food. Even as we try to save environmental tracts by designating local and national parks, every park has absolute boundaries, which means nature is still broken up as if it were a puzzle.

We know that we do not know the ultimate effect of our actions. The "edge effect" discovered by conservation biologist Thomas E. Lovejoy shows that in forests plowed through with roads, the greatest diversity and life will be found in the center of an intact tract of forest, whereas diversity peters out as one nears the road.[18] And the smaller the tract of land, the more rapid the loss of diversity in trees and wildlife. Moreover, the differentiated species of our planet are disappearing with alarming regularity. Extinctions of many types of plants and animals are accelerating. Even the animals we protect in man-made environments do not fare well. Animals who live in boxed-in conditions get the same diseases we get: Domestic pets, such as dogs and cats, get cancer, and animals in zoos can get cancer and psychoses, replete with autistic symptoms, such as pacing, bouncing, and repetitive behaviors.[19] Added to the mix are definite concerns about climate, in what I call "global storming"—a rising incidence of extreme weather as a consequence of climate change.

Humans have treated nature as separate from ourselves and have assumed that all parts are separate from one another. The war on disease and the war on the environment echo the wars that have plagued humanity for all time, in the saddest of ways. The human conviction that nature is designed as a puzzle made it seem natural for us to set up different parts in opposition to one another, even if the parts were ourselves. If, however, nature is somehow a continuous "only one," these strategies lose their hold as the only options. There may be a way to weave together what we perceive to be parts of nature in a harmonious way that fosters emergent complexity and stability.

IN THIS AND EARLIER CHAPTERS you have seen why science took the path of treating nature as a puzzle, a whole made of parts. And you have seen that, tenacious though this assumption has remained through the years, exploration of nature using this lens has revealed nature to be something else.

What, then, is the big picture? Do we have a law that captures the stability behind all of nature—what makes nature be what it is—or are we going to get one soon? Why are our laws transcendental and fragmented? No one in science can tell you. Laws have not brought us to ultimate enlightenment. We do not have a law uniting everything. Moreover, many of our laws and theories are actually opposites. We cannot explain the emergence of complexity and what holds it together. Our explanation of why things fall apart is simply an observation that they do, in a closed system, and is phrased in terms of energy, whose identity is unknown.

And though science was meant to explain nature, we find ourselves in the curious position where the actual stuff of nature is less real to us, in terms of nature itself, than the laws that describe it. This is not said as a critique of science; I do not mean to get entangled with whether, and whether it is significant that, science no longer corresponds to nature as we experience it and relies on abstraction instead. But there is a point to be made that science is based on the Puzzle Hypothesis—which the human mind accepts because we believe *it* is true to nature as we experience it. In other words, we accepted that nature is a puzzle because that seemed natural, but following it through to the nth degree took us to entertain ideas that are not natural.

Yet neither have we come up empty-handed in our researches. We first developed laws in order to codify regularities, or patterns, in nature's behavior. Today's scientists continue to search for such regularities. But for all this time, we have considered only consistently repeated interactions between nature's parts, as we perceive them.

The regularities themselves, however prized, have been underlings to our almost tyrannical focus on parts as primary; we turn our eye to parts first and hunt for regularities of those parts second.

Here is the opening to a new way. If regularities are so important, perhaps it is time we consider them in their own right. Now that we know that we have always seen nature through the lens of the Puzzle Hypothesis, and that we do not have all the answers we expected from it, we can set it aside to look at nature the other possible way: as an indivisible "one," not made of parts. We can turn our attention to the rhythms we hold in such regard, the rhythms that are the foundation of the scientific laws of the universe. To the extent that they seem to embody the essence of nature as we know it, it might be worth thinking of the regularities independent of the parts through which we observe them.

We have, after all, discovered a unique regularity that permeates everything, a commonality that appears on all scales: waves.

PART 4

The Discovery of the Nature of Nature

Physicists do not yet know if there really exist ultimate laws that express the final conditions of all existence. . . . Conceivably, life might be able to change those laws of physics that today seem to imply its extinction along with that of the universe. If that is so, then might not life have a more important role in cosmology than is currently envisioned? That is a problem worth thinking about. . . .

In fact, it may be the only problem worth thinking about.[1]

—Heinz R. Pagels, physicist

To my mind there must be, at the bottom of it all, not an equation, but an utterly simple idea. And to me that idea, when we finally discover it, will be so compelling, so inevitable, that we will say to one another, "Oh, how could it have been otherwise?"[2]

—John Archibald Wheeler, physicist

Prelude: A Final Commonality—Waves

Historically in science, people have focused on the extremes of the micro and macro material universe. They have worked in various disciplines and specialties—physics, chemistry, engineering, mathematics—and used advanced technology, in the hopes of finding a hidden secret: an invariant that does not change. I came from an entirely different background. I was an internationally competitive sprinter in track and field, a medical physician, and head of the United States Olympic Sports Medicine Council, as it was called at the time. My insights came from direct, personal, everyday experience, not hidden and not found in a laboratory experiment. Life itself broadcast its secrets when I came at it from the opposite direction. I focused on the motion of this most complex thing, of life, rather than matter.

In this section of the book, I introduce the discovery that gives us an entirely new understanding of nature—of the nature of nature—based on my personal experience with a biological organism and, in particular, its rhythms. Rhythm is motion that pervades the universe. I begin with briefly describing my personal history, which gave me the underlying capacity for this discovery. I then take you through the new understanding of rhythms and waves that shows the true nature of nature, of waves that have shaped human experience since our beginnings and have shaped the universe for all time—for waves are what we ourselves and the universe ultimately are.

NO ONE IN THE SCIENTIFIC WORLD has known what waves are until now. They can be dealt with mathematically very well, of course, with Fourier transforms and such, but that treatment sidesteps questions of what they *are*.

It is worth taking a moment to ponder waves. Whatever they may be, they surely are, to be blunt, an oddity in terms of the Puzzle Hypothesis. They lack the materialness that helps us identify things as things. They are motion, but, as opposed to motion such as the flight of a baseball, they don't seem to be of, or "belong to," an object. A wave in the ocean, for example, is not the water itself, and it is not the motion "of" the water—wave after wave will ripple the ocean, but the water stays put. That is to say, a wave moving toward the shore is not motion of water traveling, or else the whole massive ocean would migrate to the beach. Nor is the wave the forward motion of the surface of the water, even as the wave moves forward. An object floating on water will bob up and down as waves travel through, but the item is not propelled onto shore.

To a layperson, the idea of a wave is often nebulous. It seems when we say "an ocean wave" that we are referring to the water. But we also know the water is not the wave; the water is wav*ing*. Waves seem to simply exist in nature, but inasmuch as they have no inherent boundary, it is difficult to classify them in our parts-and-wholes schema.

Somehow, free from the restraints of the world, no matter *what* is waving, a wave is just motion that is in fact not confined to an object. It appears to be "pure" motion, spreading out in an up-and-down pattern.

To deal with waves, science defines them as a "disturbance in a medium" accompanied by a "transfer of energy." Unlike tangible matter, which seems to assert the "real," waves that ripple through that matter are seen as a pattern or a form that sweeps along in concert with energy (whose identity also escapes us). We calculate them to our satisfaction, right down to the quantum world, without knowing what they are or why motion should take such a pattern throughout nature.

Despite their oddness with regard to the puzzle of nature, the ubiquity and longevity of waves throughout nature is indeed well appreciated. When reflecting on the fact that the atoms of the human brain are replaced every few years, Richard Feynman commented that "the thing which I call my individuality is only a pattern or a dance. . . . The atoms come into my brain, dance a dance, and then go out—there are always new atoms, but always doing the same dance, remembering what the dance was yesterday."[3]

Feynman's dance, while delightful, also highlights the fantastically important role of rhythmic, cyclic motion. In the words of K. C. Cole in her book *First You Build a Cloud*, "The 'dance' is more real than the atoms. The 'abstract' patterns of the physical universe are more concrete than the things you can feel or touch."[4] Cole continues that we all look for patterns—physicists as well as parents, psychologists, economists, and so on—but that people make "a common mistake to dismiss the pattern as something of no real substance. . . . Patterns are more real than rocks or atoms or black holes. Patterns are us." By that she means that were it not for patterns, all would cease to exist as we know it.

The "dance" of the world's atoms—the patterns our very "building blocks" follow, when you come at it from the Puzzle Hypothesis—end up perpetuating the reality of the puzzle itself. Said again, because it is so important: The patterns the objects of the material universe follow are more essential for maintaining nature, as we know it, than are the objects themselves. It is hard to imagine a more profound indicator that the patterns themselves may be worthy of study, independent of the material that follows them.

Cole describes waves (which she differentiates from "splashes"—a one-time affair of the same variety as a wave) with riveting charm.

A wave is a lot larger than itself. It can separate itself . . . and carry information far from its source, bending around corners, going right through things, sometimes capsizing people or even whole countries in the process. . . . A wave can do all these things because it is not made of "stuff"; it is a movement of information. . . . But once set in motion, the wave moves quite independently of the people caught up in it. . . . The people stay where they are; only the wave spreads.[5]

Waves arise here again as a powerful but indefinable participant in nature. We are confronted with a reality in which waves are potent purveyors of energy and information—both of which are abstractions—somehow causing significant movement and changes even though we perceive them as immaterial. We simply do not know what a wave is.

And yet, as strange as a wave may be when we try to pin it down, waves are, after all, very natural and comfortable to most people in an experiential way. No one stares at ocean waves perplexed or questions that there is a rhythm to their own sleeping and waking. These waves are normal, familiar, and expected.

And somehow waves are a surprising universal outcome of our studies of matter in motion. Matter has dematerialized into energy, motion has demotionalized into energy, and energy—an abstraction, to be sure—is carried on the mysterious phenomenon of waves. While we have not found any "thing" fundamental yet—not the basic particle or the theory of everything—we keep on coming up with evidence that in every part of life, in every aspect of it, there are waves. Not things, but waves, or, more accurately, wave motion.

Everyone in science looks for commonalities, for things that cross the barriers of science, and wave motion twinkles with possibility. Stephen Hawking said of the uncertainty principle, "Maybe that is our mistake: maybe there are no particle positions and velocities, but only waves. It is just that we try to fit the waves to our preconceived ideas of positions and velocities. The resulting mismatch is the cause of the apparent unpredictability."[6] In the same vein, perhaps the way "we try to fit the waves to our preconceived ideas" of a puzzle is the cause of the gaps in our knowledge. Perhaps waves and the Puzzle Hypothesis are, to use Hawking's term, the ultimate "mismatch."

I BEGAN THE STORY of the Puzzle Hypothesis in the time before history, and we have traveled together to where we are today. In sum, we have no theory of everything. What we thought was simple became complex. What we thought was logical became illogical. What we thought should be natural became unnatural and even transcendental. We have many technological advances but are equally saddled with problems, and have difficulties in grasping the stability of life.

And those questions the Puzzle Hypothesis set us up with? While we answer them with science the best we can, at every level they stubbornly reemerge. What *is* everything, how does it work, and what accounts for order and chaos in nature? We simply can't nail it.

We have never evaluated the Puzzle Hypothesis. When we adopted it, it seemed universal, simple, logical, natural, and useful. We are so sure that it is true to nature that the mounting evidence that it does not fulfill its promise to simply and logically furnish natural and useful answers is lost on us. We don't even know where else we can turn. Think about this: *We will sooner say that reality is literally made of differential equations and statistics, or is consciousness-created, than say it is not a puzzle.*

Unlike the traveler in Robert Frost's poem "The Road Not Taken," who laments his choice of one path over another, saying, "Oh, I kept the first for another day! / Yet knowing how way leads on to way, / I doubted if I should ever come back," we did *not* save our first choice. We did not even recognize that we ever had it. We went full throttle toward our choice that nature is *by its nature* divisible and threw away the possibility that nature might by its nature be "one." And we also threw away all the insights that might come with it.

We must face the fact that nature as a puzzle—as science has portrayed it—is not, after all, the universal reality. And we must consider looking to the mysteries that perplex us—to life, above all—for a new angle of understanding nature as it exists.

We must recognize the pervasiveness of waves.

CHAPTER 9

My Story

I USED TO RUN by the ocean and relax on the beach when I was a child. On the Atlantic shore near my childhood home, waves sparkled everywhere. They rippled and rolled, fell forward and receded. Always in motion, waves played an endless changing rhythm on the sand.

I warmed myself in the sun. I dove in the waves to cool off. Like any child, I ran as hard as I could until I dropped—breathless and laughing—on the sand. Then I would get up and sprint again. This is what children do: It was my natural way.

Later in life, I came to realize that my childhood life by the ocean was no different from the rhythmic way people have always lived when living with and within nature. I naturally alternated what I did as the sun, clouds, and moonlight alternately swept the shore.

My body's rhythms likewise shifted. They changed, not in isolation but as the natural cycles of heat and cold and the sun and the moon changed and as I made my own cycles within those changes. My senses and mind were imbued with the experiences around me: the endless sight of the ocean waves, the rhythmic sound of their rise and crash, their misty smell, their tingling taste, and the rippling, pulsing feel of them. The temperatures always shifted—hot and cold—in the water, air, and sand. It was an open environment. No streets, buildings, or other boundaries confined me or separated me from my environment's natural rhythms, as they would have in a city or other human-built place. In those formative years, my rhythms concurred with and nested within nature's.

I continued my running in high school and in college. Instead of running naturally on sand by the ocean, however, I shifted to sprinting on a flat surface. My course was a straight line on an artificial cinder track. I was timed with a stopwatch, for specific distances—100, 220, or 440 yards at that time. Though I still thought in terms of the natural stop-and-go cycles of running from my early youth, I adjusted to this different style of running, and in my senior year of high school, I became captain of the track team. But I always trained in cycles—go/stop, go/stop and so on.

In high school and at the University of Pennsylvania, studying premed, I took courses in biology, chemistry, and physics, and for the first time, I learned about waves in depth, classroom science–style. The waves that I had made and experienced as a child, so very real, though undefined, were taken apart for me and defined decisively.

I was taught that waves have no material substance. I was taught that waves are a "form that moves up and down." They were also defined as a "disturbance in a medium"—a disturbance, as if a wave is not right and the proper way to be is linear and invariant.

Matter makes waves, I was taught, in that order: A rock dropped in a pool will make waves in the water, but waves do not make a rock or any other kind of matter. According to science, they cannot, in our universe. A wave is not a material object. It therefore cannot be assigned a mass.

Not only were waves nonmaterial, I was taught, but waves carry the quantity we call energy—a purely mathematical abstraction of their capacity to do work. That is, waves carry energy but do not carry matter. This can be seen in that the matter disturbed by a wave does not move; while a bottle floating on the ocean will bob up and down as a wave moves forward, for example, the entirety of the ocean's vast waters does not travel onto the beach when the wave does. Only energy moves forward, and up and down, with the wave.

And so, I was taught that waves are not stuff in the sense of the real stuff of the universe—precisely because waves are not believed to be material. For science, this was a given.

The scientific perspective demands that waves be understood through mathematical analysis. This, I learned, was made possible by treating them with lines. I was taught that the wave moves in a straight line—a far cry from the constantly moving and undulating waves I had experienced all my life. I was taught the "proper way" to comprehend a wave: to formulate two independent features, a frequency and an amplitude, to describe it. The way to understand the wave was to plot its speed, even if the speed changes, as a frequency averaged out along a straight line of time. The height of the wave, the amplitude, also must go along a straight dimension.

These waves were understood as if they were contained by boundaries. All waves were considered separate from one another. They were contained also as a concept: an "item" of sorts, with boundaries that separate it, conceptually, from the material that is waving. Because they were separate, they could be superimposed over one another, and they could be added and subtracted from one another.

In biology, too, I learned to see waves through the lens of technology and mathematical analysis—again, isolated from context and from the experience of nature itself. An interesting fact caught my attention in biology. I learned that all our senses are transmitted as electromagnetic waves in the nervous system. The very senses through which we learn about nature are waves—though yet again, this fact was stated in the language of linear, disembodied mathematical waves.

The classroom courses in which I was being taught about waves marked a personal transition in which I went from accepting nature's rhythms to studying waves the way science construes them. Before, I had seen and experienced waves as they simply *were*, without the imperative to abstract and analyze them along lines. Now I was seeing waves through the lens of mechanics and mathematical analysis, isolated in context from the experience of nature itself.

However, at the same time that I was studying waves in the classroom environment, I physically continued to make waves—as a sprinter doing interval training and, eventually, as captain of Penn's track team. But, again, the waves I was making were not the natural waves—seamlessly blending with nature's rhythms—that I had enjoyed as a child. These waves were measured by a stopwatch and averaged out on a timeline, true to the standards of a Newtonian clockwork universe. When I was doing this interval training, I was always told to keep moving: sprint, then jog or walk, sprint, then jog or walk. The interval training was running and then slower running, but I was never permitted to sit down. This is still the way of the sprinter today. The sessions of recovery that I had naturally done as a child, sitting and cooling down between bouts of running as all little kids do, were completely absent from my formal training.

I focused on increasing my speed when I burst into a run, and in time, I competed in national and international competitions. Running remained integral to my life.

AND THEN I WENT to medical school. As was the case with everyone in medical school at that time, there were no introductory classes about life or about health. We students started off with cadavers.

It seemed odd, if not disturbing, to me that to learn about life, I dissected dead bodies. But anatomy was taught early in medical school. I studied pathology early on, too, as well as histology and chemistry. Where were health and performance? Nowhere to be found. The fundamental stuff to study was the anatomic material of the body, piece by piece.

Nor was there a chronobiology course to study the rhythmic phenomena of a living biological system. Such a course did not exist. And anything regarding waves was just a repeat of my college education, where we learned about the rhythmic patterns of the cardiovascular, neurological, and muscular systems, as graphed by the electrocardiogram, electroencephalogram, and electromyogram—through which electrical waves of the heart, brain, and muscles, respectively, are charted along a line. All the

graphs intimated that a *linear* understanding of biological change is a true understanding. In the same way, we learned mechanical play-by-plays of signal transduction pathways and feedback and feedforward cybernetic loops, which were all ways
of understanding sweeping natural rhythms through linear, local cause-and-effect
interactions. Even the loops of feedback and feedforward—though the word *loop*
would indicate a curve—were separated out as two independent directions, moving
in straight lines opposed to each other, up and down. I was likewise taught that
homeostasis and balance were linear ideals toward which the body strives.

For 2 years, the courses had nothing at all to do with live patients. I learned to
focus on molecular mechanisms as the cause of disease. And when I finally did meet
with living people, they were all sick.

I realized that learning to be a doctor was learning about the parts of the body
and about disease. If a body was sick, our puzzle way of seeing things (which I was
not questioning at the time) determined we must find the *part* that was causative.
What about *motion*—isn't motion one of the hallmarks (indeed, when organized
and synchronous, perhaps *the* hallmark) of life? What about the rhythms that were
somehow always present in, even indicative of, physical life? No. Parts, and relationships between parts, were the unwavering point of study. Interrelated repetitive
motions—of the body as a whole—were not up for consideration.

But I did not think about this at that time. I went on to specialize in surgery. And
as a surgeon, I looked at blood vessels, because, like every surgeon, pieces of the body
were my focus. The pulsating heart and the pulsating blood vessels, so fragile yet so
vibrant, were under the surgeon's knife. The pulsations themselves were not only
largely ignored, but I would clamp blood vessels to stop the pulsations in order to
work on them. It was the same thing as in med school. Literally, we took people
apart, with the intent to heal them.

Together with my brother Herbert, I helped create the specialty of vascular surgery for vascular disease. We developed the Dardik Biograft, a method for salvaging
limbs with deteriorated arteries by surgically grafting glutaraldehyde-tanned human
umbilical veins (here I explored the value of living tissue for restoring health instead
of Dacron, an artificial synthetic graft popular at the time). My brother and I won
the 1976 Hektoen Gold Medal from the American Medical Association for the
Biograft. But still, even as a doctor, I felt medicine was far from really, truly creating
health in any patient. The approach of medicine, in my experience, was a war on
disease, but never to cause health in those lacking it.

My interest in sports and vitality was undeterred, however. I was asked by the
United States Olympic Committee (USOC) to be a physician at the Olympic trials in
1972, and in 1976, I was asked to be a physician at the winter and summer games. In
1978, I drafted the bylaws for Congress to create the Sports Medicine Council of the
USOC, in the Amateur Sports Act—which established the USOC as the governing

body for all amateur sports. I participated in establishing the US Olympic Training Center for potential Olympic and Olympic athletes in Colorado Springs. As founding chairman of the USOC's Sports Medicine Council, my focus was on enhancing the performance of the best. That, to my satisfaction, was all about life.

I had come full circle: from a childhood of experiencing natural waves, through different levels of academia where waves were flattened and abstracted, finally back to thinking about and trying to understand cyclic processes—in healthy organisms, natural and in action. At the Olympics, I was working with the same rhythmic phenomena that I had known when I was a child. I was back in the context of life, health, and performance.

It was obvious to me that rhythms demanded some attention here. The energy metabolism responsible for not only athletic performance but, in fact, all of life involved rhythmic processes. The ATP and ADP processes through which cells get their energy were cyclic. The Krebs cycle, was, as evident in its name, cyclic. The cell cycle itself was the means for cell division and growth. Every hormone in the body was secreted in a pulsatile way. The list was endless.

By now I was sufficiently convinced of the importance of rhythms in human life and performance, and I brought in world-renowned chronobiologists Chuck Czeisler, Frank Sulzman, and Chuck Fuller to work with the Olympics. Our objective was to focus on the rhythmic phenomena of human behavior—to explore how the circadian rhythm, hormonal and physiological cycles, jet lag, and a host of other rhythms work and, more specifically, contribute to and affect performance, good or bad.

We established an Elite Athlete Project at the time of the Olympic Cold War with the Iron Curtain countries—Russia, East Germany, and others. We were aware they were taking drugs for enhancing performance, and we had no idea what was going on in the United States. So we brought in five elite teams—track and field, weight lifting, cycling, women's volleyball, and fencing. We had enough information on the first three teams that we suspected they might be taking drugs to enhance performance, and then we brought in the other two, which were benign. We tested urine and blood regularly and discovered drug use. This was the beginning of testing for drugs in athletes in the United States.

Controversy aside, I recognized that the bottom line was that when athletes take drugs to enhance performance, they are really using chemicals to influence chemical processes already at work in their bodies. The drugs involved were ones that affect rhythmic processes, inasmuch as they mimic and/or enhance natural hormones that are secreted rhythmically. I knew there had to be a way to enhance human physiology and performance—our own chemistry, our own drugs naturally present in our bodies—without taking in drugs, artificially, from the outside. How to enhance natural cycles for optimal performance was of great interest to me.

Studying chronobiological activities stood in contrast, however, to the standard practice of medicine and medical and athletic research, which focused on the structural, material parts of the body. Biomechanics, exercise physiology, sports psychology, and nutrition—every specialist had a say that his or her *own* field determined excellence. But even when all the specialists sat down to analyze films of performances piece by piece, each chiming in with particular insights, the whole of it never explained what put great performances over the top. Something was unifying a great performance that no specialist could claim for his or her field. It was apparent to me that there was more to a record-breaking performance than tinkering with the particulars.

The concurrent rhythms that intrigued *me* were simply not considered, a world away from this piece-by-piece approach. Only pieces, and piece-by-piece interactions, were on the table for discussion.

I wanted to put together all that I believed to be important. There was disease and dying on one hand, taking people apart. On the other hand, with the Olympics, we were looking at the performance of the whole organism. One dealt with survival and acute chronic disease and problems; the other, with health and performance. My focus was on what pulled it all together to be as one. I began to wonder if rhythms were in some way responsible for synchronizing everything toward high performance—in health as well as in athletics.

It always concerned me that everything I was doing, in my whole medical training and practice, was looking at parts. If the whole body is affected by a disease, are we certain enough that the affliction arises in a pinpoint that we look only under a microscope? Rhythms, cycles, waves, oscillations, fluctuations—all names for the same thing—stayed in my mind. I wondered how synchronized rhythms impact performance and health.

CHAPTER 10

Tragedy Yields Discovery

JACK KELLY, brother of Grace Kelly (Princess of Monaco) and a close friend of mine, died unexpectedly in 1985, at age 57. Jack's death was both unexpected and shocking to me because he was a four-time Olympic oarsman and in top physical condition. Yet he died of a heart attack while taking a long run, something he did frequently. Shortly before his death, Jack had been elected president of the United States Olympic Committee and had reappointed me as chairman of the Olympic Sports Medicine Council. And then he was gone.

It just didn't make sense to me. Jack was a well-trained athlete. The best of the best. His resting heart rate was about 40. I had measured it when we used to run together: I would sprint ahead and recover while he caught up, running steadily. It just didn't make sense to me that Jack, an Olympian who could run many miles, would die. He should have adapted to it. He should have been able to handle it. He should have survived.

But he didn't survive. Isn't survival what we're talking about? Isn't all of athletics about being in top physical shape, to perform and survive under extreme conditions?

Yet it was not only Jack. Over 2,500 years ago, Pheidippides of ancient Greece ran 26 miles after the Battle of Marathon to alert his people of victory in the battle and the threat of the approaching Persians. "Victory is ours!" he is said to have cried when he got to Athens—and then, as the legend goes, upon stopping his run he collapsed and died. We now run marathons in his honor. We run to celebrate a physical feat of great endurance—which ended in sudden death. After Jack's death, I was struck by the poignant irony of this.

Then there was Jim Fixx, who wrote the revolutionary *The Complete Book of Running* in 1977. He died under a tree, having stopped to tie his shoelaces during a run.

Jack, Pheidippides, Jim Fixx, countless other athletes who have died after extreme bouts of prolonged exercise—all dying during the *recovery* from exercise. You would think, if anything, the heart should give out in the extreme peak of exercise, not in the recovery afterward. Something didn't make sense.

In the famous book *Running without Fear,* which he wrote after Fixx died, Dr. Kenneth Cooper—considered to be the father of aerobics—talked about the "great cool down danger." The book suggested you keep jogging and walking after you complete a long run because of the risk of dying afterward. This advice was itself based on an explanation that appeared in an article in *JAMA: The Journal of the American Medical Association.* It was known that intermittent leg muscle contractions, during running and walking, cyclically compress the veins and thereby push venous blood back to the heart. The authors of the study reasoned that blood wouldn't pool in the legs during exercise because of these contractions—but could afterward, while the runner was standing still or sitting, in the absence of those muscle contractions.[1] The subsequent lack of blood-flow back to the heart, this model suggested, could drop blood pressure, causing an arrhythmia, potentially a heart attack, and ultimately sudden death. Dr. Cooper's suggestion to continue jogging and walking during recovery was supposed to keep the muscles pumping the blood back to the heart as you slowed down, thereby preventing it from pooling in the legs.

I found the need for the idea of running without fear to be odd, to say the least. Why was it necessary to write a book called *Running without Fear*—with a chapter including "the Great Cool Down Danger" in its title, no less—when endurance running is supposed to be great for you? To me that rang a serious alarm bell. If it is healthful to run in the first place, it doesn't make sense that you have to be afraid of dying—especially during recovery. Running is a natural human activity. And these athletes should have acclimated to their routines. Shouldn't running have enhanced their cardiovascular health, performance, and longevity, instead of exposing them to the possibility of sudden death? Why were their heart rates plummeting after exercise?

I began to think about recovery—since that is when people were dying. Something else must be occurring during recovery that could cause sudden death.

The link between recovery and sudden death demanded answering a different question first, however. The question, a most basic one for me, was: What *is* recovery? Recovery had always been treated as an "interval" and not as its own phenomenon.

It occurred to me then that recovery from exercise has defining qualities. I realized that it is a physiological phenomenon in its own right, one that is physiologically identical to meditation—a practice whose value was just coming to medical attention at that time. No one had ever understood that before.

Meditation induces physiologically favorable changes in response to stress. Specifically, the heart rate slows down, blood pressure drops, stress hormones come

down, breathing slows down, and the muscles relax. Today the benefits of meditation for mental stress reduction are well known in medicine.

I recognized that these mind-induced changes during meditation are identical to
what happens during recovery from physical exercise: The heart rate slows down,
blood pressure drops, stress hormones come down, breathing slows down, and the
muscles relax.

I saw that recovery from exercise is an *active* process! There is an active physiological recovery response going on, as active and dynamic as the stress response of
exercise itself. This was not a passive event like turning off an engine.

I had never even considered that. I had certainly never seen it discussed in literature on medicine or athletics or in fitness journals. It seems obvious now when I say
it, but recognition of this fact had somehow been missed. Recovery from different
forms of exertion marks real, significant changes and processes in the body. Good
changes—changes that we bring about purposefully through meditation. This was
new. (Could we train recovery? I put that question aside for later.)

I wrote a paper in the journal *Advances* that outlined how the meditation response
and recovery from exercise are identical. The paper described both the idea and how
to make use of it:

> *In this paper, I propose a new method to induce physiological relaxation—through
> exercise . . . a program founded upon the premise that exercise and recovery from
> exercise involve physiological processes that, respectively, are synonymous with stress
> and relaxation.*
>
> *The program uses heart rate controlled exercise as a stressor, and the comple-
> mentary recovery period is exploited as a process of "active" relaxation.*[2]

The program tapped into the anti-inflammatory, antioxidant benefits of recovery. It even surpassed meditation in that respect, in that it took a person through a
fuller range of recovery. It dropped a heart rate from, for example, 180 to 60 instead
of 70 to 60, as meditation did. This to me was crucial: being able to induce a full
range of anti-inflammatory and antioxidant responses.

At the same time, I was astonished to realize that the unity of those physiological
processes—the single identity that characterizes both exercise/recovery and anxiety/
meditation—provided a bridge between the physical and mental realms. One
process—a physiological, documentable stress-relaxation process—is shared by both
mind and body. One rise and fall is done by both. Right there was the mind-body
connection that so many scientists had been searching for. No one had ever been
able to define a mind-body connection, and as we will explore later, here, suddenly,
was an explicit commonality, of a physiological stress-relaxation process, that the
two realms share.

IN MY EXPERIENCE at the Olympics, we had taken recovery for granted because, though you had to go through it at some point, it was more of a necessary evil. It had to be done because the body demanded it, but overall, the focus of training was on exercise. Endurance was—and still is—the ideal in most sports, as well as in fitness training for the general public, and that necessitated minimizing the need for recovery, which in any case was considered best if you kept jogging and walking through it. The idea of training recovery—to monitor its progress as an essential process just as we do with exercise, and to establish ideal patterns just as we do with exercise— would have been an oxymoron. No one even considered training it.

What an epiphany!

Would you ever tell anyone who is meditating to keep being anxious while he relaxes? Would you tell someone who had not slept in 24 hours that it is best to stay perky and alert when she finally falls asleep? Well, that is what we were doing when we told people to keep jogging during recovery after a run. I saw that the remedy for the "great cool down danger" did not make sense. Why would one stress oneself during recovery? It effectively prevents the systems of the body from going through normal, healthy, and expected restorative processes. Jogging and walking through recovery must, to some degree, compromise the anti-inflammatory, antioxidant, antistress relaxation process and the benefits that come with it.

Recovery, no question, is *good* for you.

The big question then stepped forward: If recovery is good for you, how could it be a danger? How was it that people were dying during recovery?

As an experiment, I took a few triathletes and distance runners and checked their heart rates. All had low resting rates, about 40 beats per minute, like Jack. I had the athletes run up several flights of stairs to raise their heart rates to a peak, around 170, after which I had them immediately sit down. As I watched, their heart rates rapidly dropped down to the low 30s and even the high 20s. It was a shocking drop. It was perhaps all the more shocking because a rapid recovery after exercise was supposed to be good for you. But I realized that such plummeting, low heart rates could easily cause an arrhythmia or even a total loss of the heartbeat— and sudden death.

The American Medical Association's idea of blood pooling in the legs had missed the mark. The prolonged stress of exercise preceded a significantly sharp descent into recovery, and the athlete's recovery physiology could not pull out of that dive. I realized that training recovery, so that a person does not experience such a rapid drop in heart rate, must be vital for preventing arrhythmia and cardiac arrest.

The training we were doing was all wrong. The endurance athletes trained in the spirit of the popular phrase "No pain, no gain." It glorified the idea of the burn; in fact, it was all burn, with no chance to pull in a deep draw of refreshing recovery on a periodic basis. Yet hours' worth of straight exercise compared to a mere 5 or 10 minutes of recovery was like a chronic panic attack: all go, go, go; stress, stress, stress—exactly what meditation was meant to counter in the mental realm.

For some reason, these athletes' recovery was forced to happen so fast that the heart rate shot down (and along with it, blood pressure, of course) and went *below* the already-low resting heart rate. That exaggerated, sharp descent meant it might keep dropping, and unless the athlete continued jogging and walking to counter the dive— until the heart rate could swing back up and stabilize—the heartbeat could crash.

My endurance runners had demonstrated it through sprinting up the stairs and then immediately sitting. The cause was not blood-pooling itself but what I now saw as an inadequately trained physiological recovery response. The combination of a trained, low baseline (resting) heart rate and a rapid descent into recovery—both of which were supposed to be good for you!—was a perilous danger.

I CONSIDERED THE BEHAVIOR of an endurance runner in contrast to that of a sprinter, who runs more like an animal in the wild.* A sprinter is all stop-and-go behavior, bursts of activity alternating with recovery. Unlike endurance athletes, sprinters do not push off recovery endlessly; rather, they avail themselves of the physiological changes of recovery between sprints. Sprinters in general are not predisposed to sudden death after exercise.

I tested this on myself and some other healthy people, none of whom were distance runners with low resting heart rates.

How did our heart rates respond to a strong sprint and immediately sitting down? All of our heart rates shot up and then naturally dropped, going down perhaps two or three beats below the baseline (where we had started)—and then, unlike the endurance runners' heart rates, went back up.

So when I started with a heart rate of 60, for example, took a quick sprint so my heart rate went to 180, and then sat down, my heart rate went down and then bounced back up to around the baseline instead of crashing. From 180, it dropped to 174, 163, 151, 137, 128, 116, for example; whatever the specific numbers were, over the next few minutes, it eventually went down to the low 60s and then climbed back up and settled in the low 70s.

My heart rate didn't go down as fast, was certainly not as low, and didn't drop too far below my original baseline heart rate before climbing back up, even ending up slightly higher than the original baseline. This was far better, the opposite even, of what had happened with the long-distance runners.

---

* In Chapter 1, I described that before civilization, when we lived in nature, we surely did not run and run and never rest as long-distance runners do. To survive, we hunted and gathered cyclically, in daily cycles, as seasonal cycles allowed, alternating between activity and recovery. Only the heightened necessity of pursuit (us pursuing prey or being pursued as prey!) kept us going, but that happened in cycles, too. The rhythmic tasks of living meant we would not have run and pushed endurance, day in and day out, for years.

I realized that marathoners do not train recovery in a healthy way. They are all about endless burn, whereas sprinters incorporate recovery into their training. Training recovery properly is crucial. It seemed to me that exercise physiologists should be trained as exercise-recovery physiologists if their ultimate goal is to improve performance and health.

I saw a definite similarity between these endurance athletes and people who succumb to drug addiction or suffer from post-traumatic stress disorder. Drug addicts who overdose die after falling into a coma—not while high but while in a crash after the high. Post-traumatic stress disorder, as the name clearly underscores, doesn't kick in during the stress that is experienced, but afterward. Sometimes people cannot recover *from* the recovery that naturally follows their upswings. They cannot pick back up again. From this perspective, the key was that whatever the cause of Jack Kelly's death, it somehow involved the overall process of exercise followed by recovery. For me, blood-pooling was, definitively, not the complete answer. If one pointed the finger at the pooling of blood, it did not lead to understanding the primary problem to begin with. It still endorsed chronic stress. It led to the idea of jogging through the important phase of recovery.* It barred recovery from taking its rightful position as a necessary complement to the stresses of exercise.

AND THEN IT CLICKED. It was an idea that resonated with my lifelong awareness of nature's cycles and rhythms. It was a novel idea whose simplicity belied its significance.

One half of the idea was that aerobics, or training exercise, is *pro*-oxidation, *proin*-flammatory: It elicits a stress response.[3] The second half was what I had just realized, that recovery is an *anti*oxidant, *anti*-inflammatory, *anti*stress, relaxation response. The "click" was seeing their connection. There is a relationship that connects exercise with recovery, the same that binds stress with meditation. Pro-oxidation then antioxidation, proinflammatory then anti-inflammatory, stress response then relaxation response, over and over again; no process persists alone, forever. The changes always give way to one another. They are all natural, indivisible cycles.

Exercise and recovery are not isolated behaviors. Exercise and recovery is *a wave*.

Our behavior itself is the best witness to this truth. You cannot exercise without recovery and you cannot recover without exercise. They clearly cycle and are recognized as cycling, back and forth. Yet even though they must cycle back and forth, we have always treated exercise and recovery independently, indeed, even as opposites. In truth they are a *continuum*. That was the key.

---

* Of course, it was indeed necessary for the overtrained endurance athletes to jog and walk through recovery, but that was only to counter the problem created by ignoring recovery during training. One has to properly train recovery in the context of endurance training. This topic will be covered later in the book.

The conventional way of thinking had shunted attention away from the reality of this continuity. The Puzzle Hypothesis, whose influence and scope we have spent time exploring in earlier chapters, has told us to regard and treat exercise and recovery as two separate entities. It also promotes seeing any relationship they might have as a connection—as if there is a relationship between "parts" of the puzzle of nature.

But in truth, the back-and-forth flow from one to the other means more than that exercise-recovery are connected: They are not *dis*connected. They have never been disconnected. Neither has an identity independent of the other. The natural flow of exercise into recovery and back into exercise belies the separate way we have regarded them. Treating them as a puzzle is not warranted, and it serves no benefit. It is far more instructive to contemplate the changes of recovery as a process continuous with, and descending down from, the peak of the exercise it follows.

The nature of this wave was a far cry from the popular notion of exercise through workouts or fitness classes that people schedule for a half hour or an hour several times a week. This was a reality of unboundaried, cyclic changes that we all experience, wholly different from a discrete dose of something to do in a punctuated segment of time. An aerobics class is over and you go home. Exercise and recovery cycles are an experience of change that flows back and forth throughout a person's lifetime.

The wave does not "end" after you recover, and this seemed important. The heart rate picks right back up when you get up and walk after recovering. It will drop again when you sit down to change your shoes. It rises when you run down the stairs, drops when you sit in the car, and so forth throughout the day. This is true to the nature of waves, which exhibit no natural boundaries.

Indeed, the origin of such waves stretches back in time, to the beginning of our lives, and they keep going forward as we age. This is the reality. Waves are a pattern of *repetition*. They move not just "up" and "down," so to speak, but cycle, repeating forward through us, as regularities of nature so stellar that they persist regardless of the details in our lives.

WHAT I REALIZED is that the pattern of their inherent connection provides a unique vantage point from which to conceptually understand exercise and recovery as a wave. The wave is natural and real, like the waves I had been aware of since childhood. Science also addresses waves, defining a wave as a disturbance in a medium or a pattern or form that moves up and down. Given either perspective, a wave of exercise and recovery is quite noticeable. It is a physiological stress-relaxation process that certainly qualifies as a disturbance in a medium or a pattern that moves up and down, which naturally repeats as it travels through our bodies. It is an observable wave.

My recognition of this wave, which was a relatively small but significant act of identifying an otherwise unrecognized natural phenomenon, was, in fact, the first of what would become a cascade of recognizing heretofore unrecognized natural phenomena. The reason it was so significant was that it asked a different primary question of nature than we have ever asked before.

Our first question of the natural world has always been "what is it?" and only after we identify whatever part of nature we are focusing on, do we ask "how does it work?" Here I asked how the working happens—not how "it" works, for I brought no particular "it" into the picture, but just how motion works. The way I was thinking focused on motion first and, at this point, motion alone. I saw wave motion that I did not tie to matter.

How and why I took this stance is hard to say, but the best I can say is that I just saw it. I saw that in life, this wave motion is real. I saw an upswing and downswing of something shaping our physical behavior: some natural wave compulsion that makes change.

It was almost as if some "thing" flows through us—something we burn and recover, some surge, that peaks during exercise and stress and dips during recovery and relaxation. I would almost have said that a wave of energy compels us, but it was not energy as science defines it—for science's energy is a mathematical abstraction with no tangible reality. I saw this as closer to the popular notion of energy as a genuine thing in nature. I couldn't find an exact way to describe this with already-known ideas, however, which made sense because it was new.

This novel focus on a perceptible cycle of motion *not* secondary to matter allowed me to see it as it was. I saw a wave of motion, one easily recognized if you can let go of identifying matter first (i.e., if you let go of needing to say what matter is moving and look at motion alone). It was a "one" wave of something like energy, in whose context exercise and recovery play out for the entire body. Ultimately, I was recognizing a powerful pattern of undivided motion in its own right.

To refer to it, I called it an exercise-recovery wave.

THE RECOGNITION OF MOTION in this pattern offered me a different perspective of how we change. I realized that the human body is subject to many similar exercise-recovery waves. Regardless of the material systems through which we see them, they are all the same. And they all repeat in cycles.

Stress and the physiological changes that accompany meditation, I had already seen, wave through us in the same sort of wave: We feel anxiety and we relax, cyclically. In the same way, we experience hunger and we eat, over and over—a person awakens with a slight gnawing in her stomach, gets up, gets washed, and goes to the kitchen; after a cup of coffee and a roll, she feels satisfied; at some point in the midmorning, at work, a creeping feeling of hunger starts to gather in her belly; she waits and works as it intensifies; she finally gets up and gets a sandwich and a drink and the wave subsides in satisfaction; and so on, throughout the day. So too we wake and we sleep: We all experience an arc of being awake during the day and sleeping at night, day in and day out, for all our lives. And, of course, we exercise and we recover. Though the puzzle mind-set would predispose us to think of these behaviors as pairs of opposites, in fact, none are separate entities and should not be treated as if they are independent from one another. They naturally cycle back and forth.

That is to say, anxiety, hunger, wakefulness, and exercise are all ways we "burn energy"—or to use a term not appropriated by science, they are all exercise—and they give way to a recovery of energy, or in plainer terms, to recovery. They are all forms of exercise-recovery.

Everything we do, all our behaviors, are waves of this type. This was an important point, which would come into play again soon.

What captivated me, beyond the existence of a continuity that departs so sharply from the way we've treated behaviors as a puzzle, was the commonality, the wave, that all the behaviors share. Each is a demonstrable upswing of activity or exertion that courses through us, naturally followed by a downswing of recovery. Though we call each type of wave behavior by a different name, and see them expressed in different physical systems, we are all almost passive in their common wave path.

It was, I saw, a common-denominator language: a language common to all our systems at once.

I FOUND THE WAVES fascinating, but at the same time, I also sensed a tug pulling me from our standard framework. I must have realized, on some level, that acknowledging them meant reversing direction from the reasoned steps of the Puzzle Hypothesis. I was seeing unity in movements that had been regarded as separate. And I was seeing waves through that physical human body whose changes had been regarded as a disjointed set of individual physiological steps. I was seeing motion in its own right, not tied to matter.

At the time I was recognizing all this, I was a practicing vascular surgeon with a medal of recognition from the American Medical Association and founder and chairman of the US Olympic Sports Medicine Council. I was in a fulfilling career with plenty of professional recognition. I was not trying to be a renegade—but I *was* looking for answers that the standard framework was not providing. So I looked at what was natural—at what was naturally happening.

And somehow, through some insight, and without meaning to challenge our usual way of thinking, I refrained from taking the step to isolate these waves' flow from one to the other, from graphing them, or from construing them in any way as objects with parts and wholes that relate with one another. This stance maintained the integrity of motion.

Indeed, the natural nature of waves supported this approach. There simply *was* no discrete unit of motion to warrant treating the waves as a sum of parts and wholes. The idea of parts and wholes and the practical techniques we derive from it are children of a long-forgotten marriage that ever remains pertinent: the marriage between how we physically perceive material objects and how we intellectually relate to them. We think in terms of edges and boundaries because we sense edges and boundaries in *matter* and physically treat matter that way. Yet here I was, looking at motion. The wave motion itself exhibited no such quality.

I felt clear that in order to understand that motion, I could not move into abstractions—I had to pay attention to what was actually happening.

Whatever the waves were, they were something real that we experience directly. They rose and fell as naturally as waves in the ocean. Indeed, we personally experience them every day, feeling them pass through *us* the way waves lift and lower water.

I'll restate this point because it was so important: Somehow, as I was recognizing these waves of exercise and recovery we experience in all our behaviors, I refrained from framing them as a "part" of nature as a puzzle. I did not treat them as an isolatable entity one could measure in an absolute way. They were not a "thing" with a proper place. They were simply a repetitive process that endlessly happens—an up and down of something akin to energy that flows through us all. They were a natural pattern of motion, easily recognized if one could get past the insistent sense that we must think of matter first. Hard though it may be, if one can desist from feeling one's body is primary, there is no reason to outright reject the idea of exercise-recovery waves passing through a person.

The purpose of all the opening chapters is finally coming to fruition. I hope you too can now hold back the impulse to assign edges and boundaries and refrain from framing waves as "parts" of nature as we explore them further. It's going to be quite a ride. We are about to see how waves wave within waves, without boundaries and with an inherent natural continuity that changes everything.

The wave aspect of exercise and recovery opened up for me the whole area of what happens when a person runs and then stops to recover. Apparently, when the people I tested exercised and recovered, the dip into recovery had everything to do with what happened during that peak exertion. Their heart rates always plummeted when recovery followed a bout of vigorous exercise.

And then it happened: The most extraordinary recognition hit me. I was confronted with a realization whose enormity is difficult to describe.

By necessity, it has to be explained step by step—that is, after all, how explanations must proceed. But as we go through the ideas, bear in mind that the wholeness of the concept, particularly in contrast to our age-old part-by-part approach of the Puzzle Hypothesis, set it apart in an astounding, even beautiful, way. In its totality, it was one idea with many implications.

Here we go.

A bout of exercise is always followed by recovery, and during my experiments, the athletes' heart rates plummeted during that phase. That meant: The body was not the only entity descending into recovery. The heartbeat was, as well—at the same time. In other words, as the athletes' exercise-recovery wave descended into recovery, their heartbeats also took a dive.

But, I realized further, it was not an averaged-out heart "rate" that was falling. It was actual motion. And not just any motion. The heartbeat is the same sort of wave motion that rises and falls during exercise-recovery. The heart muscle, like the body, "exercises" and "recovers" with every beat of systole (contraction) and diastole (relaxation).

The waves of the body and the waves of the heart were the same "stuff"—waves of exercise-recovery.

And the wave of the heart's motion fell just as the exercise-recovery wave fell.

Here were two waves of the same sort waving *simultaneously*. A spark of excitement went through me with the notion of simultaneity.

The waves change together, with no boundary. On what we have been calling two different scales—the scale of the body and the scale of the heart—waves rise and fall together simultaneously, as one.

A representative sketch can help clarify the inherent relationship. The heartbeat's surge changes in two ways: The heartbeats themselves are waves of exercise-recovery, and they accelerate and decelerate when the body as a whole does. That is, they rise in frequency and intensify in amplitude during the upswing of exercise, right as the body surges with accelerating, intensifying exercise. And their deceleration flows right down with the body's dip into recovery.* (Note that the diagram is a representation of a pattern, or relationship between waves. The waves that unfold through us are, of course, not confined to flat, dimensional lines.) When we refrain from breaking up motion in accordance with matter, we can see that these human body wave motions naturally change together, on two scales at once.

Everyone experiences this. Your heart beats faster and harder during a bout of exercise and slows and beats less intensely during recovery. (The blood vessels pulsate with this beat, as well, weaving it through all the organs and limbs). This changing heartbeat itself constitutes a larger wave that rises and falls while your body as a whole experiences a wave that rises and falls.

---

* Anyone can do this themselves: Run up the stairs and then immediately sit down. You will experience a wave of exercise-recovery and, simultaneously, waves speeding up and slowing down in your heart.

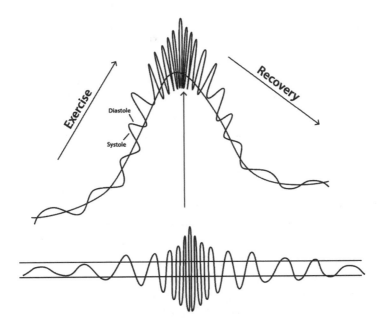

The heartwave oscillates—speeds and slows—as we exercise and recover. It climbs up and down, so to speak, a bodywave of exercise-recovery.

It is motion we can see via matter without anchoring it on matter: the large wave motion of the body as a whole and the smaller wave motion of the heart changing within it. We all experience these two layers of change *that change synchronously.* Medical doctors may separate the material of a heart from the body by dissection, but no one can separate the heartwave from the bodywave, as I began to refer to them.

Synchronicity means neither wave motion directly causes the other: This was key. Traditionally, science seeks to determine local, linear cause and effect. Therefore it seems tempting to say that the rise of the exercise-recovery wave causes the heartbeat's wave to rise.

But that is not what actually happens. The onset of exercise surely affects the heart to beat faster and harder, but those very fast, hard heartbeats can equally be considered to bring on the physical surge of exercise itself. In fact, if the heart didn't beat faster and harder, there would be no experience of exercise. And both waves naturally come down together in the physiological recovery response. The actual body and heart waves, as they happen, and as I saw them happening, are what you might call motion interdependent—inherently happening together and not causal from one direction only.

The more traditional and therefore appealing possibility, that they are two real waves discontinuous from one another, does not hold up to scrutiny in this case. If they were two distinct entities, it would have to work this way: Somehow, when the first wave went up, it would, through some local agent, nudge the other wave to go up. But if that were how it worked, then one could at least *exist* without the other—maybe not in practice, but it would be possible. Yet it is impossible. It doesn't make sense to have a body exercise-recovery wave without the heartwave, or vice versa. Since that *doesn't even make sense*, since it's impossible to separate them, then logically they can't be separate things.

A hint of the enormity of what was to come lay in that one key idea: that even though we can see it on two scales—even though I am explaining it that way—the living organism features a *single* phenomenon, of motion changing with and within other changing motion. The pivotal point is that what we see and think of as two layers of motion can be understood as a single wave phenomenon, of waves waving together across scales. It is one entity, so to speak.

Remember—though I use the word *wave*, which sounds like a finite thing with edges and boundaries, I am talking about motion only. In lifting it from its attachment to matter, I am pushing away the boundaries we impose on it. We can think about wave motion without boundaries, even if it is a little hard to do so. We can see that when it changes, it changes as one phenomenon (but that we call two scales) of waves waving in waves simultaneously.

FOR ME, THE SIMULTANEITY was riveting. The absence of a single arrow of causality made that simultaneous change fairly sing with significance. It suggested an entirely new way change happens in nature—new to our probing eyes but as old as our existence.

It in fact pushed the unity that I had just begun to recognize onto center stage. First I saw that exercise cannot be separated from recovery and has never been separate from it. Then I saw the heartbeat is a wave of the same sort that cannot be separated from, and has never been separate from, the body's exercise-recovery wave. And then I saw that the heartbeat makes a wave—speeding and slowing, and intensifying and subsiding in intensity—in a rising and falling wave that is not separate from, and has never been separate from, the body's exercise-recovery wave. Again, setting aside the material substance and its boundaries, I saw the motion as one entity, of waves traveling in waves.

This idea of one phenomenon of waves offered a sharply different take on our notion of causality.

Because the heart and body motion unfold together—as waves that travel in waves—it allows us to see what could not be seen: Where it changes on one scale, it simultaneously changes on the other. It does that because it is not a material with boundaries between parts and wholes. It is motion, which can have its own qualities— which does have its own qualities, and apparently, cross-scale simultaneity is one of them. Motion can be in other motion and still be one. Motion can be in other motion and all *is* one—because motion is not the sort of thing that inherently has boundaries. A wave can ride another wave as one so-to-speak thing.

Believe me, I know it can be a bit of stretch to think this way when we are so accustomed to segmenting all we encounter, but the fact remains that no boundaries exist to waves. A wave riding another wave, as both change, is not deeply hard to understand. We see them as scales, but in actuality they change together. It's simply what they do. So long as we remember that our words tend to make us think of things as separate but that words such as *wave* or even *both waves* help us get to the right idea instead of actual objects, we can see that there is unity and continuity to true wave motion. We can see a nonseparateness that allows for simultaneous change, where a framework of parts and wholes does not.

The waves are one phenomenon of interrelated wave motion that we, from our human perspective, identify as waves on different scales.

It was startling but enthralling to recognize an "only one" phenomenon of motion in motion going on, the opposite of our traditional approach to understanding nature. Nature showcases many apparently different parts, and from there, we long ago drew the understanding that it is a whole made of parts, with no choice but to reject the possibility of an inherent oneness. Waves waving, surprisingly, provided me a curious way out of this predicament. A wave is easily identified as a single entity. And yet, its up-and-down pattern incorporates what one might identify as opposites. It is one, yet it houses differentiation.

Waves waving takes this inherent coexistence, of what we discern as different being one phenomenon, to a new dimension: It crosses scales seamlessly, as well. We see, as separate scales, two layers that wave together. They are one despite our differentiating between them. It is similar to how we recognize the phenomenon of a peak and a trough as a single wave, but it goes even further, such that we can recognize what seem to be two scales changing as one forward flow of motion. Oneness—in the incarnation of waves waving—unifies opposites *and* scales all in one go.

I mean this not in a mystical or transcendental way. It is the actual reality of how waves happen. The many facets of nature once compelled us to adopt a certainty that nature harbors absolute boundaries between parts, but waves waving rebuff the need to apply that conviction. They rise and fall as one, and they do so across scales. Waves waving are inherently continuous while also allowing for what *we* identify as different parts.

These ideas gained traction in my mind.

THE WAVES' INTERDEPENDENCE REPRESENTED to me a novel insight into the workings of human behavior and health. The waves surely drew from the same pool of experience as the scientific concepts of motion and energy do, yet in no way could they be construed as energy in the scientific sense. The scientific idea of energy is an abstraction—and this wave phenomenon was not an abstraction representing a "capacity" to cause change. The waves were simply change itself happening. The nature of the motion of waves waving differed fundamentally from the nature of matter, which we regard as separable, and whose boundaries we apply to abstract out energy. It was the inherent relationship between scales that caught my attention.

So even though traditional analysis of waves requires taking them apart, charting them on lines, and relating them with abstractions such as energy, that approach would not have been fruitful here. Mathematics could not, at this point of my understanding, adequately describe the cross-scale relationship.* Taking the waves apart to discuss them in terms of mathematics and energy would, in fact, inherently cancel the reality of the way the waves happen. It would eliminate the ability to see the consequence-laden simultaneity of motion that I had just recognized.

One might say that the flip side of mathematics is the living organism, with all its irreducible complexity: and that was what was showcasing the indivisible simultaneity now apparent to me. Life puts on display a fantastic wave phenomenon of simultaneous change not immediately visible in the mathematically receptive realm of nonlife. It put forth, to me, a new world to explore, a new starting place. Somehow, in the most complex phenomenon we know of, in the realm where we cannot properly apply mathematics and where we have never yet been able to figure out how parts add up to the whole (that is, in the phenomenon of life), waves clearly wave together in a singular, inherently continuous event.

I COULD HARDLY BELIEVE I was the first physician or scientist to notice it. Here was an amazing wave phenomenon, happening daily, in the body of every person as they exert themselves and recover. But because the Puzzle Hypothesis orients us to think that nature is separable parts that relate, mathematically, by either linear cause and effect or by statistics, it had blocked medical researchers from recognizing it.

The simultaneity I was seeing was not only new, it was completely without precedent in medical literature. Concurrent change of this type had been somewhat recognized but construed in a different way. Physicians have known for some time,

---

* It should go without saying that this phenomenon of a wave within a wave could not be analyzed by the traditional method of treating waves by superposition. (Superposition is based on constructive and destructive interference, how waves affect each other when they cross paths on linear trajectories; it adds and subtracts amplitudes in a linear fashion, as an electrocardiogram does.) This here was waves waving together, each uniquely visible on its own scale but nonetheless united (and therefore not to be treated as many individual waves).

for example, that the heart's rhythm naturally changes when we breathe in and then out. The technical name is sinus arrhythmia. Sinus arrhythmia is not only normal and healthy but is actually good: It decreases with age, diabetes, and cardiovascular disease.[4] It is the natural way of the heart. But look at the name science has given it: sinus arrhythmia. It sounds problematic because of the word *arrhythmia*, but that's only because we consider varying rhythms to be *not* normal—science favors *invariants*, as we have seen.

The difference in perspective was critical. Sinus arrhythmia portrays the heartbeat as traveling a straight line on an electrocardiogram (ECG) graph, by itself. The focus from my point of view was, specifically, the opposite. Changes in the oscillatory heartbeat happen along with not only breathing but also with the changes of exercise-recovery. The whole of it can be rightly represented only by curves within curves, which never appear on an ECG.

To me, with my medical background, that was astounding. The ECG is the gold standard in medicine. It is a tool of measurement whose acclaim is so expansive that its graphic image signifies the world of medicine in advertisements, articles, and books on health and disease. And even so, as an authentic window to reality, it is lacking. It offered no means to assess the reality I had just seen. The way Willem Einthoven created it,* the ECG graph severs the heartbeat's solidarity with its partner, the behavioral wave of the body. Any representation of the way the heartbeat rises and drops in synchrony with the bodywave, which significantly affected the athletes I'd tested, is patently absent from it.

Because of science's conviction that we are to construe change in a step-by-step linear format, the way the waves genuinely change together is unrecognized and, diagnostically, untapped.

The origin of this monumental oversight was a divisive issue: the chasm between true continuity and its obverse, discontinuity. Discontinuity entails both linearity and continuous divisibility. The basis of my new approach was to reject the jump to discontinuity—the a priori assumption behind the Puzzle Hypothesis and embedded in all its aftermath—and instead to embrace continuity as we naturally experience it. Continuity perforce means that the waves travel in waves, not in lines.

Conceptually, the ECG's portrait of the heart rate and the heartwave I was seeing, in fact, could hardly be more opposed. The ECG *is* lines along a line. With it, the body's behavioral beat—waves of exercise-recovery—is extirpated, leaving in its vacuum a neat line along which the heartwave is etched; the heartbeat itself is straightened out in order to measure it; systole and diastole are represented along a straight line, where each beat is portrayed as flat; and significant variations in the heartwave's frequency are averaged out to report a heart rate. The way the endurance

---

* Dutch physiologist Willem Einthoven developed the ECG in 1903, as discussed in Chapter 4.

athletes' heart rates plummeted during recovery required that I incorporate all the waves that happen at once, which resulted, instead, in waves in a wave.* The fact is that the bodywave of exercise-recovery nests with the heartwave.

THE UTTER MISMATCH BETWEEN the wave and the line stood out forcefully to me yet was not of my origination. It has been long known and accepted in science that absolutely perfect straight lines, which never change, do not exist in nature. As discussed earlier, lines are prized anyway for their incredible usefulness since they— and anything culled from its natural state and plotted on a line (including waves)—can be divided into constituent parts and points; they can then be subject to calculations that indicate linear cause-and-effect interactions in conformance with the Puzzle Hypothesis.

Lines seem an excellent way to break down nature, but here I was, and it was a simple matter of trying to understand. People, like Jack Kelly, were dying. That brought me back, inescapably, to the simple reality that the waves change together and do not change regularly as something one can break into parts and trace on a straight line. It made a difference. Regardless of whether we separate them conceptually, linearize them, and treat them with mathematics, in nature—the real deal, the nature in which and with which we live and breathe—the motion of the heart is *in* motion and never linear. We experience waves that move together simultaneously.

I had no choice but to look at waves in the ways they really travel instead of analyzing them on a graph. With all my professional background, it wasn't easy to renounce the highly developed analytical framework of scientific abstraction and return to something as simple as natural experience, but there I was. Though the waves of the body lie outside the scope of the ECG, it would be wrong to ignore them and their simultaneous tie-in with the heart's changes. The heartbeat's wave travels along those bodywaves, whether they be subtle, as when one sits waiting in a physician's office, or pronounced, when one sprints 100 meters and recovers. Never does the heartbeat's wave travel along the idealized lines of science.

I settled on calling the simultaneously changing, nested body and heartbeat waves the *bodywave* and *heartwave* to mark their sharp differentiation from an isolated heart rate. The heartwave and heart rate were opposites. (And the bodywave was absent from any association with the heart rate.) The distinction was essential, from my point of view.

---

* In the most elemental sense, it was more consistent and elegant to think of waves moving along waves. Waves and the waves they travel are of the same kind. Also, they are real, whereas lines as mathematicians imagine them do not exist in nature.

I VISUALIZED EINTHOVEN'S ECG and translated it. Where the ECG graph lifted wave motion from reality and forcibly straightened it onto a chart, I restored the ECG to being dynamic waves as I now understood them. It was not a static picture; it required I acknowledge a real event in which heartbeats, as a forward-moving wave, change together with another wave.

It required I let go of the boundaries that make a closed anatomical system and accept the openness of simultaneous wave motion.

It was a simple shift but profoundly different. It gave a new insight into the age-old question of how the body works. It works, at least in this instance, through simultaneous wave motion. It is, in fact, a way the body's motion has always happened, but we have never seen it because we have thought of the body as parts and wholes. We have never seen it because we have not seen waves waving as one. We have never looked at how motion works without first carving it up and tacking it onto matter. I was on a new track: a track of scouting out "how motion works" before trying to answer "what it is."

The change in perspective is challenging. Historically, science has always put material items first. We have treated matter as the prime reality, doing the moving—and we have relegated motion to a secondary status. Traditional thinking would say that the body makes the wave or that the heart exhibits wave motion.

But here is the arresting fact: The waves exist, along with the material of the body and heart, for the entire lifespan of a human being. Do they not warrant attention equal to that of the material stuff of the body and heart? Do they not exist on equal ground? There is no reason—other than our a priori preference to think of the universe as fundamentally material—to attribute the waves *to* matter.

Yet that a priori preference runs deep, as do its consequences. It is what directed Galileo to propose that the way "*it* works" is in a straight line—a concept we still cleave to hundreds of years later. We proceeded to incorporate that linear motion into modern science, into Newton's laws of motion, Maxwell's wave equations, and beyond—not just for its practical, mathematically calculable applications but as the way we imagine motion actually exists, a way that precludes seeing simultaneous wave motion.

The simultaneity of wave motion I had just discovered was in fact antithetical to that way of seeing things. Continuity is irreconcilable with a model of wholes and parts. Galileo sought *in*variants to be those parts. Changing motion—whose structure I was open to discovering—was therefore something he tried to break down. It itself is what pushed him to move "a hair's breadth" from real motion and into idealized linearity.

And so, though Galileo acknowledged that there is nothing older in nature than motion, he never did actually look at motion alone, I realized. He truly could not see the continuity I was seeing. Instead, he treated reality as a composite tapestry, artfully separating out threads of matter, motion, and forces. Those were the players in his theory, which he tested: Having isolated those different parts, he calculated linear interactions between them. And because his calculations worked, and despite their inexactitude, we say his theory was proven.

But at what cost? Galileo's concept of linear motion also brought us to consider the heartwave as lines on a line in the ECG—and therefore to miss the phenomenon of the heartwave, which ultimately led to danger, and even death, for people from Pheidippides to Jack Kelly. Galileo's permission to depart reality in favor of the theoretical line caused our failure to see the crucial necessity of recovery. Galileo's suggestion that we abstract lines from within waves, compounded with the Puzzle Hypothesis, narrowed our focus—such that we did not see the simultaneity of waves waving. There are obviously enormous practical applications in calculating as he did, but those applications should not stand in to prove that he had actually tapped into the nature of reality. Practical applications, even civilization-changing applications, do not necessarily mean we understand the full picture. What works to track the path of an inanimate cannonball must not—does not—apply commensurately in the living world. Life in motion reveals a truth about waves that, apparently, less complex objects do not easily express.

And, as opposed to Galileo's suggestion, what I was now seeing was no theory. A theory is a potential explanation, of how parts relate, that requires testing in nature. Here—here there was nothing to prove. These were facts. Something was happening, something has always been happening, and we'd missed it. In the rich complexities of the living organism, a single wave phenomenon that exhibits simultaneous change across scales happens out in the open—but hidden from anyone who sees nature only as a sum of parts.

There is a general sense in science that Galileo permanently enlightened the world when he abstracted idealized lines from motion. But the heartwave tells a different story. Centuries of scholarly respect for Galileo have never dampened its quiet insistence to beat its rhythm in rising and falling waves. The motion of the heart does not travel along a line. It disrespects our reverence for material boundaries for everything. Instead it rides as a wave, through time, together with the waves of the body. All of science, all its theories, the entire framework built on Galileo's lines: Waves waving forcefully suggests a different way. It suggests the total opposite— waves in waves, not on lines.

What an incredible shift as far as the history of human thought goes! Galileo straightened out waves to make them understandable according to his standards; so many centuries later, waves waving together demanded I accept them as they are. Where the ECG strung the heartbeat along a line, where Galileo sat very still to suggest a line in his pulse, I turned it around. I threw that heartbeat up and down, like Galileo's cannonball, in my burst of exercise and recovery. Instead of throwing matter, I threw the motion of the heartbeat into that arc of the bodywave. I conceptually unified motion that never had in fact been separate. The shift was absolute.

IT STRUCK ME DEEPLY that Galileo's lines had set us up to misunderstand waves, such that we could not see the truth of how the heartbeat can fall dangerously during recovery. Yet I was meanwhile intrigued with how its implications went even further.

When we accepted Galileo's argument that linear motion is fundamental, we set ourselves up for unanswerable questions—specifically because we separated the two layers of waves and treated them as a composite of lines. In the case I was examining, when you straighten out the heartwave, it disconnects from the whole organism. It appears that it travels on a distinct scale, different from the scale of the body. And if all scales are distinct from one another, then there's nothing to account for their synchronous change. You actually *create* a problem of understanding how the heart "affects" other aspects of the body. Then you have to figure out what changes in the heart rate "signify." You get hierarchical levels that you can't connect—though if you had looked at motion as it happens in nature, before separating each wave according to matter and tracking them separately along lines, you would have seen the waves moving forward *as one*.

To me, that was absolutely incredible. In his 1989 Robert H. MacArthur Award lecture, ecologist Simon A. Levin stated that "the problem of relating phenomena across scales is the central problem in biology and in all of science. Cross-scale studies are critical to complement more traditional studies carried out on narrow single scales of space, time, and organizational complexity."[5] In other words, relating motion across scales of nature has been a huge challenge.

But the problem was actually created by the way we see nature—at least in the case I was now examining. The heartwave and bodywave change together, and we would be fools to separate them and then try to see how they relate across scales, as Levin explains that science needs to do. Here, I had discovered a blatant unification of motion, across scales, in a biological organism. We'd been missing it—absolutely totally overlooked it—by treating the body on distinct hierarchical levels.

AS A PRINCIPLE, the idea of waves waving in a biological organism had breathtaking potential. It could, and should, apply to the wave motion of all scales, going down to the biochemical, atomic, and subatomic realms as well as up to larger scale behaviors and cycles, not only those of a human being but also of the solar system and galaxies whose waves and cycles coexist with our own. This line of thinking I would now pursue.

I found myself crossing back over the bridge that separates formal science from nature as we experience it. I let it all wash over me, all happening at once, to see what was simply *happening*. Just the motion. "What" was waving was another wave. That distinction just fell away. The delicate all-at-once quality would be destroyed if I attempted to chart them or confine them as a property of something material. If I focused on "what is it?" "how it works" would slip through my grasp. It was one or the other. I chose to think about wave motion alone. I was ready to go down deeper.

Looking back, I can say that I was fully justified in doing so. The Puzzle Hypothesis is indeed a hypothesis, untested. There has always been another choice: that nature is not by its nature inherently divisible, but that it is an "only one." Waves waving present a crack in the Puzzle Hypothesis and offer the potential to see unity as a working force of nature. The simultaneity of waves waving—of changing motion changing synchronously with other changing motion—presents a genuine, natural phenomenon of a "one" instead of parts and wholes. It does mean leaving the ideals of formal science behind, but the natural phenomenon of synchronous waves held, for me, even with my professional perspective on the line, promise enough to do so. The waves are real. They were compelling, for me. You might call it intuition; you might call it observation; whatever it was, that was the route I took. I went with looking at how motion works in lieu of asking what it is.

It was as if someone had turned on a spotlight in a blackened room. Right there was motion of a new kind—wave motion, continuous and synchronous across scales. I was willing to follow this idea as far as it would take me. Without traveling deep into a jungle or to an extreme with a microscope or telescope or computer analysis, I had found something "new" in nature. Here all along, but totally new. A discovery about motion—of waves waving.

CHAPTER 11

The Three Principles
of Waves Waving

AT THIS POINT in my thinking, it was necessary for me to craft a new concept of motion. It was time to consider only waves—not waves *of* the body, heart, or any other material thing but wave motion itself as exhibited in a living organism.

It was time to set aside the concept of motion as an invariant, objectlike line and replace it with the concept of motion full *of* motion. It required an awareness of a dynamic process, not subject to boundaries, that shows simultaneous change across what we call scales.

Just to entertain this new concept, I had to bear, as you, too, will for a short while, a particular discomfort, namely, of not being able to gratify the old habit of identifying a discrete "thing." The way to understand waves waving is to open the mind to the unfamiliar reality that motion has no natural parts. It means accepting motion as something that is not actually a thing—but is, instead, what you might call change itself, happening in nature. It means seeing wave motion carrying other wave motion as an event. It means starting with this motion instead of with matter.

The parts-and-wholes perspective will not allow a person to recognize changing wave motion unified across scales. It was time to go for a new understanding. It meant starting not with the nonliving world that science begins with but with the dynamics of the most complex phenomenon we know of—life—through which waves waving are uniquely observable.

This was something I felt comfortable doing given my background as a physician and with the Olympics. But it is something anyone with a truth-searching mind can do.

To see a universe that hosts the continuity of waves waving, the entire Puzzle

Hypothesis needs to be replaced with an explanation—of everything we know and experience in nature—that allows for that continuity. Matter, motion, space, time, order, chaos, energy—the continuous cross-scale motion shown by waves waving must account for it all.

It has to because it dismisses our explanation of nature as a discontinuous puzzle. The facts we have collected over centuries of scientific experience must find a new home, a new context, that makes sense of them while also allowing for continuous waves, if one is to entertain continuity at all. Indeed, the new context must explain not only all the data we obtained while working with nature as parts and wholes, but also how we seemed able to divide nature if it is in fact ultimately indivisible. This is what we have to determine: whether waves waving can provide the ultimate alternative to the Puzzle Hypothesis, an "only one" that accounts for it all.

I SET OUT TO EXPLORE the waves waving, going down on deeper scales to cellular and molecular motion, and eventually hit the quantum realm. As we know, waves make a surprise appearance in subatomic particles in wave-particle duality. The amazing thing is that by the time I got to examining the quantum, the model I had developed explained that weird realm perfectly.

It was a remarkable journey, and one that I will take you on now. It was a journey that ended with waves waving—what I came to call SuperWaves—replacing the Puzzle Hypothesis as the explanation of nature as we experience it.

It was therefore, at the same time, a journey that propelled me to a new understanding of what it means to understand. I did not design new experiments to test waves waving. The experiments have already been done. The task before me was to reunderstand the findings of science in this new context. Every fact—both facts we can scientifically account for and those we cannot—means something different in this model, and understanding those facts is not an exercise in putting a puzzle together. There are no truly independent parts to piece together. Understanding means going through the facts we know and reunderstanding them—as wave motion inherently connected with other wave motion in action, instead of trying to put small, separate ideas together as a puzzle of parts and wholes.

The journey is exciting and swift, but it requires that I begin by describing the characteristics of waves waving within waves.

First, I used the heartwave and bodywave to derive the properties of how waves wave. I captured the pattern I'd observed in their simultaneous motion with three essential principles. The pattern, as clarified through these three principles, became my model.

From there, I proceeded to make a hypothesis, which is that all natural cyclic motion works according to these principles. This was not a hypothesis in the scientific sense, of course; a scientific hypothesis is a possible explanation of phenomena or relations that then gets tested. Here the explanation was already known, put forth by nature itself. My so-called hypothesis was that the rule naturally extends beyond the particular instance in which I'd observed it (in the body).

Therefore, the test was different, as well. A scientific controlled experiment attempts to get rid of all influences and parts except for what is being tested—an approach that makes sense only if nature is composed of parts. The test I embarked upon was to show that motion across all scales is inherently connected, so nothing could be gotten rid of. The test was to look at motion elsewhere in nature to see if it conforms to the rules—the rules of motion I'd derived from nature itself.

So I took the three principles to the level of the molecules within the cells of the heart and the body, to see whether they described motion on that level. I knew that all cellular and molecular motion is cyclic. My question was whether those cycles cycle up and down in sync with what I had discovered—in continuum with the heart's and body's exercise-recovery waves. This would indicate further cross-scale simultaneity, of the sort I'd observed on the scales of the bodywave and heartwave.

And that was what I saw. I saw that waves on the cellular and molecular level do indeed behave according to those same principles. This stood as the first test, an initial proof.

Then I went further, exploring wave motion in atoms, in the nucleus, in electrons, and in other particles. I saw that the quantum world fits in with waves waving so spectacularly that it offers even more stunning, and incontrovertible, proof. Through the quantum world, waves at last revealed their fundamental role in nature. I saw how they create matter as we know it, as well as space and time, organization and complexity, and relative disorder and disorganization on and across scales. This understanding set me up to explore waves waving back on the macro scale, incorporating gravity, the galaxies, and the entire universe.

This is the journey on which you will join me in the next several chapters.

WE BEGIN WITH the core description of the discovery.

In many ways, this chapter from here on is the academic paper explaining the discovery of SuperWaves. It details its characteristics and how those characteristics affect the entire natural universe. It is the chapter you can return to at any time to review the properties of SuperWaves. It describes my first step in exploring the discovery, in which I spelled out the qualities of waves waving.

I derived these characteristics from what I'd observed of the simultaneously changing waves within waves. It was tantamount to deriving and stating that 1+1=2 in a parts-and-wholes universe. Mathematics, closed systems, locality, linearity, invariance, symmetry, building blocks—these are principles one derives from the Puzzle Hypothesis. I was deriving the principles for a different model. I had to lay out how the wave "parts" of nature relate.

Of paramount importance, of course, is that in the continuous phenomenon of waves, there are no absolute parts. The waves are motion in motion, not boundaried objects.

To keep this fact in mind, we must continue to resist our natural inclination to put absolute boundaries around everything and recognize that motion has none. Nothing in nature is ever truly still—motion is literally everywhere, on every scale at the same time. When motion moves together in waves, there are principles to the way it does so, and while we refer to nested waves as if there are discrete parts in order to discuss the principles, we must understand that nested waves cannot be fully and fundamentally divided. We must retain the recognition that *we* identify peaks, troughs, and other features as "separate parts" of what is a "one" phenomenon and that *we* identify "separate scales" in nested, simultaneously changing waves within waves. We may draw an edge beyond which we temporarily ignore related aspects and nested motion, but we must keep in mind that the motion itself is ultimately indivisible in the way it happens in nature. It is a "one," an "only one," that varies because it *is* motion. It is a true continuity unlike anything we could imagine when we accepted that nature is discrete.

This unique property, of ultimately indivisible characteristics that can be identified as if they are separate in order to discuss them, allows one to examine and clarify the principles behind how waves are shaped and how they move. The principles are what one might call the rules of waves waving. They are the rules through which we can understand and describe natural motion. The rules describe the different ways motion works—the components, if you will, though they are not discrete, of waves waving within waves.

This is a new set of rules, different from other, long-accepted ones, such as discreteness, invariance, closed systems, locality, linearity, and building blocks. For me, it was like picking up a diamond and appreciating its different facets.

I saw three main facets. These are the three principles of waves waving.

1. The continuum of frequencies and amplitudes is nonlinear by nature.

2. The processes of attraction and repulsion come about through the inherent relationship between carrier waves and inner waves, such that the process of attraction is greatest for inner waves as carrier waves come to a peak and repulsion is greatest as carrier waves come to a trough.

3. Scales influence each other simultaneously as waves wave forward, such that there is an inherently continuous relationship between what science has been referring to as separate phenomena of action at a distance and causality.

As I surveyed these principles, I saw that all together, they entailed that waves waving are, by their nature, an *inherently continuous fractal*.

I refer to the principles as facets to bring to mind one single gem, which can be appreciated from different angles. All the facets ultimately reflect the reality of an indivisible continuum of waves. They are the new rules of motion that I discovered when I looked at motion alone.

For better or worse, we will discuss the rules in the abstract so that we can then see them in action on scales further down. Bear with me through this chapter, because spelling out the principles is essential for recognizing them elsewhere in nature. It is good to have an ultimate statement of the rules of waves waving within waves. Abstract though they may seem, the principles are explanations of what happens with natural wave motion. These explanations will come to life in the next chapters, when we see the rules manifest in the cells, biochemistry, and molecular biology of the living organism; in the quantum realm; in the solar system and galaxies; and more. The story will resume after the principles are spelled out.

I will now address each one in turn.

THE FIRST PRINCIPLE of the continuum of pure wave motion is that frequencies and amplitudes are inherently nonlinear, as I will now show.

In order to discuss this facet, I have to adjust the way we use the terms *frequency* and *amplitude*.* Though science measures and defines frequencies and amplitudes as single-scale measurements along a line,† I use the words in their nonscientific, general sense. We all know that waves go faster and slower, and that they can be more or less intense. So I made new definitions for frequency and amplitude based on that.

The pattern of waves waving affirms that waves exhibit nonabsolute, *relative* frequencies and amplitudes. Waves speed up and slow down: For this quality, I use the term *frequency*. They become more and less intense: For this quality, I use the

---

* Simply borrowing these words, which we use to describe waves in the wrong way, will not do, yet I cannot just coin a term either—words, by their nature, capture circumscribed objects or events (rock, tree, jump, or run, for example). An entirely new word might wrongly indicate the existence of a discrete new "thing." So I have made a compromise: Since I am describing a natural phenomenon that we are all already familiar with, I take words we already use and adjust their parameters.

† Scientists measure how fast a wave oscillates, or how quickly a cycle repeats, and then average it out over units of linear time to get a frequency (e.g., 20 cycles per second or a heart rate of 72 beats per minute). Amplitude, the intensity of a wave, is measured as the linear height of the wave on a graph.

term *amplitude*. Frequency and amplitude are qualities of a wave *that are always changing*.

The terms should bring to mind active motion—of an accelerating and decelerating, intensifying and deintensifying wave traveling in other waves that are also changing—and never a static measurement averaged out over linear time.

The inherent nonlinearity of frequency and amplitude derives from what we have seen: that waves wave together across scales.

The paradigm I will use to explore them is an athlete who trains cyclically.

Having gone through his regular day of intermittently walking around and sitting, suppose an athlete now does a set of more pronounced cycles. He sprints and recovers, sprints and recovers, sprints and recovers.

Two scales of waves change together during this behavior, as described earlier. There is the bodywave of exercise-recovery: Its frequency and amplitude rises and falls with every sprint and recovery. And nested within, there are heartbeats, themselves waves of exercise and recovery, whose frequency-amplitude rises and falls with the arcs and dips of the bodywave. This had been my basic observation of how waves naturally wave.

The concurrence shows, among other things, that as frequencies and amplitudes change, they relate across scales inherently. That is to say, though we have separated out so-called scales in the past, in reality, frequencies and amplitudes on these supposedly different scales do not change independently from one another.

This relationship is on display in the way the waves unfold. (Remember not to get stuck in semantics: The frequency does not actually belong to the heart or body; those words indicate the "layer" of wave we're referring to.) As it happens, and as we will understand in relation to inner scales later, heartbeats that speed up are also heartbeats that pound harder and harder: They are both high frequency and high amplitude.* The exercise-recovery bodywave is also rising in frequency-amplitude. As the exercise-recovery bodywave comes to a peak, so do the inner, increasing frequency-amplitude heartwaves. And they all come down together.

In other words, when waves change substantially in frequency and amplitude, so do those with which they nest. The more dramatic the sprint, the more dramatic the rise of the inner heartbeats' frequency and amplitude. This means, the sharper the slope of the carrier wave as it rises, the sharper the slope of the rising nested waves within; these become higher frequency-amplitude in their own right.

The reverse happens on the recovery side of the carrier wave. When the sprinter sits down and exercise gives way to recovery, the heartbeats decrease in frequency and in amplitude as sharply as the bodywave does. It is the way the waves happen.

---

* Physicians define this relationship as the ejection fraction. When the heart beats faster, it also beats more intensely, with a stronger ejection fraction, meaning it expels more blood from the heart, and vice versa.

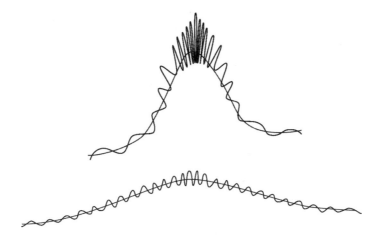

Above, a sharply sloped wave with more sharply sloped frequency-amplitude waves climbing it and decreasingly sharp waves descending it. Below, a less sharply sloped wave with less sharp frequency-amplitude waves climbing and descending it.

Both diagrams show that the sharper the slope of the carrier wave, the sharper the slopes of the inner waves (or alternatively, the flatter the slope of the carrier wave, the flatter the slopes of the inner waves).

I'll say it again because it is foundational. The sharper the slope of the carrier wave (the bodywave in this case), the sharper the slope of the inner waves' (the heart-wave in this case) increasing frequency and amplitude as they climb it. And as sharply sloped as the carrier wave is in its descent, the inner waves sharply decrease in frequency and amplitude as they descend it. This concurrence depicts the inherent relationship that exists between frequencies and amplitudes. The key idea is that in the complexity of a living human organism, frequencies and amplitudes change together, as a continuum, across scales.* The continuum of frequency and amplitude of a wave is *inherently nonlinear*.

This is a so-to-speak rule—a fact about what we have observed. The nature of the waves is that one wave climbs up and down another wave that has its own changing frequency and amplitude. Therefore it will never be linear.

The two sides of the inner wave do not travel a line. They travel a curving carrier wave that *itself* keeps changing. We see this by looking at the waves themselves. The upswing of systole and the downswing of diastole are never identical from heartbeat

* It is important to keep in mind that this continuum of frequency and amplitude is an expansive phenomenon—a real one we experience. It is easy to slip into the idea of waves as either abstractions or as being as flat and thin as the lines we draw to represent them. Real waves come from all directions, spiraling and spreading out, pulsating throughout the body. Heartbeats rest on no linear plane, though we straighten them to print an approximation on an electrocardiogram; the pumping pushes the blood in all directions, which means the waves travel in all directions. One can almost picture mounds of waves, very frequent and intense, coming and going in all directions when a person runs fast.

to heartbeat. Even as we breathe, the heartbeat speeds up and slows down with each breath. The changes are even more pronounced during the body's exercise-recovery.

This feature of the wave, stated as a principle, is that the repeating waves are inherently nonlinear.

Said in reverse, if frequencies and amplitudes were identical from heartbeat to heartbeat, the waves would travel in a line, the way Galileo and Maxwell portrayed them.* But it is not so in a living organism.

Though waves waving are nonlinear, they are not nonlinear in the usual scientific sense. Nonlinearity, the scientific term, does not capture the idea. It is the closest expression we have, but it automatically implies lines are a status quo from which the nonlinear entity is a departure. It implies that invariant lines are an anchor, an essential part of reality, and that waves that stray from them are errant.[†]

The continuity of waves waving, in contrast, simply has nothing to do with lines. The waves travel a wave. There is no need to idealize an invariant line for them to cling to because they are not required to be unchanging parts of a puzzle. In the very way the waves exist, lines do not offer any substructure. Change in waves is fine—it is, indeed, natural.

In the same way I modified the terms *frequency* and *amplitude* to mean, specifically, changing frequency and amplitude, I adapted the word *nonlinear*. I called the continuum inherently nonlinear. By this I mean that the waves, by their nature, are never exactly the same, on the upswing and the downswing, from one to the next. Said again: Inherent nonlinearity means that by its nature, any given wave, as it rises and falls, is never exactly the same from one cycle to the next because it climbs and descends another wave that is itself not linear. This is the nature of wave motion in motion.

(I know it is cumbersome to keep using the word *inherent* to describe the qualities of waves, but there is little choice. *Inherent* specifies several important qualifications: that the relationship between frequencies and amplitudes is natural and irreducible, rather than one between parts, and that the nonlinearity of frequencies and amplitudes is natural and not a deviation from linearity. We are deriving ways to identify how waves wave within waves, and these properties necessarily—inherently—emerge from the way they wave.)

I am explaining this principle as a rule by which waves wave, with the intent to show you that it happens this way on and across other scales, but let me pause to say that I found it incredibly exciting when I first saw it. "Every major unsolved problem in science, from consciousness to cancer to the collective craziness of the economy, is nonlinear," writes Steven Strogatz in *Sync*. "For the next few centuries, science is

---

* In every heartbeat, systole is normally of shorter duration than diastole—the heart contracts quickly but takes longer to refill with blood—and is thus asymmetrical within each beat, but I am speaking of a comparison between each beat and the next. As a sprinter accelerates, for example, the systole-diastole of each beat will be changing from one to the next—because it is landing on a wave that is itself changing.

† I am here referring to the plain scientific meaning of the word *nonlinear* and not incorporating the sense of emergent complexity, which we will address soon.

going to be slogging away at nonlinear problems."[1] Here I was, leaving matter out and seeing that motion is naturally nonlinear. I had a flash of intuition that this nonlinearity, as a rule of motion, would be vital to clearing up these unsolved problems.

Thus I established the first facet of how waves wave. The changing frequency and amplitude of the inner wave, climbing up and down the carrier wave, is never waves on a line. Built right into the phenomenon is inherent nonlinearity. This is how the motion works.

FURTHER OBSERVATION OF WAVES waving brings us to the second facet. As the heartwave rises to a peak during a burst of exercise, one finds a high density of fast, forceful heartbeats. This we have seen. But when we abstract out the motion of waves waving from the material body and heart and look only at the process, we see a particular pattern.

As the smaller inner wave climbs the carrier wave, those inner waves speed up and are clustered, swarming, synchronous, and dense. And with this increase in motion and compression comes an increase in temperature—it's warmer. (Think warming up with exercise.) In turn, when the waves come down, they slow down, spread out, and flatten. In this trough, the temperature is cooler.

Those inner waves do not all cluster together on their own, as we have seen. The upswing of the larger carrier wave pulls them together. And the downswing of the carrier wave pulls the inner waves apart.

This is the key to the second principle. The exercise-recovery carrier wave, as it crests before the downswing, is like a force pulling the inner waves together toward the peak. It is like a force through which the waves are attracted to one another—a force through which they pull together.

Contrast this with the heartwave when the body is recovering. As the heartbeat slows, everything cools down; the waves are not as intense, and their frequency has decelerated to a lower level. This relatively spread-out motion is the way waves wave in the trough. The wave motion is still nested within the inherent continuum but is not as highly compressed. The inner waves are relatively flat and elongated, what one would call repelled, because the trough of the carrier wave spreads them out.

Science defines a force operationally as that which is responsible for a pull and a push. Science states what a force does, but not what it is. This second principle, of attraction and repulsion, shows such forces happening as a natural feature of the way waves wave. That is to say, attraction and repulsion naturally happen between waves because wave motion clusters and disperses as it moves up and down other wave motion.

This is facet number two. Attraction and repulsion occur naturally in the peaks and troughs of waves, respectively, because inner waves are influenced by carrier waves that they climb and descend, as all scales move forward together.

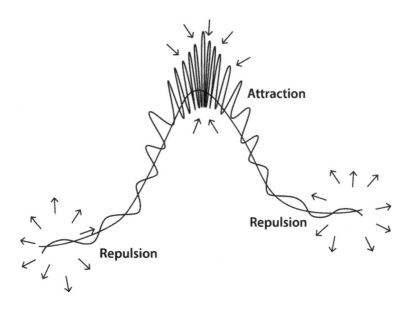

Attraction naturally happens as carrier waves compress inner waves in a peak. Repulsion, or dispersion, of inner waves naturally happens as carrier waves spread out in a trough.

I'll say that again to help drive it home. The principle is that the rising peak of the continuum of waves causes what we recognize as attraction or compression of inner waves and that the descent into the trough causes repulsion or spreading out. The carrier wave acts as a force that compels the inner waves to group together and spread apart. It is a natural account of the force—an organic explanation of how it exists—instead of a superadded power that somehow pushes things together and pulls them apart.

In this way, attraction and repulsion necessarily emerge from the continuous relationship of waves across scales, instead of appearing as unexplained phenomena that require a theory to account for them. They are plain to see.

Again I experienced a thrilling intuition. Here was attraction and repulsion visible in motion, not particularly related to matter but presenting a pattern of clustering and spreading that we are very familiar with. I imagined the continuum of motion I had discovered—motion without boundaries, in the form of waves nested in waves—permeating the natural world, like an endless lake shimmering and pulsing with ripples upon ripples upon ripples that crest and spread and ride one another in all directions, naturally influencing one another in peaks and troughs, such that we see what science, without knowledge of cross-scale wave influence, has called attraction and repulsion. There would be no "force" of attraction and repulsion as science has called it. Clusters and dispersal of motion would emerge as natural qualities of the continuum.

Such naturally occurring attraction and repulsion held nearly unthinkable potential for explaining natural phenomena. I could hardly wait to examine patterns of it on deeper scales to confirm that, on these further scales, attraction of inner waves occurs in the peaks of carrier waves and dispersal occurs in the troughs. It would invigorate our understanding of nature even further to see compression and dispersal as natural aspects of motion throughout the universe.

For the organization of the waves in this way, in contrast to the puzzle-based interpretation, is at once simple and breathtaking. The traditional scientific perspective, stemming as it does from the bottom-up Puzzle Hypothesis, attributes clustering and swarming of anything and everything to the localized level of that thing itself. It requires a superadded explanation about interactions between parts of nature to explain why they attract. The perspective I am offering here shows inner waves held together and spread apart by a carrier wave. There is no need to superadd any explanations for their attraction and repulsion. It simply happens in the way waves wave.

A RIVETING REALITY ACCOMPANIES the existence of that top-down influence: We have control over the waves.

We can raise and lower the frequency-amplitude of an exercise-recovery wave with a quick sprint followed by sitting down and thereby direct the heartbeat's wave. A person, in other words, can directly influence the heartwave with the bodywave.

We have always understood the human body and life from the perspective of the Puzzle Hypothesis, which says that changes happen from the bottom up, as building-block interactions on a given level. Waves waving shows a top-down influence of larger waves on smaller waves. It represents a top-down direction of control that happens at the same time as the bottom-up does.

What an astounding overhaul of our idea of change! The top-down direction of control presents an electrifying complement to the traditional direction in which we have understood change to happen.

The potential for this means of effectuating changes in the body is staggering, in fact. Not only can the bodywave influence the heartwave as they wave forward together, but within every wave of the heartbeat lie many smaller biochemical waves: cycles that are open to influence through this continuum. This was of extreme interest to me given my background in the progression of disease in medicine and in health and performance in sports.

Electrifying as the practical applications are, however, and though we will proceed to them soon, for now, let me clarify what the coexistence of a top-down and bottom-up direction of control means in principle. It means that causality differs from what we have always thought. It means that no causality is absolute and in one direction only.

This is the third principle of waves waving. As one layer changes as it moves forward, the inherent continuity determines that another layer changes simultaneously with it, no matter the direction of influence (bottom up or top down).* That is to say, the very nature of waves determines that forward-moving change occurs on every so-to-speak scale. Should one layer shift as it moves forward, other layers that inherently nest with it will change too as they move forward. This means that, whereas cause and effect has always been seen as local, linear interactions between parts, change on what we see as one scale inherently "causes" change simultaneously on other scales. Change is simultaneously causal in both directions.

I call this way of changing simulcausality. Again: It is an observation that if one wave changes shape as it moves forward, so does the other, such that the waves simultaneously shift on two different scales together. If for some reason one scale waves forward in a different pattern, the change automatically is present in the other scale as the waves move forward, because they ride one another.

Simulcausality challenges us with a fairly radical idea of how change happens. But we must not lose sight of the fact that it was something I simply observed from experience. I started with motion as its own phenomenon and saw it follows different rules from what we derive when we start with matter. And when I had endurance athletes run up a flight of stairs and then sit down, there was no doubt that two layers of wave motion changed simultaneously, each affecting the other. This is simulcausality.

Simulcausality, the third facet of this idea, is, frankly, enough to stop anyone in their tracks. It means that local cause and effect—a pillar of science—is not the sole modus operandi of nature. While local cause and effect does happen, it does not happen in isolation; at the same time, there also exists the top-down factor of carrier waves influencing the motion of waves inside. Grasping such simulcausality requires an enormous shift in our thinking.

Though it discredits the accepted view that local cause and effect is the exclusive method of change (which, though hard to comprehend, gets easier as we tie in more scales and see how they naturally change together), it doesn't suggest an organizational force that precisely dictates all changes from above either.

Instead, it presents a harmonious union of both. Human behaviors show an

---

* The inner wave, the heartwave, rises and falls with every beat, and it does so while the bodywave—which is the carrier wave—rises and falls through exercise and recovery. Each seems to exhibit a degree of independence, rising and falling on its own so-to-speak level. Yet the heartbeat rising also can be said to account for the rise of exercise-recovery in a bottom-up direction (and, certainly, a person could take some drug or injection and speed up the heartbeat—that too would be a bottom-up change), while the bodywave organizes the inner heartbeat wave: That was the discovery that was new and exciting.

actual, clear relationship of simultaneous changes across scales that incorporate both local cause and effect and top-down synchronization into the process of change. It is a relationship in motion, rather than a static arrow, that comes into play as our waves progress across scales. Science casts causality and synchrony as opposites. In waves that wave, they coexist as a continuum. The oneness of the phenomenon allows for it.

The existence of simulcausality's top-down organizer, seen in the body and heartwaves, is a phenomenal discovery. For all time, people have recognized order and organization pervading the natural universe, but we have had no way to account for this naturally through the Puzzle Hypothesis. Indeed, one of the corollaries to the concept that nature takes the design of a discontinuous puzzle is that there can be no top-down organizer within nature itself. This way of thinking led to the second law of thermodynamics, which predicts progressive disorder for the universe. But an open continuum of waves is a reality that happens nonetheless. And it showcases a choreography of order and organization built right into the natural progression of motion.

The work of Dutch physicist Christiaan Huygens can help one visualize this simulcausality of waves. In 1665, Huygens observed two pendulum clocks attached to a single wall oscillating in synchrony. He further found that if he disturbed the swing of one pendulum, it would soon return to swing in beat with the other clock. Even if he mounted several asynchronous clocks to a wall, in time they would all beat in synchrony. How did this happen?

Physicists break apart these waves and attribute their synchrony to a phenomenon called sympathetic resonance, or coupled oscillations, a piece-by-piece explanation of wave behavior that you can posit if you treat the waves as discrete. I hope at this point it will be redundant to say that I was looking at the genuine behavior of waves, suspending the tendency to break them apart and analyze them piece by piece, to see how they naturally wave together—because waves are not discrete.

So to see how these waves wave simulcausally, picture several pendulum clocks mounted and swinging on a wall, as Huygens's were. All have a beat. Now think of it this way: The clocks, together, lend the wall its own pulse—a greater, larger beat. The wall is now throbbing with its own larger rhythm. That large, strong beat, in turn, shapes and regulates the inner beat of each individual clock.* Now that they are happening together, if one clock gets knocked out of synchrony, the wall's powerful beat realigns it. Because the inner rhythms lend their beat to the wall that in turn organizes them together, the whole process is nonlinear, working in both directions at once, as it moves forward. If either layer changes, it is naturally reflected as changes in the other layer. This is a simulcausal relationship.

---

* So too is the air around the whole setup vibrating with sound waves and treating all clocks to the same rhythmic environment.

Huygens's clocks show layers of waves happening at once and influencing each other at once. In the waving waves that I was seeing, however, there was a pivotal difference: Unlike the clocks example, where the clocks' inner waves create the wall's carrier wave, the waves of the heart do not create the bodywave that in turn coordinates them. The waves co-occur from the beginning of every human life. Like the clocks and the wall that were already in motion, life's waves all happen together.

WE HAVE, OF COURSE, been seeing simulcausality during this whole discussion of waves, but now that I have articulated what the principle itself entails, we can see that it shatters our old worldview. Motion is change, and change has been, from the scientific perspective, supposed to happen only from local interactions between parts. Here is motion causing a change in motion—that is, motion causing a change in other motion without reference, at this time, to material parts. I'll say it once more: *Motion can and does orchestrate motion **in** other motion, from the top down as well as the bottom up, without overt mediation through material parts*—though we will soon move on to see how the material ties in, as well.

As revolutionary as this idea is, it is hard to see only because we have become accustomed to thinking in one particular way: to thinking that nature is by its nature material, made of discrete building blocks. When we started with nature as primarily a material phenomenon—when we asked "what is it?" before getting to "how it works"—we severed natural cross-scale relationships between motion and other motion and blinded ourselves to the top-down organization of motion by other motion. We could not see the simultaneous causality of motion across scales because of that choice.

Making the shift to accepting that motion is its own phenomenon, with its own rules, opens many doors. It suggests that many scientific problems may not be problems at all. As incredible as it seems, the problems of complexity and action at a distance are issues specifically because of how we have understood nature to be. Complexity, for example, is a problem because things change in ways that seem organized beyond what is possible through local and linear interaction. Action at a distance, in the scientific sense, is alarming because it suggest that local parts "know" how to be coordinated without a messenger through space and time. But motion changing in synchrony with other motion tells us that, at least in the model we are drawing from, motion is organized collectively in tandem with causality (i.e., simulcausally). Even the mysterious nature of gravity—for which Newton had no suggested mechanism, and for which Einstein developed his theory of general relativity because of the scientific problems with action at a distance—is open to fresh analysis on the basis of this new understanding of motion.

The playing field is wide open.

We will now step into the game, but as we do, let's look over our shoulders at the whole of the ever-changing wave pattern itself.

A UNIQUE, ALL-PERVASIVE PATTERN characterizes waves waving within waves.

The principles I derived show that the singular, cross-scale phenomenon of waves is self-similar across scales. That is to say, rather than waving randomly on different scales, the waves' pattern is like a fractal (i.e., it is self-similar).*

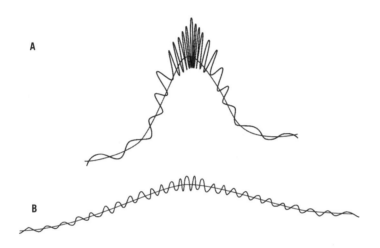

The waves are self-similar. (A) During cycles of sprinting and recovery, the bodywave and heartwave exhibit the same sharp slopes shooting up to intensely high, clustered peaks and the same steep downward slopes to recovery. (B) During a slow run—a more prolonged, low bodywave of exertion-recovery—a person's heart rate wave rises moderately, with no pronounced clusters of high frequency, high amplitude waves in the peak, and gently descends into recovery. The bodywave and heartwave are self-similar.

When someone runs, whether in a sharp sprint or in a slow jog down the block, the shape of her heartwaves and the shape of her bodywaves mirror each other. I realized that the two continuous scales assume the same pattern.

Yet though the pattern is like a fractal, it is not a fractal in the way they are usually defined. The pattern of waves waving is like a fractal in that it is self-similar across scales, but it differs from a fractal in that it has nothing to do with points, lines, or dimensions. Nor can it be broken into parts—fractured, like a fractal—to see the self-similarity.

Indeed, one cannot precisely measure this fractal because, as an inherent continuum

---

* Fractals, you may recall, are the mathematical class of complex geometric shapes, in which any scale can be magnified to resemble the whole.

of motion, it is always changing and therefore has no lines, no edges, and no boundaries. There are no standard units to count with ultimate precision.

The pattern of the heartwave is organized as what I called an inherently continuous fractal (IC fractal). Again, no term in our language can really capture an inherent continuity. It is fractal-like but free of boundaried, independent scales. It is totally whole, and seamless. It is similar to matryoshka dolls—the self-similar dolls that nest within one another—but also unlike them, for the dolls separate from each other on different scales whereas the waves do not. An *IC fractal* is the best term we have.

At this time, my intuition, which would soon be borne out by the data, was that this fractal relationship extended far beyond the scales of the heartwave and bodywave. But even on those two scales, it is possible to conceptually recognize the IC fractal relationship.

All the principles of waves waving—all the facets of this one gemlike phenomenon—come together to create this fractal-like pattern. Indeed, the inherent interrelationship between all the rules becomes clear within its framework. The nonlinearity and nonlocality of the nested continuum of frequencies and amplitudes means scales change together as waves move forward. The attraction in the peaks and repulsion in the troughs likewise ensure that the peaks are self-similar and the troughs are self-similar. Simulcausality ensures that the changes in one layer are reflected in the other. Though I have stated them as separate rules, one can easily see that they overlap because they are all facets of a single phenomenon. And they come together to represent the IC fractal.

I mentioned in Chapter 8 that fractal patterns abound in nature. There are fractal patterns in branching trees, arteries, human lungs, snowflakes, coastlines, broccoli, clouds, the gaps in Saturn's rings. Mathematicians and scientists who work with fractals marvel at the similarity between the patterns they mathematically find and the patterns in nature.

Yet the correspondence between the math and the real world is not exact, and in reverse, no one knows why fractal-like patterns should exist in nature at all. In a *New Scientist* essay titled "Don't Mention the F Word," science writer Amanda Gefter states the commonly accepted idea that "cosmology is founded on the assumption that when you look at the universe at the vastest scales, matter is spread more or less evenly throughout space." On the other hand, she says, new work on fractals by, for example, Italian physicist Luciano Pietronero suggests an alternative view, "one in which the irregular distribution of matter that we see around us never evens out into a smooth structure, but repeats itself at ever grander scales." She continues, summing up the scientific dilemma, "The most profound question in physics today is how to unify the really small with the really big—and when it comes to matters of scale, fractals may turn out to be a key ingredient."[2]

Here, in waves waving, a fractal pattern necessarily emerges from the way that motion is. There is no trouble unifying scales because as they change, the scales are not separate in the first place. And order and organization necessarily emerge from these cross-scale changes. I was blown away. Like the scientists who study fractals in the material parts of nature, I wondered whether fractals hold the key to the cross-scale organization of the universe. But here I was compelled by the natural fractal of motion in motion rather than by mathematical fractals. I was coming from the peak of complexity, from the intensely organized phenomenon of life. We have always thought the design is a puzzle; here was a different pattern vying to explain it all. Could it be that waves waving somehow organizes matter into different fractal shapes, which are flung throughout the many scales of nature and which science dissects as if separate?

It was time to go down to smaller scales of nature and see.

CHAPTER 12

Cell Cycles

WHEN I FIRST DISCOVERED the phenomenon of waves changing together across scales, I saw that it nullifies the premise behind the scientific approach. It shows an openness utterly incompatible with closed systems. And, with almost exuberant flair, it exhibits natural, fractal-like organization across scales: a feature potent enough to suggest that it may be the source of stability and permanence in the universe, just as the tangibility of matter suggested, eons ago, that *it* was the source of stability in the universe. There could be no better rallying cry to look here instead for the source of order in nature.

Recall that it was in the living person that I had first seen the simultaneous changes of heartwaves and bodywaves. My next step was to gather information about what happens on the cellular, molecular, and biochemical levels within and then to evaluate whether the principles I had derived about nested wave behavior properly account for those scientific observations of said cellular, molecular, and biochemical motion.

I realized, from both my medical experience and my experience in Olympic sports medicine, that everything in the body is pulsatile and rhythmic. The cell cycle, as its name indicates, is a rhythmic process through which a cell divides. This fundamental cycle spans the life of a human being; from the time of conception, in the fetus, during childhood, and throughout a person's life, cellular division and growth goes through intermittent cyclic patterns of division, with built-in pauses in the division. This is true for every type of cell in the body, from bone and skin to heart and brain cells.

Within the cells themselves, the double helix of deoxyribonucleic acid coils upon itself—is supercoiled, even, in fractal-shaped coils upon coils—and oscillates. While generations of students have been routinely presented with DNA's double-helix structure as its defining characteristic, recent research has revealed that the motion of the molecule is key to its function, says Helen Pearson in her article "Beyond the Double Helix" in the journal *Nature*.

The double helix . . . regularly morphs into alternative shapes and weaves itself into knots. Cell biologists, meanwhile, are exposing the surprisingly dynamic life of the molecule after it crumples up into chromosomes. The latter form fleeting liaisons with proteins, jiggle around impatiently and shoot out exploratory arms. . . . Some researchers believe that these mysterious movements may be just as important as the genetic sequence itself in deciding which genes are switched on and off.[1]

The movement of the DNA is rhythmic: Some researchers use the expression "as if it is breathing" while others describe "dancing DNA"—both of which underscore the rhythmicity of DNA's motion. RNA is also folded on itself in this dramatic way and pulsates dramatically like DNA. And proteins, which are believed to be manufactured by the motion of the DNA and RNA, also do what is called protein folding. In protein folding, the proteins pulsate in a twisting manner. "Pulsing has been observed in many types of proteins, from alternative bacterial sigma factors to mammalian tumor suppressors such as p53, and has been shown to function in diverse processes from stress response to signaling to differentiation," note researchers Joe H. Levine, Yihan Lin, and Michael B. Elowitz in an article in the journal *Science*, adding that "many genetic circuits actively and spontaneously generate dynamic pulses in the activity of key regulators, and these pulses temporally organize critical cellular functions."[2] Scientists are well aware of, but not able to account for, all this rhythmic, pulsatile, organized behavior.

Metabolism is usually spoken of as if it were an engine running the cell, but adenosine triphosphate (ATP) and adenosine diphosphate (ADP) cycle back and forth from ATP to ADP to provide energy at the cellular level. It would be wrong to imagine cellular energy coming from a one-way source like coal in a steam engine or gasoline in a car, poured in and burned continuously. Rather, the metabolic process is cyclic. Indeed, scientists view the metabolic cycle, this cyclic switch from ATP to ADP and back and forth, as cycles of energy expenditure and recovery: like the cycles of exercise-recovery I described earlier. In the same way, the ATP itself is replenished through what is known as the citric acid cycle, or the Krebs cycle. And all the chemicals, the minerals, and the electrolytes oscillate, in rhythmic motion; calcium oscillations are one well-known example in this realm.

Hormones are secreted in a pulsatile way. And surprisingly, their cyclic release affects the expression of genes. A recent study from the National Institutes of Health explains:

The release of hormones by the body's glands can occur in an episodic, or ultradian, pattern, which consists of repeated periods of release that take place throughout a 24-hour, or circadian, period. Glucocorticoid[s] . . . are steroid hormones secreted by the adrenal glands. . . . Glucocorticoids act through the glucocorticoid receptor, which is expressed in almost every cell in the body and regulates genes controlling development, metabolism, and immune response. . . . In addition to being released from the adrenal glands in a 24-hour circadian pattern, these hormones are also released in a pulsing mode, cycling approximately every hour, in what is referred to as ultradian cycling. In this new study, the researchers demonstrate that ultradian hormone stimulation induces the pulsed expression of genes (known as gene pulsing).[3]

Thus, a hormone that is cyclically secreted, every hour, also impacts our genes, which are themselves pulsating.

THE TAKEAWAY from all these facts is that from the cell cycle down to chemicals in the cell, everything is in a constant state of cyclic motion. Even that part of the cell that orients us to think of nature as a puzzle—that is, the cell membrane—vibrates up to a thousand times a second. The truth is that everything is in rhythmic, vibratory motion. The motion goes by all different names, be it vibrations, oscillations, fluctuations, waves, rhythms, or cycles. But whatever one calls it, this cellular motion is so essentially variable that it is not the least bit linear, and is of a repetitive wave nature. Everything in the cell that is essential to life is doing the same thing: cycling.

Waves are a language common to all this motion—a common denominator language, common to what appear to be different scales of life. I had already spelled out the basic phrases of that language and had accepted that that motion follows rules unrecognized by science. I then assumed that down in the microscopic level of the cell, waves follow the same pattern and the same principles. They, too, I assumed, are examples of synchronous wave motion across scales. I would examine this realm to see if these assumptions were borne out.

The scientific view of such matters rests on different assumptions that have been ably examined. Biochemist Erwin Chargaff articulated it well: "Our understanding of the world is built up of innumerable layers. Each layer is worth exploring, as long as we do not forget that it is one of many. Knowing all there is to know about one layer—a most unlikely event—would not teach us much about the rest."[4] In the same spirit, biomedical researcher Zoltán N. Oltvai and physicist Albert-László Barabási proclaim in an article in the journal *Science* that "the basic dogma of molecular biology [says] DNA is the ultimate repository of biological complexity. Indeed, it is generally accepted that information storage, information processing, and the execution of various cellular programs reside in distinct levels of organization: the cell's genome, transcriptome, proteome and metabolome."[5]

And more, as I mentioned in an earlier chapter, Paul Davies poses the very question I would now explore in his book *The Fifth Miracle*, but he could not resolve it. "Could life be like this: apparently complex but actually very simple, like a fractal, and therefore the product of a simple, lawlike process?" Without the framework of waves waving to work with, and bound by the scientific belief in truly separate layers, Davies concludes: "Personally I do not believe it, not least because it demands a view of nature that is incredibly contrived. To claim that there really is 'a code within the code,' generating living creatures on demand from simple formulae, is just too far-fetched."[6]

What I saw is that the synchronized motion of waves waving is simple in a different way from the "simple formulae" of science. It thus holds the potential to account for, without contrivance and for the first time in our history, the organized motion of life. I had already seen a relationship of inherently cross-scale motion in the living organism that is indeed, to use Davies's words, "apparently complex but actually very simple, like a fractal."

So now I replaced the scientific view of separate layers with the reverse. I accepted that all of these so-called layers—from the genome to the transcriptome to the proteome to the metabolome and beyond—are *not* unbridgeable or hierarchically separate with "distinct levels of organization," as Oltvai and Barabási called them, but rather, function through waves that relate as an inherent continuum across scales.

We cannot follow Chargaff's lead to "know all there is to know" about them in the sense of finding a fixed invariant, but we can see patterns that repeat with regularity. I assumed that all the cellular, biochemical, and molecular biological waves relate, not in hierarchical levels going up like a pyramid, but in dynamically twirling, clustering, and dispersing spirals, nested inside a fractal-like wave continuum as they move forward.

I followed through with this assumption by posing it as a question: What would reality be if waves waving were "it"? What if this model—and not the model of local, linear causality—is the one that portrays how our cells and molecular biology work? What if shifting the model—from seeing everything as separate parts to seeing a continuity that unites wave motion—solves our problems? Science sets up cellular motion as a complex network of molecular interactions and even still cannot account for the organized motion of life. But the waves waving model determines that wave motion is synchronized like an orchestra: many different instruments, all of them playing their own parts but also guided by a conductor (a carrier wave) who adjusts as they change. How would waves waving explain what is happening if waves waving, and not the Puzzle Hypothesis, accurately reflects the nature of motion?

I filled in the specifics with the behavior of the cells, the biochemistry, and the molecular biology of the heart.

The millions of cells of the human heart—and the molecules within those cells—oscillate with every heartbeat. The chemistry is activated during systole, meaning it works harder, and it is less active during diastole. The same goes for cellular metabolism. But while these facts hold no special significance for science, I realized they show something about waves. *The cells and molecules within the heart exercise and recover just like the heart does, and they do it in accordance with what the heart does.*

That is to say, we saw earlier that the heart exercises with the contraction of systole and recovers during diastole—and now we see that in that same span of time, the cells and molecules within *also* exercise and recover: They oscillate faster and harder with systole and recover during diastole.

The cells and molecules, in other words, make exercise-recovery waves as the heart itself does, and all those waves of the cardiac cells and molecules, within the heart, nest fractally, as a continuum, within the heartwave—just as the heartwave itself nests within the bodywave of exercise-recovery.

That was my first application of the heartwave model. It followed through with the nested character of wave motion, detailing how cellular motion in the heart is continuous, if you will, with the wave continuum I had previously identified. In a way, it was nothing new; it was simply applying the same pattern, going down. But it was brilliantly consistent.

Then I worked in another fact. As blood leaves the heart, it spirals through our arteries, going forward and then a bit backward, forward and a bit backward, in a most rhythmic, pervasive cycle that gently jostles every organ and tissue. Surely, then, the cells and molecules within all our other organs partake of the same rhythmicity that the heart does.

I therefore ventured that beyond the heart itself, cellular motion within the other organs and the blood vessels also wave. The cells are also more active during systole and the accompanying contraction of blood vessels and are relatively slower during diastole. I was encouraged by the well-documented fact that the blood vessels themselves—going to the heart and other organ systems, including the kidneys, the liver, the lungs, the brain, the skin, and further throughout the body—divide into smaller and smaller blood vessels in a fractal shape, like a branching tree.

And last, I posited that all molecular oscillations, vibrations, fluctuations, and rhythms within—including the movements of DNA, genes, and proteins—nest in the same way. This was the broadest stretch of the model, for unlike cellular rhythms in the heart, which are known to change with every heartbeat (and, as I saw, furthermore, with the body's wave of exercise-recovery), the movement of DNA, genes, and proteins had never been scientifically linked with our larger-scale rhythms.

As far as further innovation about waves goes, however, this last step was not much. The idea was simply that molecular wave motion is no different from the other so-called scales'. This was a reasonable step because none of the waves happen

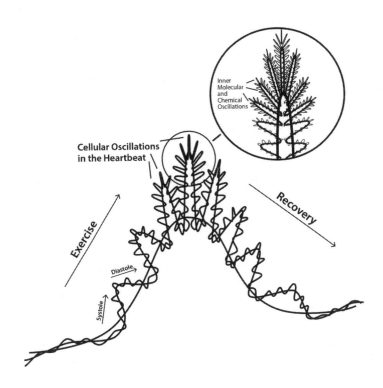

Nested scales of waves waving within waves. The heart "exercises" with every systole (contraction of the muscle) and "recovers" with every diastole (relaxation of the muscle) and does so more vigorously on the upswing of bodily exercise and less vigorously on the way down to recovery. Within it, the cells and molecules "exercise and recover" more vigorously during systole and less vigorously during diastole.

To the extent that the heartwave becomes more and less pronounced with the changes in the bodywave, the inner scales of the cellular and molecular waves change as well. They altogether work harder and recover more dramatically during sharply sloped heartwaves and altogether do everything less variably during gently sloped heartwaves.

in isolation. The nature of wave motion determines that the continuum naturally extends from the macro to the micro, wherever wave motion exists. In contrast to traditional closed systems, it is overtly open. This is what necessitated that the wave motions of DNA, genes, and proteins nest in the fractal, as well.

These assumptions may sound as if I were coming up with other new ideas, so let me reiterate that I was just considering the one idea, applied further. The idea is one new model of how motion works: that wave motion nests with other wave motion as it unfolds and that all so-to-speak scales of nested waves influence each other as they move forward. It reflects a natural reality, rather than presenting a mathematical model for how parts relate. The inherently continuous relation of waves across scales is unrecognized by science (which focuses on separate instances of motion construed as straight lines) but is nevertheless recognizable by any person, scientist or not, who thinks honestly.

So when I say I assumed that the vibratory motion of the DNA, for example, travels forward with and within the cyclic motion of the cell within which one finds that DNA, and with and within the cycles of the tissue in which that cell resides, and with and within the cycles of the organ in which that tissue is found, and with and within the cycles of the entire body that includes that organ, I mean that I assumed that all wave motion is the same nested, cyclic process of activation and relaxation, or what we are calling exercise and recovery. That's it. The idea is a simple one.

It is, indeed, startlingly simple when one considers how deeply complex cellular function seems to be when treated as wholes made of parts. The motion of the cell's seemingly complex material structures is nevertheless comprehensible as a pattern of motion alone—just as I had understood the motion of the heartwave and bodywave as a pattern of motion alone. When seen as waves waving instead of a mathematically related assemblage of lines, it is overtly understandable. It is understandable as a simple, natural reality.

Therefore, we can revamp our understanding of what is actually happening. As an individual exercises and recovers; as the heart beats faster and slower; as the blood travels with the heart's rhythm, faster and slower, in a branching fractal through our internal organs, limbs, brain, and skin; as every cell and organelle within the cell vibrates and oscillates faster and slower, not in isolation but in the context of the wave motion of our tissues, our organs, and our very selves—everything experiences the same IC fractal rhythm of motion in motion in motion.

The simulcausality works, in this case, in the following way: All the billions of vibrations, oscillations, cycles, and so forth are nested continuously with and within one another. All those nest in the cycles of the heartwave (and other organs), and those nest in the cycles of the bodywave. Together, these minute cycles constitute the cycles of the different organs and tissues of the body, and these constitute the cycles of the bodywave, while *at the same time*, the bodywave affects the cycles of all the organs within, and the cycles of the organs affect the cycles and oscillations of all the cells and molecules within.* This is simulcausality: so-to-speak different layers all changing somewhat on their own while also building up to larger-scale changes, and yet, while also being shaped and organized by those larger-scale changes.

---

* This can be understood, again, through a parallel to Christiaan Huygens's clocks on the wall. Imagine thousands of clocks on the wall, each made of thousands of tiny, inner oscillating mechanisms. The collective oscillations of the tiny mechanisms make up the beat of each small clock, and the collective beats of the small clocks make up the wall and its rhythm. Meanwhile, the wall's powerful beat contributes to and shapes the rhythms of the clocks, whose own strong beats in turn shape the rhythms of the tiny inner mechanisms.

All the motion of a living human organism, I began to realize—of the inner scales as well as the larger scales—naturally relates through waving waves' inherent openness and asymmetry. It is a continuum of thousands, millions, even billions of biochemical interactions on many different levels, going right up through into the macro scale. The fact that none of these nested cycles is entirely even—their nonlinearity— allows for the variability that is a defining feature of living cells, while the common language of wave motion nevertheless unites them across scales.

This was the first application of the model. I called it LifeWaves.

IN ASSUMING ALL LIVING motion nests in this IC fractal, one facet of waves waving stood out to me. The rhythmic motion of the carrier waves helps shape clusters of inner waves as they unfold. This top-down angle of change—the facet of attraction and repulsion—commandingly asserts that the behaviors of the living organism participate in ordering its inner changes.

Specifically, the model determines that the motion found in the molecular realm, highly variable as it is, is also guided by carrier waves, which I also called attractor waves. It states that organized, rhythmic motions—such as genes collectively working together, molecules making proteins, proteins folding to create evermore complex molecules like hormones—all nest in carrier, or attractor, waves, which guide them as a collective. On the other end of the spectrum, seemingly random wave motion—what science calls fluctuations, in the cytoplasm, for example—nests in the troughs of the carrier waves. As such, it is not random at all; the carrier wave dictates, in its simulcausal way, that these low-frequency, low-amplitude waves must disperse and be flatter than waves nested in peaks.

I accepted this explanation of how motion works on these scales. My hypothesis had been that the rules of motion I'd discovered would hold true on deeper scales, and thus far, the facts already known matched the pattern exquisitely.

A great beauty of this understanding is that this one phenomenon houses tremendous organizational intricacy and relative disorganization all at once. We have always wanted to know "what holds everything together and why do things fall apart?" and here was an uncontestable answer for, at least, the order of wave motion. I would not come to the stability and instability of matter (as opposed to motion) until I hit the quantum realm, but when I matched the waves waving model to cellular and molecular motion, I accepted that the relative stability of such motion— the hallmark of life—is due to its simulcausal, fractal, nested wave continuity.

The waves waving model directs us to understand that there is a top-down influence, not just a bottom-up one, on all our cellular and molecular changes; whereas the accepted scientific view is that all such changes—specifically, Brownian motion and Darwinian mutations—are only bottom-up and therefore random. The former worldview is open and continuous; the latter is closed and discrete. But only one can be true. One can almost see smoke rising as these two worldviews collide.

Brownian motion was first observed by Scottish botanist Robert Brown as the perpetual jiggling of pollen or dust in a drop of water. The clash arises in that scientists have long presumed that the jiggling of the cytoplasm and other molecules within the cell are also examples of Brownian motion (i.e., supposedly random). And further, from there, scientists have worked hard to understand the organized motion of complex molecules—such as proteins, RNA, DNA, ATP, and ADP—as "somehow" emerging from that random motion.

But when we step back and add to those theories the context in which they emerged, we see that Brown's observation took place within a closed system.* Even scientific histories of the era include this condition as an essential component of Brownian motion. The scientific perspective a priori assumes isolationism: that there is no way for oscillations to be organized from the top down. The idea of irregular motion being somehow under a natural top-down influence is inconceivable when you begin with the assumption that nature is a whole made of discrete parts.

That is why traditional cause-and-effect cannot explain why and how ordered motion arises in the cell; recall Davies's question: "As a simple-minded physicist, when I think about life at the molecular level, the question I keep asking is: how do all these mindless atoms know what to do?"[7] Waves waving, applied to that very same motion that stymies standard scientific explanation, shows exactly how they know what to do. The molecules composed of "mindless atoms," as they move about in their characteristic constant motion, do not travel in isolation; they nest within other moving molecules, cells, and organs, as well as the organism itself. And the motion itself, by its nature—and not by the nature of the matter through which we observe it—nests and is continuous with other constant motion. The motion of the "mindless atoms" is guided by its inherent continuity with other wave motion. The very nature of motion is that no motion is wholly distinct.

Waves waving thus negates random fluctuation by placing so-called cellular Brownian motion squarely in the midst of billions of coordinated, nested waves of motion that span scales all the way up through the entire organism. Just as the open continuum of the heartwave and bodywave stands opposite to the linear propositions made by Galileo, Newton, Maxwell, and Einthoven, the same phenomenon—waves waving—shows a top-down influence that stands opposite to Brown's proposition of bottom-up randomness in a closed system. In that, it stands opposite to Darwin's proposition, as well.

Similar to Brownian motion, evolution by natural selection asserts that the root

---

* Einstein's statistical account of Brownian motion was, necessarily, also a result of calculating it within a closed system.

of all change comes about by random mutation of genes. (As it happens, Darwin was Brown's friend and visited him before going to the Galápagos Islands on the *Beagle* in 1831; Brown may have given Darwin the idea that all change happens randomly, from the bottom up.) But again, this new understanding of how motion works determines that change in DNA is guided by other waves of motion.

Mutation is, as is well known, achieved through movement of DNA's atoms and molecules, and the nature of waving waves reveals that this motion cannot be purely random. The movement of the DNA's atoms and molecules happens right in the midst of other ordered, cyclic motion. And the nature of motion is that it is inherently continuous across scales. Therefore it cannot be an exclusively bottom-up process. A top-down, outside-in influence—carrier, or attractor, waves—orchestrates the motion while it changes.

The implications on this scale of change are enormous. The inherent relationship of inner waves with larger-scale behavioral waves validates that there exists a guidance to molecular mutations that have seemed random when coming from a bottom-up, isolationist perspective. It is a different mechanism. Change is guided through simul-causality, and though it is inherently nonlinear (and therefore not predictable with exactitude), neither is it random.

Now, as I was realizing all this, I knew that such a proposal is all but heresy in the scientific world. Evolution by natural selection is a cherished, fiercely defended theory that is considered one of the most well-proven ideas in the history of science. Top-down guidance has been outright rejected on the basis of our isolationist, puzzle worldview. And since I had not yet followed through with the idea all the way down through the quantum, I had not yet resolved how electromagnetic waves, for example, might act in continuum with the waves of the oscillating DNA to cause genetic mutations. But even so, how could I dismiss the reality of ordered motion in motion? Waves waving in waves is the way motion works in a living organism. I was applying the model, and this was the implication.

As much as I knew that I was taking on an enormous challenge by opposing the idea of fundamental randomness with the idea of organization being a natural aspect of motion, the groundwork for such a reversal had already been recognized by many scientific thinkers. In his book *Synchronicity*, F. David Peat says:

> *Science has always treated its laws as mathematical abstractions. . . . But suppose that this orthodox view of science is strictly limited; suppose that, at some level, a formative and ordering principle does indeed operate within the universe. Such a principle would act to generate the novel forms and structures of nature and would be the motivating force behind all patterns and conjunctions. Clearly, a formative principle must be very different from what is normally meant by "a law of nature," which is the abstraction and generalization of scientific experience. But*

what could it be, this principle of animation and generation that differs in such a radical way from the conventional laws of physics?[8]

I appreciated the beauty of what Peat recognized.

The "formative principle" of wave motion is indeed "very different from what is normally meant by a 'law of nature,'" as Peat says. And it is especially so in that it is not an ordering principle per se—it is an observation of how order *naturally* emerges in the way motion really works. I believed I had before me the "principle of animation and generation that differs in such a radical way from the conventional laws of physics." I believed it was waves waving within waves.

It was time to go to the quantum realm.

CHAPTER 13

The Quantum Explained: The Resolution of Wave-Particle Duality

WHEN I TOOK THE model of the inherent continuity of waves waving and turned to see how it applied on the scale of particles and the quantum realm, the revelation was instant and total. I saw that what the Greeks had said and what scientists cling to today—some form of the atomic hypothesis—actually points, instead, to waves waving within waves. The facts scientists awkwardly try to piece together as discrete parts and wholes make perfect sense within the wave continuum.

By going down the inherently continuous fractal, tracking the same three principles of the single-wave phenomenon at work, I saw the reason for all the weirdness, mysteries, and paradoxes that science has found. Science has been trying to understand nature with a model that is not true to its nature. The discrete quantum that Planck created as an "act of desperation," the paradoxes that Einstein and Bohr could not resolve, the findings that drive the best physicists to say that something is wrong with our explanations—all of it dramatically was resolved in one go with a single change, from the puzzle model of discrete parts and wholes to the continuity of waves waving within waves. It changed the head-scratching nature of quantum findings into an awe-inspiring proof for the fundamental nature of nature.

This perspective cleared up everything, as we will soon see. It showed that waves waving within waves compress into what we have been calling a particle and that that has been the source of what science describes as wave-particle duality as well as what science describes as a particle appearing in the peak of a statistical wave function. It also showed that the way waves cluster into a particle automatically creates Heisenberg's uncertainty principle and more. The perspective of waves waving within waves made it clear that what scientists cannot explain evades them specifically because they treat as parts and wholes that which is ultimately not discrete.

It was its own kind of proof. It was logical. It was simple. It was proof by Occam's razor. John Archibald Wheeler said, "To my mind, there must be at the bottom of it all, not an equation, but an utterly simple idea. And to me that idea, when we finally discover it, will be so compelling, so inevitable, that we will say to one another, 'Oh, how could it have been otherwise?'" That happy place of understanding is where uniting quantum findings with waves waving delivers us all. In one quick stroke, it reworks our troubling findings to make a lot of sense. Everything falls into place. How could it be otherwise?

I will now show you how it all works. I will show how in the peak of carrier (attractor) waves, at the height of the compression of spiraling and swarming waves within waves, waves waving condenses into what we call matter. I will show how the world of matter, motion, space and time, and order and chaos are all expressions of waves waving within waves. I will show that that one amazing unity, the other choice that we could never see, the "only one," accounts for nature as we experience it.

To do so, I will show that the core findings of quantum physics—the very regularities that make science what it is—all arise from waves waving within waves. No new model is necessary: The model is the same one I had already discovered at higher so-to-speak scales. It is the same wave continuum that I identified as the heartwave and bodywave. It is the same continuum that accounts for organized motion and relatively disorganized changes in cells and molecules. We will simply travel that one last step down to the quantum to watch it in action.

And we will understand the nature of nature.

Like a person walking for the first time on the moon, we will now walk through our very own world and see it for the first time as it truly is. To play on Neil Armstrong's words, what was one small step for a man, will now take us on one giant leap for humankind.

I WILL FIRST lay out the basic problems of quantum physics. I will then show how a shift in perspective cleanly eliminates the problems.

From the vantage point we have been taking—the position of accepting the waves waving model and how its three facets work in action—we will see that quantum happenings are all easily understood as expressions of those facets at work. We will see again and again that quantum weirdnesses arise because scientists try to explain reality by way of the Puzzle Hypothesis.

Several specific mysteries and paradoxes stood out to me in the quantum world. There was the elemental discreteness in whose terms—quanta—Planck first described wave energy in the quantum realm, and there was the related phenomenon of quantum jumps, through which particles seem to jump from one point to another without passing in between. There was wave-particle duality, through which both elementary particles and light exhibit particle-like and wavelike properties. There was the Schrödinger wave equation, which is used to formulate where one might find a particle and which relies on statistical probability rather than direct knowledge. There was quantum field theory, in which invisible fields or lines of force somehow knot up or bunch up to be a particle or a virtual particle. There was the Heisenberg uncertainty principle, according to which one may measure a particle's momentum or location but never both at the same time. There was the measurement problem, where the act of observation seems to affect the particle's existence through the so-called collapse of the wave function, such that physicists have proposed that a particle occupies all paths until it is measured. And there was what Einstein called spooky action at a distance, in which particles separated by distances instantaneously act in synchrony (which I touched on in an earlier chapter).

In all of these cases, as we have seen, science substitutes the ability to calculate and predict for the ability to understand the nature of existence. Though quantum predictions are extremely successful, what these predictions represent in the real world has never been determined. Every suggestion is difficult, to say the least: Whether it is Niels Bohr rejecting the possibility of understanding any more than the mathematical equations, Max Born saying mathematics is the reality, or others saying consciousness creates matter or that multiple universes exist to account for all quantum possibilities, every explanation is hard to swallow.

All these theories are what they are because of the context in which they arose: the assumption that nature is designed as a puzzle. They are attempts to make sense of findings that don't make sense on that platform.

First there was Planck, the father of the quantum, who, in an attempt to account for the organized spectrum of light radiated by blackbodies as they are heated up, said that light energy is exchanged in discrete packets. Planck himself called that move an "act of desperation" because making waves discrete made no sense. And yet, he proceeded. The scientific idealization of the Puzzle Hypothesis, which Planck implicitly accepted, meant that it was all right to apply edges and boundaries even then.

Then Einstein took the technique further. He said that light itself—seemingly proven to be waves—exists in discrete units, as particles called photons. These discrete wave segments, whatever that might mean, were understood to be particles that have no mass.

And de Broglie took it the whole way through. He said that all particles—including electrons, atoms, and other matter—are waves, which again committed waves to conform to the discreteness of material objects. Edges and boundaries thus achieved total dominance. The scientific ideal that discreteness be maintained at all costs—such that it can even be applied to massless particles and continuous waves—had marched, unhindered, into the quantum realm.

This historical progression shows a critical point. It was not that physicists had desired mathematical structures that do not correspond to anything in our experience or imagination. Nature itself had pushed the challenge. Scientists' own findings at the quantum level had demanded that the scientific commitment to classical physics be modified, but as scientists, they were on their own when it came to figuring out how.

Planck, the first to confront the challenge, went with discreteness over normalcy. Einstein and de Broglie then carried through with his approach. They all followed their a priori commitment to the puzzle model. But it left them no choice but to come up with what they themselves called quantum weirdness.

The Puzzle Hypothesis had reached its limit in the quantum, I realized. It allowed physicists to calculate formulas for what they saw, but it could not address the natural human sense of what makes sense—so much so that many people began to abandon the hope that we might ever make sense of nature.

But the fact is that in the time before recorded history, people had embraced the Puzzle Hypothesis because it seemed to be an accurate reflection of reality. Now, so many thousands of years later, even while physicists staunchly maintained this model, its outcome—its explanations of what exists—betrays that cause for acceptance. That the simple idea of fundamental building blocks, for example, finds its final incarnation in the idea that something can be a particle on one hand and a wave on the other—a wave which is somehow discrete instead of spreading out—is, after all, not simple but weird.

The scientists who developed quantum physics had had no other choice.

But now, there was another choice: the possibility that in the quantum, nature is displaying an open, cross-scale continuum of inherently connected wave motion.

So I turned it around. I rejected taking edges and boundaries for granted. Because I had started with motion instead of matter and had seen a cross-scale continuity of wave motion in a living organism, I now had a different model to take down to the quantum. Mine was a model of motion in motion, a model in motion, in fact. It was replete with a sensible, natural account of phenomena such as order and density and relative disorder and dispersion. I knew that the realm of the quantum was a world awash with waves of all kind—waves almost begging to be recast in the inherently continuous wave fractal instead of as a pastiche of lines. Those waves, which so puzzle physicists, are no different from any other wave. There was no reason they would not follow all the principles I had seen on other scales. The insights had been so great, so far, that I expected the new model to account for what we had discovered in the quantum realm—even for matter and its edges and boundaries.

So I extended it, going down.

I let go of science's interpretations and let the waves waving model replace its model. I looked at the quantum world as representing deeper so-to-speak scales of that IC fractal.

And with that one maneuver granted, everything fell into place.

RIGHT OFF THE BAT, the very creation of quantum physics stepped forward for reinterpretation. Planck had seen consistent, organized, stable emissions of light waves radiating from a heated-up blackbody. The emissions seemed to happen in units whose existence he could not explain—what he called quanta.

Those apparent units make perfect sense in the self-similar continuum of waves waving within waves. What happens is that carrier waves create, in their peaks, relatively stable clusters of highly peaking inner waves. This happens through the second facet, the principle of attraction and repulsion.

Planck, recognizing these stable clusters of waves but not recognizing the carrier wave that compresses them, figured out an equation to describe them. And with no other way to account for the clusters' stability, he proposed that energy waves are emitted as discrete quanta. The quantum was his attempt to explain a reality behind the mathematics. I saw right away that there are no truly discrete quanta. Planck's so-called quanta are, in reality, peak after peak after peak, nested with, and held together by, other waves. What Planck imagined as a boundaried wave unit is this: a compressed stability of groups of waves, caused by the relatively sharp slopes of the carrier waves within which those apparent clusters (of inner waves) nest.

Those compressed peaks are, of course, in continuum with untold layers of waves. On any given layer, there are spread-out waves on their own scale; up a scale, there is the carrier wave, in which both the compressed peaks and spread-out troughs nest; and up a further scale still, there is a yet larger carrier wave that the inner carrier wave itself climbs up and down and that simulcausally influences the shape of that inner carrier wave and the smaller waves within it. And of course waves extend up and down from the three layers I mention here.

Deep within, within the larger carrier wave, and within the inner carrier wave, on the level of inner waves, the wave-peak clusters are in continuum with waves that disperse. That means there is a flip side to the compressed bundle of quantum energy: what we know as a quantum jump. Between quantum units, the inexplicable space that seems to separate one energy level from the next and that somehow gets bypassed as the particle travels from level to level—what physicists have called a jump—is the natural dispersion between the clusters. To anyone with an eye fixed only on compressed wave peaks and expecting linearity, it seems like a jump through space to get to the next level. In fact, the so-called space is a natural

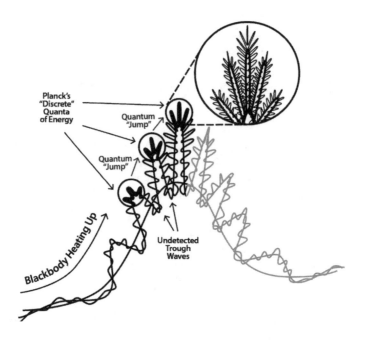

So-called quanta of energy are clustered peaks of waves as they move in waves, climbing up a large carrier wave. The "jumps"—what seem to be boundaried limits between them that they traverse without going "through space"—are simply the troughs spreading out between peaks.

dispersion that is in continuum with, and is the dip between, the clusters. (More on this to come.)

Planck had no choice but to describe clusters of wave peaks as if they are indeed separate. He did not know about the continuity of waves waving within waves. Focused on a single scale, he could not detect the continuous troughs, inner waves, and carrier waves that inherently connect with the peak-clusters and perpetuate their stability. He thought, in desperation, that he was perverting the idea of a wave by applying discreteness, when in fact, the discreteness was an illusion created by our ignorance of how waves wave in nature. Planck effectively chopped off the top of wave peaks, able only to describe their behavior through mathematics, without knowing what he was describing.

Physicists came to call Planck's chopped-off wave peaks wave packets or wavicles. Now I saw how these waves could present as discrete mathematical bundles with jumps between. It was no mystery. The math describes intense clusters of waves within waves, compressed in the peak, as circumscribed units, but does not reference the carrier waves that compress them.

THE CONTINUITY OF PLANCK'S supposed quanta with the rest of the waves waving fractal was significant in innumerable ways. It explained, most directly, other phenomena whose discoverers used Planck's technique.

Einstein used Planck's quanta on light. He described relatively stable clusters of light waves as if they are discrete particles, or photons.

But even if Einstein did not know of the inherently continuous fractal pattern through which waves wave, its cross-scale effect has everything to do with the stability of the photon on the quantum level, I realized. It is responsible for holding together the bundles of waves that science construes as discrete. It is responsible for squeezing and maintaining knots in the waves that come from all around. Einstein's application to photons showed that to be the case—and then revealed even more.

The insight was as wondrous as it was simple:

Waves waving in peaks is what physicists identify as the particle.

Whether it be a photon, electron, or any other particle, compressed wave peaks are what physicists call particles. They seem to be independent particles in the same way the wave peaks seemed like independent quanta to Planck. Waving waves had already been shown to cluster up in the peaks of carrier waves, and so, I realized, they must be materializing into particles on the quantum scale. They also, of course, spread out in flatter waves and in troughs, where they are not as concentrated—and that is when they appear to us as waves.

This is the origin of wave-particle duality. Waves waving are waves. They also, in peaks, can coalesce into what we call particles. I realized that there is a fluidity between the two forms, a fluidity that puzzles scientists, because, at this minute scale, waves can shift between being more and less bunched up as they repeat forward—between being what we recognize as material and being what we recognize as waves.

This was big—nearly inconceivable yet utterly beautiful. When physicists said their experiments show that particles are waves, they had explained their data backward. It is not that the particle exhibits wave characteristics. The waves were there already. Under the right circumstances, as carrier waves compress them, *they* exhibit *particle* characteristics. Waves waving compress in the peak to become matter. Waves waving actually is what science calls the particle.

This completely resolved the problem of wave-particle duality. Einstein proposed a duality of photons and light waves because he had no natural source to account for both. And de Broglie likewise proposed a duality of matter and waves because he too had no natural source, nothing that could generate both matter and waves. Bohr said we must settle on complementarity for the same reason. And here, at last, it was. Here was the natural source, the single generative phenomenon that eliminates the need for a duality. Here was this inherent continuum, of waves waving within waves across scales, whose peaks of waves compressed within other waves appear as what physicists call the particle. Yet when less compressed, they appear as waves. The particles and waves alternately appear primarily thanks to the second facet, of attraction and repulsion, though, of course, all facets work together, as we will continue to see.

THE IMPLICATION OF THIS solution is that a particle is not a discrete particle as science would portray it. It has no absolute boundaries. We call a bundle of concentrated waves a "particle," but by its nature, it is continuous with other waving waves. *Particle* is a genuine misnomer.

The shift in perspective here requires thinking several steps away from the old standard. I would like to say that one could picture the materialized particle like a knot on top of a string that is bobbing up and down. Except, the knot is not on a straight line of a string—a knot in the middle of a bedsheet is a better analogy. But even a bedsheet, knotted in the middle and bobbing up and down, indicates a flat, isolated layer and as such will never convey the factor that makes all the difference in the world: the nonlocalized continuum of nested waves within waves within waves, in the midst of which that knotted bunch in the middle materializes.

It is that continuum across scales that gives us both matter and motion as science recognizes them. Matter is a materialized cluster of waves waving within waves, and motion—well, the waves are motion, so the motion remains in continuum with layers that are clustered into particles and as such is still waving. That, I recognized, is the source of the perpetual motion of particles as science recognizes it. The fact is, for all our attempts to pin down nature into a stable, unchanging, invariant puzzle, we never could get rid of the motion; waves are always there.

Also consider what that means about the waves that are continuous with the so-called particle—those waves that do not compress into particles but that spread out around them, due to the continuous nature of waves. We identify them as waves in empty space. Yet I saw that there was no reason to frame the waves as secondary to space. Space was originally conceived as an arena in which matter exists, and matter turns out to be a compression in a continuum of waving waves. A turnaround was in order. Just as waves compress to be a particle, waves spread out to be space. They are space on the level of the quantum particle, in what has been called a quantum jump. They are space on the level in the macro world, what we call empty space—empty of matter, but always filled with waves. We will return to this extremely important point soon.

For now, however, we will focus on the materialization of what science calls the particle.

Here we can see that whether waving waves compress into a particle or not has

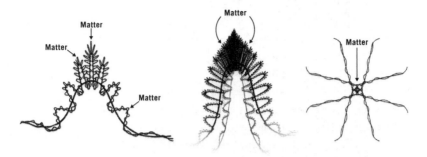

Waves compressing in the peak of nested waves is the particle.
In this illustration, three drawings are used to suggest how waves coming from "all directions" are what we have identified as a particle. Waves always wave coming from all directions in this way. Even beyond what is shown here, waves come from above, from the sides, and so forth. (Other illustrations have omitted these alternative representations in order to keep illustrations of the cross-scale relationships as simple as possible.)

everything to do with the carrier wave. And I knew that scientists had already shown this to be the case. Physicists call it the Schrödinger wave equation.

The match was exceptionally clear. By applying the facets of waves waving, I understood what is happening. An electron is a bundle of waves waving that may or may not cluster into a particle, depending, in large part, on the carrier waves that guide it. The carrier waves can squeeze a concentrated, knotted-up bundle of waving waves into what we recognize as a particle. However, the bundle may remain what we recognize as waves, if the carrier waves are relatively flat. The carrier waves themselves climb up and down still larger carrier waves that simulcausally influence their shape, so they are never precisely the same one cycle to the next.

Thus in either case, whether the carrier waves are more or less sharply peaked, the process involves untold layers of waves waving within waves moving forward together—making it inherently nonlinear—so the future of the particle is not precisely determinable. We just know that the greatest compression is in the peaks—and that therefore the peak of the carrier wave is where particles appear. In the approach to a trough and in the trough, there will be fewer clusters, and further apart.

The Schrödinger wave equation, I saw then, is science's way of mapping how the carrier wave compresses inner waves to become particles. To someone tracking wave-bundle particles mathematically, the peak seems to be the place with the greatest probability of finding a so-to-speak particle. Only, it is the peak of a real wave, not a statistical one. The statistics mathematically describe the compressing effect of a real, and variable, carrier wave.

The Schrödinger wave equation explicitly shows that the sharper and higher the slope of the mathematical wave function, the higher the probability of finding a particle. Note the consistency: If this sounds almost identical to the language I have used to describe the heartwave nested in the bodywave, it's because it is. The higher the slope (of the heartwave as it climbs the bodywave), the more pronounced the cluster of waves within the peak. *The waves are all the same.* Inner waves cluster to a peak, with great density, in high frequency-amplitude carrier waves, within further waves still, on and across scales.

Unexpectedly, a strange marriage had taken place: between the mathematical findings of quantum physicists and the bottommost levels of the waves waving fractal I had discovered. The Schrödinger wave function is a linear way of accounting for waves waving (and a good one at that). It is a perfect example of what happens when you try to track nature mathematically on one scale only so that you ignore the other scales of waves: You cannot tell what is causing the cluster. You cannot see that a real, inherently connected, nonlinear waving wave is causing the materialization of particles. You can only describe it as a probability.

Scientists were extremely close to recognizing the influence of carrier waves on inner waves through the Schrödinger wave function. It was as close as they could come. But the true nature of waves—their inherent continuity, which is inherently nonlinear—prevented the breakthrough. Because scientists construed waves on separate linear scales, they saw only disjointed waves, that they labeled a probability wave and wave packets.

Indeed, sticking with the Puzzle Hypothesis at all costs was the source of the bizarre idea that probability dictates where a particle appears, I realized. Born, the first to suggest that Schrödinger's equation meant particles appear from mathematical probability, effectively had no choice but to take that stark position.* He had no model to explain why electrons appear in a wave pattern. The carrier wave's effect seemed to be a separate phenomenon from the appearance of the electron itself.

Without a sense of the inherent connection between the scales of waves he had before him, the only wave Born could feasibly imagine to fit the facts was a wave of probability. Two waves, two scales, and yet no connection: Schrödinger and Born were in the immediate vicinity of the discovery, but ultimately could not fathom how it works. The preconditions of discontinuity and linearity effectively hid from them the inherent connection of waves across scales.

I have to say, I was delighted to realize that the model of waves waving showed that the Schrödinger wave function is a real wave. There was a deep satisfaction in

---

* Schrödinger himself had thought—and hoped—that his equation described a scientifically recognized physical wave when he developed it. But then Born showed that to explain anything beyond a simple hydrogen atom, Schrödinger's waves would need to occupy a mounting number of dimensions. This scientific absurdity effectively forced Schrödinger away from his hope. It was the reason Born said Schrödinger's equation shows electrons appear because of probability waves. As a scientific absurdity, it grew directly from treating nature as a sum of discrete parts.

knowing matter does not appear from a cluster of mathematical probability. A single electron is not a product of statistics, on the one hand, but on the other, neither is it mystical. There is a third option: an option made available by recognizing the existence of inherently continuous, cross-scale, nested wave motion. The electron is a cluster of waves within waves and behaves as such. The bizarre corner into which quantum findings had forced rational thinkers has a doorway out. No scientist has to take a stand on whether the universe is made of mathematics. The math tracks real waves. Physicists have not known what they are mapping only because no one has recognized waves changing together across scales.

THERE WERE SO MANY aspects to recognize that it was as if a cascade of insights was enveloping me. And yet, that was not nearly all. But I must say, before I go on, something that is as surprising as it is necessary to keep in mind.

For thousands of years we have wanted to know what nature is made of. We therefore naturally expect that any answer to quantum problems should settle the nature of nature's building blocks once and for all. Thus, even as waves waving resolves all these quantum-level problems, the relief it offers comes at us from a peculiar angle. The shocker is that there is no answer to what nature is built of, because nature is not designed as a puzzle. One could say it is made of waves waving, but even better, one would say it *is* waves waving.

The answers I present about the quantum, therefore, solve quantum mysteries in an unexpected way. They show that the problems were problems not with nature itself but with the model. The answers show, in fact, that quantum weirdnesses are really not a big deal. They are simply generated by applying the wrong model to nature. They are to be expected, even. All the aspects of quantum weirdness are tied together and make easy enough sense once you understand the new perspective.

For this reason, I will not spend a lot of time going through the ideas of quantum physics in great detail. I will instead touch on the biggest remaining mysteries and show how they are outgrowths of waves waving within waves. We will see that they are uncomplicated phenomena from that perspective but were made complicated when we treated them as a discrete puzzle.

WE START OFF WITH the Heisenberg uncertainty principle, together with the measurement problem and action at a distance. All fell right into place.

At the quantum level, waves waving are relatively unraveled, so to speak, and only coalesce into particles when climbing high frequency-amplitude peaks. Waves waving thus shows that there *is* no so-to-speak particle unless the waves cluster into one. That means that once the waves cluster into what science recognizes as a particle, the wave motion seems to disappear—for it is now a particle. The Heisenberg uncertainty principle had approached the truth on the level of a single so-to-speak scale. A particular level of waving waves—never truly independent, of course—can be momentum or a particle, but not both as science measures them, depending on the carrier wave's compression.

Of course, no level is ever truly isolated from the continuum. But if we focus on just one, as science does, we can interpret it as a materialized particle of waves waving or simply as motion—just, never both at once.

The uncertainty principle interpreted this reality as it did because science assumed, as per the Puzzle Hypothesis, that the particle is a discrete particle that should also have an independent momentum. When you see that the particle is a cluster of waves waving that materialize as they travel into the peaks of carrier waves, however, you can understand that the waves themselves can fluidly come in and out of materialization—being either what we call motion or what we call matter. They may appear to us as particles; they may appear to us as motion; but, on a relatively isolated single level, they may not appear to us as both.

Physicists furthermore have found that measuring particles affects particle behavior: This is the measurement problem. When they use the Schrödinger wave function to predict possibilities of where a particle may be found, physicists consider all possibilities to be equally viable: The actual qualities of the particle are considered undetermined. They seem to become determined only when measured, causing a "collapse" of the wave function. To scientists, the act of measurement itself seems to causally determine the characteristics that have been measured.

In the same way we clarified the problems of Planck, Einstein, de Broglie, Schrödinger, Born, and Heisenberg, we can see that the weirdness of the measurement problem arises from trying to frame it in terms of a puzzle. First, several related assumptions lurk behind the act of measurement itself: that a discrete entity (the particle) exists that ought to have invariant properties such as mass and momentum; that the observer and her measuring apparatus are truly separate from the particle because nature is a puzzle with absolute edges and boundaries; and that particles and the means physicists use to measure them, such as photons or gamma rays, are fundamentally different stuff and absolutely separate from one another.

The fact is, they are all waves waving and inherently continuous with one another. All possibilities for a particle are indeed open because the gathering cluster of waving waves has not yet condensed into what science calls the particle. Shoot a photon or a gamma ray, which are themselves waves waving, at that less-dense, non-materialized cluster of waves waving and that will send it spiking right up a peak—so the particle materializes. Its spread out but still moderately condensed identity as a wave, which it had had the moment before, will disappear in favor of its tightly bundled identity as what science calls a particle.

As for action at a distance, it is spooky (Einstein's description) only if nature is composed of parts and if changes can transpire only through local cause and effect. The third principle of waves waving within waves, however, simulcausality, shows that waves—including the wave-peak clusters that we call particles—change together.

It is a little tricky to explain because we are accustomed both to local cause and effect being the best mode of explaining change in nature and to the idea that those causes effectuate changes in parts of nature that are isolated from one another. As I understood it, and continue to understand it, however, this is simply not how nature works. Local cause and effect does, certainly, happen, but not in isolation: It is always in the context of larger carrier waves influencing everything (while those carrier waves, themselves, are simultaneously influenced by those inner waves). The synchrony of waves waving together allows for coordination across what we perceive as distance between objects even while local changes transpire. Inner waves nest in larger waves, and the expression of the larger waves' influence—which all inner waves experience simultaneously, even as they go through what we have called local cause and effect—has seemed to us like action at a distance.

THIS BRINGS US BACK to discussing the nature of space. We have resolved the existence of quanta, quantum jumps, wave-particle duality, the Schrödinger wave function, the Heisenberg uncertainty principle, the measurement problem, and action at a distance. Let's step back and see what waves waving tells us about the universe we live in.

So far, we have seen that waves waving is what science has recognized as matter, as well as motion. I realized that it also, then, accounts for what science identifies as space.

As a bundle of waving waves compresses into a particle, the so-to-speak particle is still inherently continuous with what we identify as the region around it. Waves continue to spread all around. That spreading-out aspect is what we traditionally have called space. Relative to the density of matter, space is that aspect of the waves waving continuum that is spread out.

Scientists have long recognized that what looks like unoccupied space is filled with all sorts of waves in the form of what science calls electromagnetic radiation. Even so-called empty space, a vacuum, does not exist. There are always what are called zero-vacuum fluctuations.

A reversal of perspective properly frames the reality. It is not that space is filled with waves. The waves are always there because they are what we perceive as space. Just as matter *is* clusters of waves waving, space *is* extended waves waving, in all directions. Matter is the clustered peaks; space is the spreading troughs—of motion *in* motion across scales, of waves waving within waves waving within waves. The cross-scale synchronous motion of waves across scales is what we perceive as matter and space, as well as, of course, motion.

Because of our traditional perspective, where waves do not materialize, to us, it has looked like there is "nothing" in between bits of matter. Because we have focused on where the peaks are concentrated, we feel that we see a particle "in" space. Nevertheless, they are continuous with one another. I gave this aspect of waves waving a name to indicate its inherent continuity. I called it matterspace.

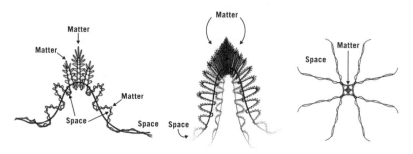

Waves waving within waves is the underlying reality behind matter and space. Again, three illustrations are used to help suggest the cross-scale, continuous relationship between matter and space.

Scientists have long recognized that (what they call) a particle's presence affects the region of space around it. They call this influence a field. The so-called effect of a field, in other words, is the physicist's way of explaining an unexplained continuity: why an apparently discrete particle seems to exert an influence through space. The facet of attraction and repulsion shows why this happens. Carrier waves squeeze and disperse inner waves—they push and pull—through what science has called a force. In other words, the selfsame waves that bring particles to materialize also pull in what appears to us as the region around them. The "field" is a mathematical expression of these influences. Electromagnetism, the strong force, the weak force— these are all different ways science describes the effects of carrier waves in the so-called region surrounding a so-called particle.

The field has always been, for scientists, an operational definition. Now we can see what it represents in reality. Both the object and the field are manifestations of the single IC fractal of waves waving, rather than separately influencing one another.[*]

---

[*] Even the gravitational field for a massive object, which means, to science, something that exhibits an influence "from" that object, is an exhibition of wave continuity—the continuity between the space around the object and the object. A gravitational field *of* an object is the compressing effect of the carrier wave. That is gravity; more to come on that soon.

Space and motion have never sat comfortably together within the Puzzle Hypothesis.

The trouble began 2,500 years ago when the Greek philosopher Zeno of Elea articulated several problems with motion, all generated by thinking about nature as a puzzle, as I discussed in Chapter 5. When a person divides motion into an infinitely small number of steps, the forward flow of motion itself seems to vanish. Scientists recognize that the challenges posed by Zeno's paradoxes are more than theoretical, as Briggs and Peat explain in *The Looking Glass Universe*.

Zeno showed that the kind of apparently continuous motion we see every day is impossible. . . . No matter how small you make the divisions, there is still the problem of how to get from one point to the next without some discontinuous (what theorists later called quantum) jump. . . . With the advent of quantum theory, the notion of a smooth, continuous path suddenly disappeared and Zeno's paradoxes were viable again. How do things get from one point to the next?[1]

Here is the answer. It is an extraordinary answer, and a summation of the points I have been making about motion and space.

We expect space to be divisible along a line. We expect motion to likewise be divisible into corresponding segments. This was Zeno's challenge: How can motion be divisible when it also appears to be continuous? The answer is that Zeno was right in posing the paradox, although his proposed resolution—that motion is an illusion—sided with the wrong half of it. Motion is real, but absolute *divisibility* is not. Neither space nor motion possesses ultimately true boundaries. What they do feature is waves within waves within waves, which, at certain levels, descend to troughs that *we* construe as absolute edges. The source of the paradox is our projection onto nature of our sensory perception—ultimately false—of absolute boundaries. The thing about space is that it is not a divisible line: Space is extended waves (within waves). We can and do break the waves up, in a way that seems absolutely broken, at dips to a trough. Therefore space and motion seemed truly divisible to Zeno. But they are not.

Our tendency to break up waves at dips to a trough is most noticeable at the quantum level. As waves concentrate in the peaks of carrier waves at the subatomic realm, on *our* level we see jumps, or gaps, in between those concentrated packets. It stands to reason that this is, indeed, the basis of our sense that nature is a whole made of parts: Between the concentrated clusters that we identify as matter are relatively unraveled waves—extending not "through" space but *as* space—which seem, to us, to be large scale jumps between solid pieces of matter. The edges seem absolute because we have not understood that the nature of nature involves troughs that are continuous with the clustered peaks we are more inclined to notice.

Nevertheless, the continuity is there. The quantum jumps that seemed to resurrect Zeno's paradoxes in fact resolve them. They demonstrate the inherent continuity between matter and space that exists throughout waves waving. Near those bottommost layers that we can detect, only the peaks of clustered waves are visible. As waves spread out in waves, beyond a certain level, the dips in the troughs are beyond detection. Quantum jumps and zero-vacuum fluctuations are as far down as we have gone, but the continuum surely extends below what we have investigated thus far.

I AM ABOUT TO move on to explaining the emergence of complexity, but before I do that, I want to clarify a few things about our basic concepts of matter in space and of time.*

Everything discussed thus far should indicate that when I say that waves peak with tremendous compression through which they materialize as particles, it is far more than a peak we draw on paper, not just a little peak, and not just coming from one direction. Strange as it seems, it comes from all around, and not from any linear dimensions at all. We may learn to imagine, from a young age, matter moving in space: that space is an empty, linear arena in which matter zings around in straight lines. Democritus called it "the void," Descartes conceptualized it as a linear grid, and most people fundamentally relate to space in these ways.† But lines, and the dimensions they seem to entail, have nothing to do with it. Waves don't occupy space in a point. They do not travel in lines—lines do not even exist as anything other than an idealization. It is the other way around. Waves waving are the stuff of nature. What we perceive as matter moving in space is a function of the peaking and spreading of waves waving within waves in all so-called directions—really, not in any classical "direction" at all.

This is a dynamic, even fun, redesign of our understanding—a buzzing, teeming reality in which matter and space and motion pull together and spread out as waves wave forward. Imagining a coffee cup occupying a linear grid of space simply isn't right. That static picture does not capture the reality of wave motion making matter and space. Every particle in that cup is an epitome of constant wave motion: a bunch of knotted-up, spiraling waves waving around, coming from all directions, in clustered clusters across scales. The coffee cup *is* a continuum of waves waving forward—what we call particles in peaks, spreading out to troughs that we identify as space in between, all in wave motion.

The continuous waves waving phenomenon is the "stuff," and from its variations come our perceptions of matter and space and motion as separate things.

Time is an essential, inseparable aspect of this continuum. When we see motion or

---

* We will discuss time when we talk about gravity.
† Einstein said that intergalactic space is curved; we will return to that point soon.

change, we experience the passage of time. (A frozen universe of no motion or change on any scale would have no passage of time.) Time, then, is that which we experience through motion and change—and motion and change themselves are waves waving within waves. Thus time is the folding and unfolding of waves waving.

Now we can understand how it is that time passes so differently on different scales. Time is not absolute. The midge fly's wings beat a thousand times a second, which is a very normal oscillation on its scale; standard atomic clocks vibrate at microwave frequencies, about 9 billion cycles per second; galaxies are billions of years old. The passage of time as an expression of waves waving is relative to the scale on which the waves are experienced. That is to say, waves mark the passage of time, but they are all happening at once. The larger arc of a large wave, for a galaxy, for instance, will see a slower passage of time (to us) than the tiny oscillations of the midge fly because change—the wave—is happening more slowly. Those waves are waving through an almost unimaginably large cycle, and so it takes a long time (from our scale) for change to pass.

This nested, forward-moving continuum also explains why time is irreversible. Time appears reversible in mathematical equations, but obviously, in our experience, it is not. The best thermodynamics could do was to state, based on the second law of thermodynamics (which we will soon address), that probability determines there is an arrow of time. Waves waving shows instead that time reversal is naturally impossible. Everything is always changing forward asymmetrically. Waves climb and descend other waves nonlinearly, such that the configuration, for lack of a better word, can never fully return to what it was before. Nested changes within changes assure that the same "moment" never happens twice.

Time is thus freed from its operational definition, that which we measure with a clock, as Einstein said. Nor is time relative to the observer, as a linear progression that differs if one is in motion. But neither is time an objective, inexorable forward march of an independent clock in the sky on any scale. Time is nothing more and nothing less than the progression of change: our perception of cross-scale waves waving as they move forward.

I called this single entity of matter in motion through space and through time, matterspacetime. It is one word, not three, because what we have called matter and space and time as if they are separate are all in continuum with one another.

ALL THESE REALIZATIONS POINTED me to a further insight that has to be included in the basics: what I called sprouting.

Sprouting is the emergence of inner scales of waves that happens in the peaking peaks of waves within waves. Matter appears when waves wave within waves to greater density, as one example, but the phenomenon of such condensed stability is of course not limited to that so-called scale alone. All sorts of organization and complexity emerge in the peaks of peaks of waving waves. We saw, on the level of life, when a human being engages in exercise-recovery waves, a climax of complexity within complexity within complexity.

As a carrier wave rises to a high frequency-amplitude peak in this way, inner waves get taller and more closely bunched. They therefore sprout their own inner waves, which had been relatively flat before.

I called this sprouting because you cannot see those inner waves when the carrier waves have a relatively gentle slope and low frequency. When carrier waves are gently waving, their inner waves are relatively subdued. But the self-similarity of the waves waving fractal determines that inner waves will rise and stand out as the carrier wave rises, too. And those inner waves will further sprout more inner waves, which, though they could not be seen before, now rise to our level of detection.

The phenomenon of sprouting leads us to an understanding of complexity in the following way. Waves waving within waves creates the phenomenon of emergence, or much coming from little. The inherently connected rise of frequency and amplitude across scales draws out inner waves from within more highly variable carrier waves. It coaxes them up and out. The inner waves rise in frequency-amplitude such that they burst through to detection on other scales—scales from which they were never separate, but in which they were never pronounced enough to be noted.

It's not that the whole is greater than the sum of the parts. In a living organism or other complex phenomena, waves waving within waves sprout levels of complexity from within to give us more than we can account for through linear addition. The process is different. The inner waves emerge, from our perspective, from no particular space at all, to create and contribute to greater levels of density, complexity, organization, and variability.

One step up from the quantum particle, at the level of the atom, we see such sprouting—what would seem to be emergent complexity—in action. What happens at lower quantum scales happens at the scale of the atom because the waves are self-similar. At the lower quantum levels, a knot of waves waving materializes as a particle; this we have seen. On the atomic level, one so-to-speak scale up, knots upon knots squeeze together to be the atom. The nucleus materializes, more solidly, in the center, as a cluster in a peak. A "cloud" of semiunraveled waves surround it but can still materialize as electrons. It is not a discrete atom, but a sprouted, relatively stable cluster of waves within waves within waves. The atom—with its stable, central nucleus and peripheral electrons flitting in and out of material/wave existence—is a beautiful confirmation of how waving waves sprout complexity.

In the framework of an atom as such, electrons span the bridge between matter and motion as we detect them. They cross freely from one form to the other. The electron can even converge with other waves, when tightly clustered into what we

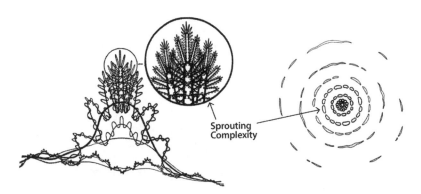

Two representations of sprouting waves waving to a peak (akin to a cross-section and an overhead view). In very high frequency-amplitude waves, inner waves oscillate very fast and hard (i.e., are also high frequency-amplitude); self-similarly, further inner waves, within those inner waves, oscillate very fast and hard, and further and further going down, such that deep levels of high frequency-amplitude waves emerge in the peaks as new levels of (what we call) complexity.

recognize as a particle, to peak even higher—for example, when a photon wave-bundle hits it. Science recognizes this as the particle absorbing energy and jumping to the next level. (It does not travel directly from level to so-called level because space is not a line. The particle doesn't travel through space—the "particle" is in continuum with the "space" around it. It coalesces with other waves and appears in the next peak.)

In 1901, physicist Walter Kaufmann demonstrated that an electron will increase in mass as its speed increases. This makes sense from the perspective of waves waving. The higher the speed, the more densely the waves compress, and therefore the greater the mass. This acceleration, rather than the Higgs boson, is what really creates mass. Mass sprouts when waves waving increase in frequency-amplitude. This acceleration is the singular reason behind all emergence: emergent mass, emergent complexity, and emergent order. On the next scale up, it is responsible for the increasingly complex atoms known as the elements.

WHEN YOU LOOK AT the periodic table of elements you are struck with its consistent pattern. It is self-similar on each scale—a reflection of the self-similar nature of waves waving. And yet, the periodic table is not perfectly symmetrical. As you add more and more electrons and protons—a greater density of wave clusters waving within waves—you get more elements in the middle of each period.

This is not an accident of atomic structure. The increasing number of elements in each row is a visible illustration of sprouting. Greater complexity and density arise in the middle of accelerating high frequency-amplitude clusters of waves waving within waves. From here comes all of what we consider our material world.

Here are a few images that show this pattern in nature.

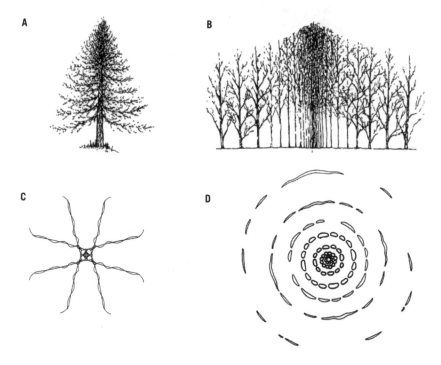

Increasing frequency of tree branches on (A) a single tree and (B) a cross-section view of clustered tree branches in a forest
(C) A cross-section-type view of waves waving from all directions
(D) A top-down view of waves waving within waves demonstrating central nucleation and peripheral dispersal

The last two we have seen before. They show the clustering tendency of waves within waves—the facet of attraction and repulsion—that accounts for central nucleation. The pattern emerges on all scales, from an atom and a cell to a hurricane and a galaxy.

Remember, these sprouting inner clusters arise because waves are dynamically in motion. Complexity and organization of all sorts arise because the waves are not just sitting there. They move, and their motion brings forth dense inner layers of complexity. Depending on what I suppose you could call the number of layers of high frequency-amplitude waves waving within each other—you can't count them, really, so I'm using the term loosely—there are different degrees of complexity and organization. High frequency-amplitude waves are nested in, and have nested within them, more layers of waves; at the same time, there is an intense squeeze of all these clusters within clusters together. Apparently that allows us to detect more: We read it as higher levels of organized variability (i.e., complexity).

A mammalian cell—with its nucleus, cytoplasm, and multitudinous organelles—therefore, is a materialization of many more layers of waves waving within waves than a steel molecule.

And it goes even further. Where complexity has emerged, carrier waves simultaneously organize that complexity. The steel molecule will be organized more "regularly"—in a flatter wave, self-similarly with the less complex molecules—so we see regular molecular crystallization. A highly dynamic sprouted wave, such as the living cell, cannot form such so-called regular crystals. Instead, the carrier wave (self-similarly with the exceedingly oscillatory waves-within-waves cell) sweeps it into incredibly vibrant, changing structures of waves within waves: a living organism.

This sprouting of emergence and organization in the waves waving pattern continues upward. As we ascend back up through the IC fractal in the coming chapters, back up through biochemistry and molecular biology, up through behavioral waves of exercise and recovery, even up through mass and gravity, we will see that the pattern of emergent, sprouted organization is found all over the universe. Scientists have long recognized that throughout nature there are jumps between patterns of order and disorder. When we discuss how simulcausality organizes sprouted complexity, we will revisit phenomena that exhibit such jumps, including self-organized criticality, phase transitions, conductivity versus superconductivity, species jumps, and more. These phenomena are the result of the inherently continuous fractal in action, as we will explore soon.

WE HAVE COVERED A lot of ground. We have seen that one model, a model of waves waving within waves that was developed before I ever considered the quantum realm, provides an elegant account of what transpires in that realm. The model shows the origin and nature of matter, motion, space, and time while resolving many scientific mysteries. It differs decisively from the scientific model of nature as a puzzle in many ways. I had much to consider, but before I moved on from the basics of the model, I turned to mathematics and energy: two ideas, fundamental to science, that approach reality but ultimately miss the mark. Our exploration of the model itself will close with this discussion of mathematics and energy.

The wrong model set out the wrong questions, as we have seen. Science sought an answer to how parts relate to make nature as we experience it but developed only partial answers because nature is not, by its nature, a whole made of parts. There is a certain amount of what looks to us like local interaction. But thanks to the facet of simulcausality, what we call local cause and effect is never fully local and isolated.

Inherently connected scales always influence one another as they unfold. The idea that local causes always precede local effects—and that our task is to discern how this transpires—is not true to nature.

This fact, extraordinary in itself, also has extraordinary implications. It explains why mathematics, as it applies to science, is often effective but never absolute and is limited in scope.

The universe is, by its nature, ordered. But no matter how much, and how extensively, we try to take it apart, it is still an inherent continuum of waving waves. This continuum is what perpetuates order as it waves forward.

Mathematics, as applied in science, sets out to track the magnificent regularities that the inherently continuous cross-scale fractal generates—but on one scale at a time, point by point. In other words, math linearly approximates one feature of waves: the repetitiveness. But it can neither capture nor report on simultaneous layers of cross-scale effects. This explains both math's efficacy and its limits in describing the natural universe.

Newton and Gottfried Wilhelm Leibnitz, with the calculus, did try to straighten out waves in order to calculate motion and change. Their method of calculation required steps that deviated "a hair's breadth" from reality, to use Galileo's term, such as dividing a curve into infinitely small points.

This deviation from reality was not a mere convenience. It was inherently necessary because points and lines cannot accurately describe nested waves within waves. Moreover, that nested motion of waves waving continues to generate the inexactitude and approximation that has always plagued even the best calculations. One can never get rid of fluctuation when modeling nature as an assemblage of points and lines, because that is not how nature exists.

What science has called the n-body problem* is caused by that same simulcausal nestedness. The n-body problem cannot be calculated because n-bodies mean multiple discrete bodies. The problem is explicitly framed in terms of puzzlelike parts interacting through local cause and effect—parts that can be very different from one another. And that is why it appears to be mind-bogglingly complex.

In reality, everything in nature interacts simulcausally. Things do not add up because they do not truly relate as discrete bodies. The whole, in fact, does not equal the sum of the parts. And that is fine. The whole is an inherently indivisible fractal of wave motion. Therefore, when waves nested within waves are sprouting, one cannot attach numbers in the same way.

Here we see the full scope of Eugene Wigner's "unreasonable effectiveness of mathematics." We see how math is so effective—as well as the unreasonable *inef-*

---

* The problem of predicting the motion of each member of a group of (more than two) objects that are freely interacting

fectiveness of math in complexity theory. What we call "nonlinearity" is waves waving strongly enough to disturb the seemingly linear progression—with new, sprouting layers of waves.

Indeed, perhaps needless to say, *nonlinearity* is a superfluous term. It is the equivalent of calling a living person "nondead."

Math works well in the quantum world because physicists put in great effort to isolate and dampen waves as much as they can. Systems are closed. The labs are made cold—very cold—to dampen waves to be as linear as possible.* Nevertheless, the waves are there.

Recall the words of Hans Christian von Baeyer: "The way the bell curve emerges is nothing short of magical. The general features of the bell curve make intuitive sense, but why it should have precisely the shape it does . . . remains a mystery. What power guides the pennies [of a coin toss] . . . to fall in such a predictable way? . . . How does this exquisite order emerge?"[2] In the same vein, in an article for *Physics World* magazine, Mark Haw explains that "Einstein himself never accepted the statistical interpretation of quantum mechanics. Statistics in a liquid of atoms was fine because you knew that you were counting real, physical atoms. But what did it mean to speak of the statistics of a single electron? What was 'hidden' behind the electron that caused it to behave statistically? This was a question that Niels Bohr's 'complementarity' simply barred you from asking, and Einstein was never satisfied with that."[3]

Waves waving exists as that reality those physicists never could discern. Whether on the macro scale of a coin toss or on the micro scale of the quantum, all motion nests in larger carrier waves whose existence scientists could recognize only through mathematics.

I realized that two scientists, de Broglie and David Bohm, both had the idea of a real wave, instead of a probability wave, somehow causing the quantum particle to be where it is. De Broglie called it a "guide wave," and Bohm called it a "pilot wave." They had been on the right track. Einstein knew it and felt something deep inside was wrong. I saw that Einstein, de Broglie, and Bohm were, so to speak, more right. All of nature has this pattern, with the highest concentration in the peak, and less in the trough, of real waves waving within waves. It is a facet of the true pattern of nature.

The beauty of nature's order is both brilliant and encouraging. In *Synchronicity*, F. David Peat spells out that mathematicians and scientists are aware of the rift between nature as it exists and the mathematical laws that describe it, but that hope is not lost.

---

* The ability to calculate so well in the cold labs of the quantum stands opposite to the inability to make calculations for variable biological systems. In biological systems, all the waves are highly clustered with simulcausal changes and cannot be fully represented mathematically. The waves aren't flat enough to be plotted as linear regularities.

The equations of physics will never take wings and fly for they are simply math-
ematical descriptions, abstractions in thought. Suppose, however, that these laws
are themselves the mathematical manifestations of something which has hitherto
only been dimly grasped. What if the laws of nature—the ones that really fly—are
not simply abstractions of experience but are the realization, within the world of
mind, of something that is creative, generative, and formative, of something that
lies beyond mathematics, language, and thought?[4]

The inherently continuous fractal is the true existence that Peat could describe
only as a possibility. The beauty of his description, and the reality that it is, can only
leave one silent in appreciation.

Might we one day develop math that effectively represents waves waving? I'm not
a mathematician, but I suppose there is an opening for a new branch of mathematics
that would be wholly different from complexity theory. It would be understood that
it cannot describe nature absolutely, insofar as mathematics works with discrete entities and numbers. But perhaps such a math might approximate a simulcausal, cross-
scale continuum as it moves forward, instead of approximating a whole made of
parts. It would never be exact, because waves within waves are ultimately nonlocal
and nonlinear, but it could be a useful way to model waves waving.

SCIENCE'S MATHEMATICAL MODELS of nature have a common finding—a finding
whose existence was itself a mystery. This commonality is energy. From the physicists' perspective, in the words of Heinz R. Pagels, from his book *The Cosmic Code*,
"the visible world is neither matter nor spirit but the invisible organization of
energy."[5] But what exactly energy is has escaped them. It is a mathematical quantity,
with no tangible existence. There has been only an operational definition: the capacity to do work. Others will add to this definition the capacity to make waste, or
entropy.

Energy, in both cases, is a measure of change, whether to different forms of order
or to decreasing order. It is a measure of motion. No one has known why there is
such a thing as energy in nature.

By now, we have fully explored what causes change. And thus we know the identity of energy.

When scientists measure the capacity to do work or dispersal as waste, what they
are attempting to do is measure the change generated by the continuum of waves
waving. The change *is* waves waving. That is to say, the question of "what is energy?"
is no question at all from this perspective. Change, including useful work and dispersal, is not an external "thing" that happens "to" material objects or discrete par-

ticles. The reverse is true. Waves waving propels change through the universe. Our concept of energy is simply a measure of that fact. You might say energy is a mathematization and, in that, an approximation of changes occurring in the waves waving fractal.

Thus, the reason for the existence of the second law of thermodynamics is exposed. In a closed system, energy always goes to waste; this is what happens when you cut off, as much as possible, the dynamic continuum of waves waving within waves. Of course, no system is truly closed. But to the extent that the influence of carrier waves is dampened, inner waves will flatten and disperse. We will return to the second law in the next chapter.

In the same stroke, this explanation resolves the puzzling existence of complexity. Scientists have not known how or why energy contributes to emergent complexity when they expect it to disperse in accordance with the second law. The simple explanation is that there is no automatic dispersion when waves wave in their inherent continuum. Both organization and dispersal exist within it. We will discuss this further soon, in relation to health and disease.

We have also seen the equivalence of energy and mass, which was a baffling outcome of the Puzzle Hypothesis. Through the famous equation $e=mc^2$, all mass seemed to be energy. Physicists treat and work with particles as abstract mathematical quantities of energy. The energy of $e=mc^2$ is, in the end, no mystery: The energy is a mathematical abstraction of wave motion. Nor is energy's relationship to mass a mystery, either: Waves waving becomes matter, as I described above. (That only accounts for the $e=m$ part of the equation, though. Waves waving also, of course, disperse as space—which we have understood through spreading electromagnetic waves—and is represented by light, or c, in this equation. The light's speed of oscillation—the squared part of c^2—is a linear approximation of time as it unfolds through waves waving. Thus, the equation $e=mc^2$ can be loosely read as waves waving=matterspacetime.)

I liked the term *wavenergy* and used it for a while to refer to the discovery, because the changes of waves waving are active and dynamic. They hoist change on everything we think of as static. It's sparklingly energetic. But this is not the energy of the physicist's dictionary. What we perceive as energy is the essential fact of change in nature.

If we were hanging on to the fabric of nature (which itself is waves waving), waving waves would be wriggling us, jiggling us—us and all of nature—as waves within waves billow through that fabric. Energy is a linear, approximate measure of that change. The concept of energy as an external phenomenon is not true to nature, not true to how change exists. Change happens naturally, and will always happen, because waves are waving.

WITH ALL THIS CHANGE always in motion, I must say that the orderliness of it was striking. It was a different kind of order. Order in nature—whether represented by laws, equations, or brute repetition—has always been of regularities. Here was an order *of change*, indeed, made *by* change. The organization was brought about by change.

I could see that the actual shifts and turns of each wave—the defiant nonlinearity that characterizes the true wave—play an active role in creating order. Rather than drawing deep to a line to find order, the waving waves themselves are and make order. No math is necessary; in fact, math would ruin the true order. It is a regularity—of *ir*regularity. It is an invariant that varies. It is a symmetry that is asymmetrical; an order that generates relative order and disorder; the invisible that is also visible.

AFTER ALL THESE REVELATIONS—at once mind-blowing and illuminating—I still saw that the idea is very simple. It makes a lot of sense. Nested waves that concurrently change might seem strange because it is a new way to think (as we have discussed at length, we're accustomed to thinking about nature piecemeal, as static parts), but it is amazingly simple once the newness wears off.

It is logical for motion to be in motion. Why would it travel in lines? There are no true lines in nature. It makes good sense for motion to be true to its own character— to be *in* motion, and not *un*changing. There is a common sense to that. And more, it is a natural phenomenon that we experience. It is real—this is nature as we experience it. Rhythmic motion is universal and not an abstraction from nature.

As mentioned earlier, Richard Feynman once said that "if, in some cataclysm, all of scientific knowledge were to be destroyed, and only one sentence passed on to the next generation of creatures," the most densely informative fact is "the atomic hypothesis (or the atomic fact, or whatever you wish to call it) that all things are made of atoms—little particles that move around in perpetual motion, attracting each other when they are a little distance apart, but repelling upon being squeezed into one another."[6]

The motion Feynman describes in his fundamental statement is none other than wave motion. It is a back and forth motion, of attraction and repulsion. And that very motion, continuous with like motion on deeper levels yet, causes waves to sprout into the material atom he describes. Stuck looking at it the way science has looked at it all along, Feynman cast it as matter in motion. But one level down, if he could have seen it, is a single phenomenon, waves waving within waves, which coalesce to create that moving atom.

The original, other choice of nature stepped forward to claim its rightful place. I saw nature not as a whole made of parts but as an "only one." I saw it is not built from waves waving: It *is* waves waving. The particle, the motion, space, and time are all manifestations of that one single phenomenon.

If one sentence were to be passed to the next generation, the most essential is this: Waves wave within waves, in an inherently continuous fractal, to create matterspacetime.

It is everything we know.

I called it SuperWaves.

IT WAS TIME to let waves be. Waves, not matter, are the bottom line. Physics has always accepted waves as a phenomenon that simply exists, although identifying them somewhat disparagingly as a "disturbance in a medium" or "a form that moves up and down." It is necessary to turn our viewpoint and embrace waves. I cannot explain waves further than to identify how they work and then—to accept. Whatever waves are, they're here. They somehow give us what we experience as the material world as well as the motion of that material, all in one. They are the stuff of the universe. Motion, in a self-similar fractal pattern, is what *is*. The incredible facts science has unearthed about nature all fit into this elegant explanation.

I remembered, on reflection, from my medical training, that our senses operate through waves. Senses are rhythmic, and they pick up oscillatory information. It was amazing to realize this. The very reason we split up nature to begin with—that our senses sense one piece of nature at a time—expires as a rationale when we get to the bottom of what senses really are sensing. Here was the very thing that made us look at nature as a puzzle—and they are waves themselves. Only, our senses cannot pick up the ultimate continuity that we encounter between matter and space at either the macro or the smallest of small scales, the quantum level. And so, we assumed everything is separate and set history on its course.

Now I could see that the universe is waves waving within waves.

It was perfect. To paraphrase Wheeler's exclamatory declaration, how could it be otherwise?

CHAPTER 14

Laws: The Quantum and Thermodynamics

I HAD REACHED a sort of trough of my own. The quantum was as far down as I could go. Beyond that, there were only waves waving within waves, far below anything any human has detected. I could not draw more knowledge by going deeper. The SuperWaves portrait of nature was complete.

And yet, though I had journeyed as deep as current knowledge would allow, the journey was not over.

This may come as surprising and even counterintuitive considering where we've come from. We thought that going as deep as possible would tell us everything we wanted to know about nature. The Puzzle Hypothesis had pointed to the very, very small as the source of all answers—for if the nature of nature is that it is made of building blocks, the smallest levels should reveal what those building blocks are and how they relate. Just knowing their nature should, from that reductionist perspective, give us a complete understanding of all of nature.

But tracking SuperWaves down led to a surprise discovery, confirmed by the very small. Particles are not, after all, discrete building blocks. They are by nature continuous with, and nest in, an inherent fractal continuum of waves waving.

In this way, SuperWaves turned out to crack open a secret we did not know nature had been holding. It is more than a discovery of what we had been yearning to uncover, the reality of what nature is. It also sprung on us a different discovery, of the design of nature.

This second aspect of the discovery means that along with the definitive answer to what nature is, we also have a new model that requires we reacquaint ourselves with already-known facts.

This in itself was completely unexpected, and for me, a source of pleasure and wonder. The road to understanding lies not through unifying the partial bits of knowledge that science has accrued and answering science's questions through that unification, but through reunderstanding the facts in their true and natural context such that the questions melt away. The final answer to our questions about nature does not fuse all our bits of partial knowledge into a compact whole, but rather, it takes us swinging around a bend—and brings us face to face with an unexpected, cross-scale web of motion in motion on which all those bits of knowledge are to be hung, like bits of holiday lighting, in a vast context of motion in motion that dissolves their boundaries. For thousands of years, while we thought that nature takes the design of a puzzle, we partly understood nature but were left with many, many questions. Now the questions can be resolved all in one go.

And yet, because of how science has trained us to think, this may seem too easy. This sense of "it's too easy" is a potent source of resistance to the discovery of Super-Waves, and it is important not to fall prey to it. I will now address both issues: the easy solution and why easy makes us uncomfortable.

The ease itself comes from two sources. The first source we have examined: that many of our questions were based on an incorrect assumption about nature. Adjust the assumption, and the questions go away.

The second source is more subtle. The scientific perspective has had us incor-rectly believing that the answers lie in the details. Specifically, as discussed in Chap-ter 2, the Puzzle Hypothesis—which underlies all of science—told us that in order for our answers to come from nature itself, the answers had to be within the pieces. The alternative—a top-down/outside-in, organizing force through which order arises—could only be conceived as something superapplied to nature.

In other words, we thought that if we were going to find order arising from within nature, we would have to study the properties of the pieces, piece by piece, to find that order's source and then reconnect the details in all their complicatedness— a task science has zealously undertaken.

And here we are, with SuperWaves showing that order, along with disorder, is a natural consequence of the way waves wave. Order and complexity, as well as relative disorder and chaos, are influenced by a top-down/outside-in generator from within nature itself. Science was founded on the premise that order and change come only from the bottom up, from the inside out, from the details. SuperWaves shows a simultaneous top-down/outside-in influence: the other half that science was missing. It not only complements local cause and effect but also recasts what it is entirely.

This approach, which is the second source of easy understanding, is why easy makes us uncomfortable. We have amassed a nearly unimaginable amount of data, and the new approach tells us we don't need it in the way we had thought.

Our situation is a strange if not bizarre one, something like that of a person who, while walking around and living his life normally, has only been aware of the left half of his body.

Imagine this person. He has worked hard to understand many things. He strives to know how he moves forward, for example. He collects data on how far his single foot (to his mind) moves ahead. He makes theories of how he stays balanced as he moves forward (based on his knowledge of his single foot), and so on.

But one day he comes to realize that there is a right half to his body, which has always been there, working together with the left half. And his whole picture changes. His theories about how his left half moves him forward are wrong. And the detailed data he collected can be reunderstood in the context of his normal human gait. The whole is more than the data of the left half doubled to include the right half. The whole gives new meaning to existent facts and easily explains so much more. No longer of essence are the details of the left half alone.

That situation is like ours here. We can let go of many of the details science has and still understand the big picture. Detailed facts are not useless, but neither are they essential for that understanding. The big picture is the framework of Super-Waves: the secret of how nature works, how things hold together and fall apart, and what it is. Even the greatest master of detail will never understand the sustained setup of nature without this framework. SuperWaves answers our great questions because it is the nature of nature.

And yet, somewhat amazingly, knowing the nature of nature is not a finale that renders needless all further investigation. True, it ends the chapter of our history in which we thought of nature as a sum of parts. But it also begins a new story.

We have, ahead of us, two pathways to explore. The second, which we will discuss in the final chapter, is how to adjust and refine our knowledge and technologies for the future survival of life on earth. But the first, which we will now discuss, is how to reunderstand the past theories, principles, and laws of nature as science has construed them. The dramatic but gorgeous truth is that this single shift of perspective makes nature wholly comprehensible in an uncomplicated way.

In all the phenomena we will discuss, the top-down/outside-in aspect of simul-causality is a star. It contributes to the organization of nature in a way never imagined from the perspective of local, isolationist cause and effect. The top-down/outside-in clustering and dispersal that carrier waves effectuate—the natural, simultaneous complement to the local cause and effect that we are already familiar with that fuses the two into a different kind of causality—creates organizational patterns that science knows well but cannot explain.

And that is how this leg of our journey will come to an end. We will visit a number of scientific phenomena and see top-down/outside-in, bottom-up/inside-out simulcausal organization at work.* We will see all the facets, in fact, working together to generate nature as we know it.

In this chapter, we will discuss the relationships between thermodynamics and quantum physics and between the second law of thermodynamics and its converse, emergent complexity. We will also look at what science has been calling fractal organization in nature, self-organized criticality, power scaling laws, and strange attractors, as well as phase transitions, speciation, and synchronicity.

In the chapters that follow, we will discuss the apparently extremely complex phenomena of health and disease. We will understand anew the law of gravity, whose scientific formulation led scientists to propose missing mass, dark energy, and black holes. And we will briefly go over a number of seemingly disparate phenomena dealing with the environment, from the latitudinally dependent species richness gradient and the impact of caloric restriction to the "edge effect" in species conservation and the Allais effect in a solar eclipse. We will see how all phenomena naturally present in nature are evidence of—and explained by—the way waves wave within waves.

FIRST, A WORD ABOUT simulcausality. Its effects are everywhere, and of all the principles of SuperWaves, it shines in eliminating problems science has had.

Yet it is perhaps one of the hardest concepts of the new worldview to truly understand. Simple though it is, it is the opposite of what has seemed normal to us. It is the opposite of local cause and effect—the opposite of discrete, isolated parts interacting with one another. Everything connects with everything else while it is happening, and that can be a little hard to comprehend at first.

We need not abandon our sense of a single cause, of course. Everything that happens, and everything that is, is caused by the way waves wave. In an ultimate sense, true causality is SuperWaves. The sole direction of causality *is* the forward motion of waves waving within waves.

But we can go further than that in understanding smaller causes as we experience them. This is where simulcausality comes into play and where things can get tricky if you try to hold on to, and simply modify, the old worldview. Simulcausality draws from the new understanding that wave motion proceeds with an inherent togetherness across scales. It is not a combination of bottom-up and top-down: It is an awareness of a unity—of the fact that both bottom-up and top-down influences unfold together as motion occurs. The inherent interconnectedness is

---

* From now on, whether I say top-down or outside-in instead of top-down/outside-in, I always mean top-down/outside-in and am using the shorter term for brevity's sake. The same for bottom-up or inside-out instead of bottom-up/inside-out—I always mean bottom-up/inside-out and am using a shortened version for brevity's sake.

reminiscent of a coin whose heads or tails side is always affected as you turn to the other; if you turn heads up, tails go down—not sequentially but simultaneously because it is one coin.* Wave motion is "one," in that it has no fully independent parts, so the forward motion of all waves naturally changes on different scales as any given scale changes.

In the universe, there is, of course, a lot of what we call local cause and effect going on. The key is that none is absolute. It is not absolute because there is also, always, this simulcausal guidance from carrier waves as all motion moves forward. It is a different kind of idea.†*

Simulcausality can be a little difficult to grasp in the abstract, but the reality of it becomes more comfortable as we see more examples of it. The patterns it creates are plainly visible when one knows to look for them. Let's move on to explanations of nature that show it, and all the facets of SuperWaves, in action.

OUR FIRST STOP is the inherent relationship, and the consequences of the relationship, between two scientific subjects: quantum physics and thermodynamics.

Science has treated quantum physics and thermodynamics as distinct areas of study. Indeed, as we have seen, the Puzzle Hypothesis prompted scientists to develop the two fields as if they are wholly unrelated. Sadi Carnot began the science of thermodynamics in that way: He chopped off the large-scale wave's cooling-down process and isolated it as the model movement of heat. He represented the natural downswing of a wave as a straight line, from hot to cold.

In this way, as discussed earlier, Carnot departed from reality just as Galileo did, conceptually, with his proposed law of independent velocities. Galileo construed the arcing motion of an object as straight lines of gravity and inertia going in different directions. This conceptualization culminated in Newton's laws of motion and gravity. Carnot's application of Galileo's lines were no less influential. Because of his straight-line abstraction, the natural downswing of inner waves' recovery was construed as a straight line of energy from a hot source to a cold sink.

Planck, not long after, cut up nature along an additional plane. Starting with the leftovers from Carnot's dissection—the upswing of a wave as if it were a straight line

---

* The analogy is limited; simulcausality is not compact and discrete like a coin, it is not static, and it is not restricted to dimensions. Indeed wave motion, as I have described, is behind the dimensions we perceive.

† One key to making the transition to grasping simulcausality is to turn an eye to the way we are used to understanding. The Puzzle Hypothesis conditioned us to expect invariants—parts that don't change but that do fit together. But change is too relevant to discard in favor of invariants: Change is motion. Change is the nature of nature. True invariance does not even exist. Therefore invariance, though easy to think about, cannot be our starting point. The way nature works is not separate things changing in a one-by-one fashion. The shift is hard, but we have to understand that all motion is inherently indivisible, one instance from another, and it is forward-moving. Through its forward motion is the only way we can see it. That means it is always changing and never invariant. It is the way it exists—where existence is, again, not a static state but comes to fruition through motion.

going up—he shaved off the peaks of the inner nested waves, such that it looked like discrete jumps from peak to peak.

Reality differs.

Reality presents these supposedly separate sciences in continuum with one another. You have to heat something up before it cools down. You have to introduce oil, coal, gas, or wood and, indeed, keep feeding the fire. We must identify the heating-up process for what it is: It is the heating of a blackbody, the selfsame process that Planck described to launch quantum physics.

In the same way I had seen that exercise is never separate from recovery, I saw that the upswing of a heating blackbody—from the scientific perspective, everything that heats up is a so-called blackbody, including the sun and human beings—is never separate from the downswing described by the second law of thermodynamics. It is the same thing, the same wave pattern—not an analogy, *but the same stuff of nature*, rising and falling across scales in the way that it does.

The necessary reunion of that upswing and downswing does not nearly deliver the full picture of what is happening with what science calls heat energy, however. It was just as I had experienced when reuniting the upswing of exercise with the downswing of recovery, which had not been enough to deliver the full picture of Super-Waves. Only once I'd discovered the simultaneity of the bodywave and the heartwave did I understand the compressing and dispersing effects of the carrier wave. It was this understanding that oriented me to see that the collective is organized through a means other than local cause and effect: through simulcausality.

Reuniting quantum physics with thermodynamics requires just the same line of thinking. It entails more than realizing those sciences were never separate. It means recognizing there is more than a single wave going up and going down; there are always inner waves climbing up and down carrier waves on innumerable, inherently continuous scales. The key is realizing that inner waves—including the clusters of waves within waves that we recognize as matter—are collectively organized, simulcausally, by the shape of the carrier waves in which they nest. *The sprouted, clustered, collectively organized peaks within peaks at the crest of the wave—the crest that joins quantum mechanics and thermodynamics—are what science has called **energy**, the capacity to do work.*

As it was with quantum physics, I realized, science's ignorance of this reality is the reason for the findings of thermodynamics and also of its mysteries.

Now we can see reality as it is.

TO UNDERSTAND WHY the second law of thermodynamics seemed indisputably true, we have to begin with what we just saw: that the science of thermodynamics axiomatically leaves out the upswing of a wave; that it frames the downswing as transpiring in an idealized, closed system; and that it identifies the downswing of the wave as a linear flow from hot to cold.

It meanwhile recognizes a pairing between the so-called flow from hot to cold with a dispersal of so-called energy. For science, in that flow from hot to cold, some energy is said to be siphoned off to do work, but most goes to waste. This has been stated as the second law of thermodynamics. (The second law, as discussed earlier, has always and only been crafted as an observation of what happens in a closed system. It was never a theoretical prediction. Scientists boxed in the downswing of waves, observed what always happens, and stated it as a law.)

The next step is to realize what is happening in terms of the SuperWaves fractal. A system can never be truly closed, but certain things do naturally happen in a relatively closed system. A relatively closed system—what science calls a closed system— is one that has been isolated from the variability and fluctuations of the environment as much as possible. Thus the oscillatory motion of carrier waves is highly dampened. SuperWaves shows, as we have seen, that waves flatten, tend to disperse, and become closer to equilibrium in this sort of system.

Such a flattening of waves waving, and the loss of variability that accompanies it, naturally leads to a decrease in simulcausal organization. Decreased simulcausal organization means waves going into troughs will disperse and will not be usable for work. That is to say, the simulcausal means through which inner waves are organized and held together in peaks, and relatively dispersed in troughs, is hampered in such a system. What transpires instead is an unraveling of organization with a bias toward wave dispersal. This is what science "discovered" as the second law.

I'll repeat this explanation because it is so important. Because carrier waves are densely populated with inner waves waving, which carry within them further waves still, and because every one of those so-called inner layers of waves rises and falls, changing in frequency and amplitude as it climbs and descends its carrier waves (which climb and descend further carrier waves), relative isolation, through closing a system, has a profound impact. Inner waves begin to flatten out and dissipate. Flattened out and dissipating waves result in a loss of scale, complexity, variability, and organization.

This set of consequences is what has been identified by science as a dispersal of energy, or waste. It is also known as entropy. It is the opposite of what was introduced in the previous chapter, emergent sprouting organization.

LET'S TAKE A MOMENT to consider what this really means about the so-called second law of thermodynamics. Two important ideas nearly leap out for further inspection once you understand that closed systems cause waves to flatten out, become more uniform, and dissipate.

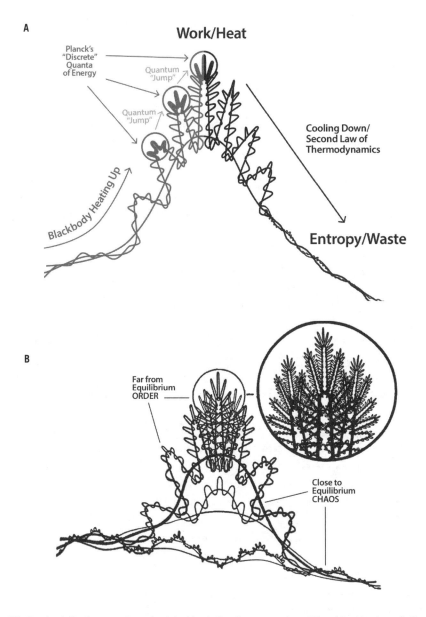

A

Planck's "Discrete" Quanta of Energy

Work/Heat

Quantum "Jump"

Quantum "Jump"

Blackbody Heating Up

Cooling Down/ Second Law of Thermodynamics

Entropy/Waste

B

Far from Equilibrium ORDER

Close to Equilibrium CHAOS

(A) On the left, the upswing of a blackbody heating up and emitting blackbody radiation (including what science had identified as quantum jumps). On the right, the downswing of a blackbody cooling down, as exemplified by the steam engine. The waves traveling the downswing, if it is boxed in as a relatively closed system, spread out. This is the true cause behind the dispersal and cooling down described in science's second law of thermodynamics.

Note that the phenomena behind science's laws of quantum mechanics and thermodynamics are in continuum with one another, as described in the beginning of the chapter.

(B) More layers of waves illustrate the relationship between order and disorder.

The first idea is that science's method of study, experiment through isolationism, actually creates environments in which a descent toward chaos is inevitable and complexity does not consistently emerge. And the second idea, the flip side of the first, is that that very act of isolationism prevents scientists from finding the source of emergent complexity.

These two ideas, which I will now elaborate on, are not new recognitions, per se. They are further instances of principles I had finalized in the quantum realm. Now we circle around and see the effects expressed on higher scales in the SuperWaves continuum.

The first of these ideas is as dramatic as it is inspiring. Scientists found a truth that reflects a fact about their method of study—and not about nature as it exists. That one deeply powerful assumption, that nature is by its nature divisible, made it impossible to see that they were not discovering a grand law of nature. They were discovering, instead, that closing a system as much as possible catalyzes a greater dispersion of waves. The second law of thermodynamics is not a law of nature: It is a "law" of relatively closed systems.

That is why an inevitable march toward entropy seemed so certain. The scientific method requires closed systems, and closing systems promotes increasing disorder. Thus every experiment's setup unwittingly but actively contributes to the mounting disorder it seems to reveal. When Sir Arthur Eddington said that "if your theory is found to be against the second law of thermodynamics I can give you no hope; there is nothing for it but to collapse in deepest humiliation,"[1] and when Einstein said that "classical thermodynamics . . . is the only physical theory of universal content concerning which I am convinced that, within the framework of applicability of its basic concepts, it will never be overthrown,"[2] they did not know that the act of closing systems affects nature profoundly. They did not know that closed systems, which science demands as a prerequisite for any study, have an impact on nature. But so it is. The scientific method itself caused the so-called discovery of the second law.

The flip side of this realization is readily apparent. If isolating systems facilitates disorganization, it also means that that isolation hampers emergent complexity and organization. There is no way a scientific experiment will ever find emergent complexity's source.

We saw, in the last section, that complexity emerges through sprouting: through the dense, swarming rise in frequency-amplitude in the peaks of waves going up peaks of waves. In highly accelerating, high frequency-amplitude waves, inner layers swarm, spiral, and sprout in synchrony—coordinated by the compressing peaks of carrier waves in which they nest.

Every one of scientists' suggested solutions for emergent complexity—be it far-from-equilibrium thermodynamics, synergetics, complex adaptive systems, nonlinear dynamical systems, self-organized criticality, spontaneous self-organization, or cybernetics—misses the essential, unified expression of waves waving within waves that creates the emergence they seek to describe. Indeed, all attempts to account for increased organization suffer from the same consequences of boxing in a system. Such ideas misconstrue nested waves as circles and cycles of energy that are "somehow" used for order. Proposals that organization increases through free energy, negentropy, dissipative structures, and so on idealize a closed system, a system which wrongfully advertises that the flow of energy only goes in one direction, from order to chaos.

And the inherent limitation of their explanations has been understood by scientists, as Paul Davies explains in an article in the magazine *New Scientist*.

Physicists are far from knowing just what it takes to create order out of chaos. They cannot point to specific characteristics in the laws of physics as "the source of creativity." It is not even clear that the whole story lies within the known laws. Some scientists suspect there are undiscovered laws, overarching principles, at work, coaxing clod-like particles of matter toward organized complexity. Sometimes the hypothesized "principle of increasing complexity" is called the fourth law of thermodynamics.[3]

This so-called fourth law of thermodynamics, scientists suppose, would reveal the secret to how complexity emerges. The true secret is that closed systems hamper the way complexity emerges.

As long as a system is closed, nested waves within waves—with all the synchrony that increases order and organization, and that accelerates and heightens inner layers of waves such that they sprout—cannot be rightly understood. Open the system, as waves waving determines we must, and you get a different outcome than entropy.

I'll say it once more because it is so essential. The open, simulcausal continuum of wave motion nested with and within other wave motion determines that a top-down/outside-in influence causes several notable effects. One is the repulsion and disorganization that science identified as entropy on the downswing and in troughs. Another is the attraction and organization that is the opposite of entropy on the upswing and in peaks. And yet another is the so-called emergent complexity of organized matter, brought forth through a compressing effect in the peaks, as deeper and deeper layers of inner waves swarm, spiral, and sprout.

This is an extraordinary way to reunderstand nature.

EMERGENT SPROUTING THUS NEGATES the first law of thermodynamics, just as it does the second law. The first law treats the universe as if it is a closed system. Its pronouncement that all its energy is conserved simply restates the Puzzle Hypothesis. But like the second law, the first law was derived from experience—and all scientific experience has been in relatively isolated, closed systems. Closed systems hamper emergence. The sprouting that defies being understood as a sum of parts cannot take place in a closed system. And so, closed systems seem to adhere to the first law.

As they do for the second law of thermodynamics, the systems in which the first law is observed actively contribute to it being found.

The first law obscures the nature of nature. Nature is neither a closed system nor a sum of parts. Phenomena come into existence from deeper, previously undetected layers of waves waving, through sprouting. Nature as we know it, through the varied expressions of the components of SuperWaves, brings forth multifaceted complexities inconceivable from the limitations of the first law.

These exquisite, cross-scale emergent patterns of organization, which are swarming, spiraling, and sprouting waves, are superabundant in nature. They are intimately familiar, as the patterns found in atoms and in cells, in the solar system and in galaxies, in hurricanes and in flocks of birds and swarms of fish. They are patterns of organized complexity, and they happen naturally, on innumerable scales, in the SuperWaves continuum that *is* nature.

The repercussions of this recognition are enormous. Swarming, sprouting, and spiraling waves bring forth unprecedented layers of inner waves, all while being compressed and dispersed by carrier waves. Matter itself and space itself, which exist indivisibly as matterspace, and organization are all natural expressions of waves wav-

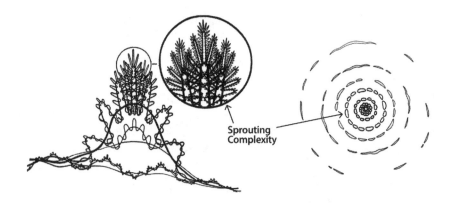

Sprouting accounts for what has been called emergent complexity and organization by science. Previously undetected inner layers of waves come forth in highly variable nested waves.

This diagram is identical to the diagram representing the sprouting of matter and stability on the quantum and atomic scales because waves waving sprout self-similarly, in the same way, on all scales of the universe.

ing within waves. Both the stuff of nature and its means of organization are entirely different from what science had thought.

WITH ALL THIS SAID, the source of the fractal organization that abounds in nature becomes clear. The nature of nature is that it is an inherently continuous fractal. When Davies says in *The Fifth Miracle*, "Could life be like this: apparently complex but actually very simple, like a fractal, and therefore the product of a simple, law-like process?"[4] the answer is an emphatic yes—but an IC fractal, of the sort he had never identified. This IC fractal organizes not only life, as Davies pondered, but also the entire universe.

The IC fractal nature of SuperWaves, moreover, provides a home to all natural phenomena. Many natural patterns and regularities are, therefore, easily recognized as exhibitions of it. I will now articulate a number of issues that the discovery resolves. I will not spend much time on them; rather, I will list them, paragraph by paragraph, to give you the highlights of how their characteristics arise from SuperWaves.

The alternating order and chaos found in what scientists call systems on the edge of chaos, or self-organized criticality, displays the pattern of the IC fractal. These are systems that exhibit a certain amount of extended stability, until suddenly they collapse into so-called chaos, after which they shift into stability again. A growing sandpile that periodically falls into an avalanche and the human brain are examples of such systems.[5] Such so-called self-organized criticality—and I say so-called because, of course, it is not in fact locally self-organizing—follows what science calls power scaling laws: Big events happen less often than smaller ones, according to consistent mathematical ratios.

The IC fractal creates systems that exhibit self-organized criticality, and as I explain how it does so, I will include how it accounts for what science calls power scaling laws, as well. Science has used the term *power scaling laws* to refer to natural relationships, across scales, in which one scale varies as a mathematical power of another. As you go up in scale through organisms from a mouse to a blue whale, for example, metabolisms, heart rates, and life spans all shift as suits the scale. (The changes happen in proportion; the size of the animals goes up step-by-step as if by addition, but their metabolisms and so forth increase by orders of magnitude, as if multiplied, in order to accommodate the step up in size.) The Richter scale of earthquakes similarly reports on a power scale of magnitude, where an earthquake of 5.0 is 10 times less intense but happens 10 times more often than one of magnitude 6.0, and one of 6.0 is 10 times less intense but happens 10 times more often than one of magnitude 7.0.

The IC fractal is, perhaps obviously, responsible for power scaling laws. Power

scaling laws display its self-similar character. But the relationship is even more revelatory. The cross-scale continuity of the IC fractal also removes the mystery that has,
for science, surrounded power laws: the mystery of why, from the scientific point of
view, power laws seem to advance up scales through jumps. To the scientist, they
make no more sense than the quantum jumps made by an electron traveling from
one atomic shell to the next.

The same mechanism that we saw neutralize the mystery of quantum jumps also
neutralizes the mystery of how power scaling laws jump from one scale to the next.

We resolved quantum jumps when we saw that carrier waves within carrier waves
compress and organize, as well as disperse and disorganize, inner waves. This simulcausal effect maintains the inner wave clusters' shape in more than one respect—
within a given so-called scale and also, inherently, across scales. This compression of
peaks in peaks, together with dispersal of troughs in troughs, is responsible for what
looks like clusters that are separated by jumps. It is true for so-called quantum
jumps; it is true for so-called power scaling laws. Power scaling laws represent a pattern of waves waving that is held fractally consistent, from one inherently continuous scale to the next, by carrier waves.

All natural phenomena that seem to alternate between stability and jumps to a
different phase of stability are organized in this same way. Systems on the edge of
chaos, which I introduced above, happen as they do through this mechanism. The
order and chaos they encompass spans an inherent continuum of scales. What we
observe, in such a system, is a mounting increase in order and organization, as happens naturally when waves climb a carrier wave. They are more concentrated and
organized as they accelerate toward the peak. Then, as they descend to a trough,
they tend to disperse and become relatively chaotic. Then they begin to climb anew
and display commensurately increasing order.

These so-called jumps from order to chaos and back again reflect the natural rise
and fall of waves inherently organized and dispersed by waves that themselves are
inherently organized and dispersed by waves. Order chaos, order chaos, order chaos
is a natural pattern in the IC fractal: exactly what one would expect from clusters of
nested waves waving within waves as they climb and descend. It is not order versus
chaos but order and chaos in a continuum.

Even brain waves during the chaotic state of a system on the edge of chaos, for
example, still retain many degrees of order relative to an idea of absolute chaos. Science has never been able to explain this type of order, using the term *strange attractors* to describe the phenomenon of order even within relative chaos. The degrees of
order retained by such a system, even in the face of chaos (brain waves, in this
example), are perpetuated because the waves are nested within other waving waves
and have waves nested within them. They will not descend to utter chaos because
they are still nested in the SuperWaves continuum.

In fact—and this is a vital point—there is no absolute chaos anywhere in nature. Even down in zero-vacuum fluctuations, there are still cycles in the waves. So-called chaos has a definitive place on the IC fractal. Chaos, which is always *relative to order* in a given system, is troughs within troughs of waves within waves.

Phase transitions, through which substances seem to suddenly jump to a different state after periods of stability, similarly illustrate the continuum that now replaces what science had construed as jumps. The phase transition of liquid water to ice, for example, demonstrates that trillions of water molecules that freely slide by one another at temperatures ranging down to 1°C will, at temperatures just below 0°C, organize into crystals of ice. This apparently sudden jump of trillions of molecules again reflects carrier waves effectuating relative coordination and stability. It is the same aspect of SuperWaves that accounted for so-called self-organized criticality.*

The jumps that have seemed to characterize self-organized criticality, power scaling laws, and phase transitions also appear in speciation. Scientists have had no way to account for why species exist as distinct groupings. Chimpanzees, as a group, are sharply distinct from gorillas, which are sharply distinct from orangutans: These are stable clusters with what seem to be jumps or gaps in between, rather than a seamless blend between individuals (no groups). "Called 'the mystery of mysteries' by Darwin, speciation is still a little-understood area of evolution," says biologist Jerry A. Coyne in an editorial in the journal *Nature*. "One of the great problems of biology is how a continuous process of evolution can produce the morphologically discontinuous groups known as species. Despite its title, *On the Origin of Species* made few inroads on this problem."[6]

Speciation stands out as a strong expression of the force of carrier waves, on a scale we have not yet examined. It is the nature of carrier waves to organize and compress stable groups, with relative dispersion and disorganization in between. Apparently, very large, grand-scale carrier waves organize the individuals that comprise a species. These are waves that take, from our perspective, many thousands and even millions of years to cycle. The troughs between the compressed clusters within are the so-called jumps between species, which have appeared as discontinuous gaps.

---

* To the extent that they are open to investigation and experimentation, the details of phase transitions make a good example of what investigation and experiment will mean in the future.

Any phase transition—even from ice to water or water to steam—is incredibly complex (as opposed to the quantum level, where the conditions under which waving waves organize to become a particle are well documented). And, as opposed to the simulcausality of the heartwave and bodywave, it is not clear here what is waving what.

In the past, scientists tried to piece together the workings of nature bit by bit. Now that we have the grand scheme, experimentation will mean understanding the details of how it works within the context of the known scheme, by trying it out. Most important will be experimenting with the influence of carrier waves. In this example of phase transitions, future experimentation with the oscillatory patterns of different substances will show us how we can simulcausally—by shaping carrier waves—effectuate phase transitions that suit our needs. This means we can learn how to organize inner molecules—which, as we have seen, are sprouted clusters of waves waving within waves—by tinkering with the carrier waves in and with which they simultaneously nest.

This is an extraordinarily powerful, and new, way to approach making changes in nature. Experiment will allow us to investigate it.

This is consistent with the idea of punctuated equilibrium, introduced by pale-ontologists Niles Eldredge and Stephen Jay Gould. Based on the fossil record, Eldredge and Gould suggested that species do not evolve gradually; rather, species are fairly stable for long periods of time and then rapidly undergo significant change through which new species branch out. This progression offers a further clarification of how speciation works: Species are apparently held stably together by carrier waves for extended periods, followed by a trough during which there is dispersal and relative disorganization, followed by rising stability and then periods of prolonged stability again. It is the same way that systems on the edge of chaos alternate between stability and chaos, with prolonged stability that periodically descends into chaos and then regroups to prolonged stability.

Again, though these explanations may sound like different ideas, they are but one idea, which reunifies like phenomena that science has treated as separate. All the organization that we call complexity, all the relative disorganization that we call chaos, and all the jumps in nature—whether they are instantaneous, like quantum jumps, or span thousands of years, like jumps of punctuated equilibrium—are expressions of the single SuperWaves continuum.

What science has presented as a splintered assemblage of mysteries are all explained by the few facets of the one inherent continuum that is nature. It is extraordinary. It is amazing that there are not myriad problems to be solved after all. Details to understand, yes; underlying perplexity, no. What science framed as dis-crete mysteries of organization and complexity are all resolved as expressions of the one true reality that is nature.

THE INHERENT INTERPLAY of the different facets, with a focus on simulcausality, is on display in the final scientific mystery that will be addressed: the mystery of synchro-nization. Synchronicity illustrates, in a most sweeping and pleasing way, how the simulcausal IC fractal pattern of SuperWaves upholds stability, and creates emergent patterns of organization, while also allowing for relative disorder and chaos.

Synchronization happens through the same simulcausal organization of inner waves by carrier waves. Electrons that couple during superconductivity, particles that synchronize in solid-state physics, pacemaker cells that fire collectively in the heart, a colony of ants working as one, a flock of geese migrating together, a school of sardines swimming as a whole, a group of fireflies firing in unison, the distinctive elements in an ecosystem working in harmony—all are simulcausally synchronized by virtue of being nested in carrier waves within carrier waves within carrier waves. Not to say that there is not local signaling going on: There is, but seemingly local signals are not isolated, and instead simulcausally feed into carrier waves that in turn simulcausally shape them and their group's behavior (on many scales going up and down).

Like speciation, synchronicity exhibits the effect of carrier waves on a scale we have not yet examined. The simulcausal process is the same but even grander. The waving waves that cluster and sprout into particles, that cluster and sprout into atoms, that cluster and sprout into molecules, that cluster and sprout into cells, that cluster and sprout into organs, that cluster and sprout into creatures, that cluster and sprout into flocks and schools—those wave clusters are squeezed together and coordinated into high levels of synchronized motion, on and across scales, by carrier waves.

Let us note an outstanding feature of synchronicity that is little discussed: the alternation between synchrony and asynchrony that truly characterizes synchronized groups. Groups whose members are synchronized also periodically desynchronize in a rhythmic pattern. Slime mold, for example, is a cluster of otherwise independent amoebas that periodically cluster into a synchronized whole (the slime mold). When food is scarce, the single-celled amoebas begin to oscillate; while oscillating, they jiggle closer and closer together. Eventually, they cohere into a new organism: a multicellular organism known as slime mold.

The slime mold exhibits greater coherence and longevity than any amoeba alone. It can travel far greater distances than a single-celled organism can. It is a paradigm of synchronicity. When the slime mold reaches food and ingests it, however, the synchronicity ends. The organism disperses into single-celled amoebas once again.

In times of food scarcity, amoebas go through great waves of rising frequency-amplitude oscillation—what we have been calling waves riding the rising upswing of a carrier, or attractor, wave. These nested, rising frequency-amplitude waves are simulcausally powerful: They enforce coherence and longevity through the simultaneously changing, top-down/outside-in carrier wave in which they nest. Their nested continuity with the carrier wave brings about the compression, emergent complexity, cooperation, and synchronization of individual amoebas, such that they organize into the form of a slime mold.

Then, when the slime mold reaches food, things change. The food allows for recovery: a downswing of the nested waves. The group of synchronized amoebas that is the slime mold dips to that trough. The carrier wave, simultaneously, becomes commensurately less powerful. And so the amoebas once again desynchronize and disperse.

Everyone marvels over the synchronicity of slime molds, but when we truly understand their behavior, we see the desynchronization is as much a part of the marvel as the synchronicity is.

Photosynthesis also exhibits synchronous organization through collectively peaking waves. During photosynthesis, scientists have found, molecules exhibit quantum entanglement—synchronously cycling through excitation and relaxation in groups even when not strongly coupled. The letter "Evidence for Wavelike Energy Transfer through Quantum Coherence in Photosynthetic Systems" in the journal *Nature* reports that instead of wandering randomly during photosynthesis, as had been previously thought, energy during photosynthesis moves in simultaneous, organized waves. Here again, we see the pattern of synchronization through waves waving.[7]

Synchronization stands out as an example of how the details science has collected can and must be reunderstood. Synchronization and its indispensable partner, desynchronization, are but another expression of order and chaos—of stability and instability—that we see in nature. There is surely much to be gleaned from the details science has catalogued about topics such as the signals emitted by ants, fireflies, amoebas, and the like, but ultimately those details still rest in the context of simulcausal carrier waves.

The signals are not linear—they are waves. And they are not exclusively local—they are simulcausal with the carrier wave that guides individual waves and wave clusters. The true nature of synchronization and desynchronization will stand as a guide for understanding what happens in all these specific cases.

Synchronicity is one of the great marvels of the natural world, and it has been considered one of the great puzzles of life. What is life, if *not* organized stability and change on many different, layered levels? The coordinated changes of life throughout a stable body are what has always lent life its air of mystery. Life might even be defined as multileveled motion in a particular pattern of stability and change.

It stands to reason that an absence of synchronization opens a person up to a relatively chaotic descent into disease, while cultivated synchronicity fosters health. In the next chapter, we turn to the synchronization of health and disease: to life.

CHAPTER 15

The Origin of Health and Disease

NOW, AT LAST, we come to what feels like a grand finale: the explanation of health and disease. It is not a true finale, in that I have yet to explain other momentous topics such as gravity. But for most, if not all, people, an understanding of health and disease is of climactic importance.

I will set out the findings of two outstanding medical mysteries and show how waves waving accounts for them. The first mystery surrounds the incredible but unexplained finding of a single risk factor for all-cause mortality: what medicine calls "low heart rate variability." The second is the utterly surprising finding that reducing bloodflow on an interval basis, in what is known as ischemic preconditioning, actually protects the heart from the damage of a heart attack, rather than increasing the damage.

As I go through these findings, I hope you will come along with me and be filled with amazement, as I was, of how SuperWaves accounts for their otherwise puzzling facts. Together, these two medical findings conclusively testify that waves waving is how our living bodies function. They affirm that the new model of waves waving within waves accounts for and makes sense of everything, where the parts-and-wholes model did not.

We begin with the first of these mysteries: a repeatedly documented strong correlation of low heart rate variability with disease and death. I will lay out the problem as science sees it and then show the powerful way that waves waving resolves it.

HEART RATE VARIABILITY (HRV) is a measure, made by physicians and researchers, of the heartbeat's beat-to-beat intervals. HRV reports a linear average of the spacing of heartbeats over a period of time. If, while one is lying down during an ECG (electrocardiogram), the heartbeat varies in an even fashion—from 60 to 62 to 61 to 61, for example—the HRV is said to be low. In this example, there is very little variation in the spacing between heartbeats. The heart rate is like a metronome. This is low HRV.

Conversely, the more variable the interbeat intervals, the greater the HRV.*

A review of medical literature reveals what to many is a surprising fact, namely that low HRV is a single risk factor for a surprisingly wide spectrum of disorders— and of mortality from all causes. It spans from in utero and infant mortality to geriatric mortality, from cancer and cardiovascular disease to autoimmune and behavioral disorders, from HIV/AIDS to drug addiction and juvenile delinquency.

Beginning in 1965, when researchers first found an association between decreased HRV and infant mortality, researchers have reported associations between:

- Low HRV and risk of heart disease[1]

- Decreasing HRV and the progression of coronary artery disease[2]

- Increased HRV and longer survival times in HIV infected and AIDS patients[3]

- Long- and short-term decreases in HRV and multiple sclerosis[4]

- Decreased HRV and increased dependence on insulin in both adult males and juveniles with diabetes[5]

- Decreased HRV and weight maintenance at higher than "usual" levels[6]

- Decreased HRV and cocaine users[7]

- Low HRV and juvenile delinquency and adult criminal behavior[8]

- Low HRV and severe brain damage[9]

Perhaps most outstandingly, as part of the renowned Framingham Heart Study, researchers reported that in a group of elderly people with a mean age of 72 ±6 years reduced HRV was found to predict mortality from all causes, including cancer, chronic disease, and cardiovascular disease.[10] This overarching study ties together, and powerfully reinforces, all the other studies listed above.

It drives home the hair-raising fact that a person's overall likelihood of dying— for any reason—grows higher as her HRV decreases.

The authors of the study say that "the biological mechanism explaining our present results remains unknown." Yet, they nevertheless conclude that "the estimation of heart rate variability by ambulatory ECG monitoring offers prognostic informa-

---

* HRV is calculated by measuring the time between successive RR intervals of QRS complexes as recorded on an electrocardiographic strip. To measure HRV and heart rates over linear time, each heartbeat cycle of systole and diastole is treated as a dimensionless point, such that the continuity of the heartwave is lost.

tion beyond that provided by the evaluation of traditional cardiovascular risk factors."[11] This is an extraordinary conclusion. It essentially suggests that heart rate variability predicts health and longevity—and, conversely, disease and dying—better than other known risk factors do.

For science, the trouble with HRV is twofold.

First, there is the basic question of why low HRV—invariability of the heart rate, of all things—is linked with death by all causes (i.e., with multiple diseases and disorders). This is almost incomprehensible from the perspective of medical science. Every disorder falls into one of the categories associated with decreased HRV. Cancer, cardiovascular disease, autoimmune disease, and neurologic and behavioral disorders—they all correlate with low HRV for reasons unknown. Why heart rate variability ties in with disease is a question of grave importance.

But even beyond this pressing, practical question is a question about our very way of understanding diseases and dying. One would think that ordered regularity would be associated with health. In the history of science, anything that is ordered, regular, and predictable is considered stable—and not likely to be a hotbed of disorder and disease. Why does a heart rate that is unchanging and predictable likely lead to disease and death?

In an article for the *Journal of Innovative Management*, Harvard physician Ary L. Goldberger explains the negative consequences of a loss of variability in a living organism.

What is the result if something terrible happens to a biological system that has this fractal "playfulness" [i.e., variability]? Perhaps the most common thing that happens is that the system loses its variability, and becomes pathologically regular. . . . Pathology correlates with a loss of fractal complexity. We have evolved the concept of disease as a decomplexification of a system. Counterintuitively, the output of many biological systems becomes more regular and more predictable with pathologic perturbations. Although we call pathology in systems disorders, many diseases manifest patterns that are highly ordered and periodic: the tremors of Parkinson's disease, manic depression, autism, obsessive-compulsive disorder, brain waves during epilepsy, and breathing patterns in heart failures. . . . The uncontrolled growth of cancer cells may be another example. . . . In fact, clinical medicine is not feasible without this paradox: without such stereotypic periodic behaviors, clinicians could not make diagnoses.

The aging process is also typified by a loss of complexity. . . .

Healthy function, on the other hand, with its broadband fractal-like variability, is much harder to characterize. It's very hard to use a single word to describe healthy behavior. We end up using words like plasticity, variability, resilience, and productivity. This type of variability enables the organism to adapt.[12]

In other words, when a person loses fractal-like variability, disease is likely. Look at what Goldberger says: Disease sets in when a "system loses its variability, and becomes pathologically regular." This is what I called the Ary Paradox in an earlier chapter: that ordered periodicity and regularity, which science regards as "order" throughout the universe, can be a mark of disease and *dis*order in the human body. On the other hand, an increase in "fractal playfulness"—an increase in variability that is self-similar across scales—associates with health and longevity.

A RETURN TO OUR roots provides the simple answer to these questions.

The discovery of SuperWaves began when I recognized the heartwave nested in the bodywave. It is an inner wave riding that carrier wave: It rises and falls, rises and falls, as it climbs with exercise and descends with recovery, in the bodywave. It is compressed and decompressed (i.e., simulcausally shaped) by the arc of that body-wave. It attains a higher temperature in the peaks and cools in the troughs.*

Then the discovery continued—that heartwave is also itself a carrier wave. It simulcausally shapes the microscopic cellular, biochemical, and molecular-biological waves that climb and descend the slopes of each heartbeat's wave of exercise-recovery.

All these scales of waves, and more, simulcausally influence one another as they progress forward.

This simulcausal relationship between inner waves and carrier waves is the key to understanding HRV. When inner waves climb a gently sloped carrier wave, they are relatively even. They have low variability, in other words. In contrast, when they climb a sharp slope, each wave changes significantly from the one before; they are increasingly variable. The two scales are, always, inextricably connected.

Yet the perspective used in science and medicine prohibits recognizing this cross-scale motion. Indeed, it not only discounts the reality of waves waving simultane-ously across scales, but it also prescribes converting waves into scales and lines. This scientific perspective compelled Galileo to tack single scales of waves onto lines. It permitted him to isolate waves as much as possible and break each into linear parts. It authorized Einthoven, the inventor of the ECG, to formalize those processes for the heartbeat. And it is the reason HRV does more of the same. HRV separates the heartwave from other waves and averages it out, in a linear scale, as a heart rate.

But from that perspective, no one has been able to explain why low HRV stands out as the single risk factor associated with all-cause mortality—whether it is a mere risk factor or, somehow, a cause. The heartwave, however, provides the answer.

The HRV mystery was an illusion: an illusion created by the Puzzle Hypothesis.

---

* This is an important point that we will return to later.

The heartbeat's changes do not occur in isolation and cannot be properly understood if we view and treat them in isolation. Instead, we must recognize that the heartbeat's changes inherently connect with inner biological and chemical cycles.

That means that what medicine calls low heart rate variability is, in reality, low heart*wave* variability.

The flatter the heartwave, the more metronome-like the heartbeat, and therefore, the lower the HRV as physicians measure it. Low HRV, limited though it is as an averaged-out measure, still reports that the heartwave is relatively flat. It therefore indicates, albeit unintentionally, that there are relatively flattened waves across scales. And we are well acquainted with the cross-scale effects of flattened waves. Over and over again, we have seen that flattened waves are the recipe for decomplexification, dispersal, and disorganization: as true for the heartwave as for any other so-called scale in nature. They are the source of relative disorder, the lack of harmony, that is the hallmark of disease.*

What is disease? Disease is the mounting disorder that grows as nested waves lose their sharp peaks, their inner organized complexity, and their high temperature and begin to flatten out, disperse, and cool down. Disease is a human-scale expression of flattening, decomplexifying, and cooling waves.

Even in our experience, after all, chronic disease presents that way. It is a recurrent, cyclic misbehavior, if you will, of different cells and molecules in the body. Each disease has its own pattern of regularity. But what all chronic diseases have in common are regular patterns of behavior uncoordinated with the health of the organism as a whole. In my experience as a physician, I also saw that everyone knows—but doesn't necessarily grant importance to the fact—that a decrease in body temperature is often associated with chronic disease.† These are all expressions of flattening waves.

One could say that the regularities and periodic patterns that characterize disease incarnate the decomplexification of waves as they flatten out. This includes those mentioned by Goldberger: the tremors of Parkinson's disease, the mood swings of

---

* I will again spell out the mechanism through which low heartwave variability allows for disease, to make sure the point is clear.

When a carrier wave becomes flattened, all inner waves simulcausally become flatter, too. They have, in other words, low variability. This low variability concurs with a loss of sprouting and swarming of the inner waves—what we would call decomplexification. The carrier waves fail to concentrate and amplify the inner waves. What happens as the waves progress, instead, is a low-complexity, relatively dispersed, fairly periodic repetition of inner waves. This is what happens with low HRV.

In contrast, high frequency-amplitude waves waving are highly variable. They change a lot, from one to the next, as they climb up and down the steep slope of the carrier wave. They are concentrated, pulled together, and amplified as they climb those steep waves, thereby squeezing and amplifying their own inner waves—and thereby unfolding as a highly organized, deeply complex forward-moving cluster of inner waves within waves. We have seen that such variable waves correspond with relatively high order and organization. A high degree of HRV, in other words, indicates highly peaking, sharply sloping waves: the source of the organized complexity that we call health.

† Flattening waves are the reason temperature drops: With greater disorganization comes what science would recognize as a lesser capacity for work (i.e., a decrease in temperature).

manic depression, the repetitive behaviors of autism and of obsessive-compulsive disorder, brain waves during epilepsy, and breathing patterns in heart failures. All such periodic patterns indicate a failure of the heartwaves and bodywaves to organize, as an emergently complex collective, the body's inner chemistry and molecular biology. If the wave behaviors of the chemistry and molecular biology were well organized, their climb up and down the sharp and variable slopes of those carrier waves would eliminate their monotonous cyclic progression.

Such is the fundamental explanation of disease. At the risk of being repetitive, I will say it once more outright. A flattened heartwave (i.e., low heartwave variability) is the simulcausal route through which cellular metabolism, biochemistry, and molecular biology lose their complex organization and cool down and fall apart.

Obviously, I am not trying to explain the specific etiology of any particular disease here. But the explanation is more profound than if I were. What I am showing is that SuperWaves resolves the traditional scientific problem of how disease happens by offering a new framework through which to understand our information. The scientific method had oriented us to try and figure out how health and disease come about—as a product of direct, exclusive, isolated causality. SuperWaves shows that diseases come about through the way waves unfold across scales together. Their top-down/bottom-up, outside-in/inside-out influence means that flattened waves on one scale not only indicate but also contribute to shaping flattened waves across all scales.

So although health and disease seem to deserve their own new principle, the principle of the universe generously offers the phenomenal insight we needed all along. SuperWaves demonstrates that low heartwave variability is the cause—not the exclusive direct cause, but the simulcause—of chronic disease. Somewhat amazingly, the well-known and well-documented link between low HRV and death from all causes stands as forceful proof that SuperWaves is the nature of nature.

The self-similar simulcausality of waves waving even directs us toward another layer of proof, no less spectacular.

Simulcausality dictates that when the disorders we recognize as disease unfold, they unfold across all scales, all at once—not "all at once" in the sense of instantaneously, but "all at once" in the sense of all scales changing together. I have emphasized the top-down angle of simulcausality because it profoundly revamps our idea of how things work. But kindred changes occur on inner scales. If the bodywaves and heartwaves, as carrier waves, are doing a poor job coordinating inner waves with the workings of the rest of the body, then the cells, chemistry, and molecular biology should also flatten and decomplexify as they devolve into patterns of disease. We have thus far discussed only the flattening of the heartwave in association with disease. But such flattening of waves must also appear on the molecular level of an afflicted person.

And in dazzling further proof—whose significance for SuperWaves is, of course, unrecognized by science—researchers have indeed observed a flattening of molecules in those suffering from disease. They have discovered that a slew of notorious diseases—including Alzheimer's, Parkinson's, ALS, and Huntington's disease—feature proteins that unfold and flatten as prions.[13]

Prions are misshapen proteins that can cause infection (most famously, mad cow disease). Whereas certain proteins normally take the form of helical twists, they can become prions by misfolding, converting into what are called flat beta-sheets. Somehow, these malformed, flattened beta-sheets self-propagate, meaning, in their presence other helical proteins flatten out as well. They are sometimes referred to as "shape-shifting" proteins. With these changes comes disease.

Though they do not recognize it, what the researchers have found is an exhibit of flattening waves across scales. Note what is detected in all the diseases: an almost eerie incidence of misshapen proteins through which helical twists are reduced to flat beta-sheets. I say eerie because observing this molecular change is like peering directly down through scales of waves waving. There is no question of what it is: It is a physical manifestation of cross-scale flattened waves. Helical twists that subside into flat beta-sheets are an overt, physical show of waves flattening out and losing sprouted complexity.

And more, scientists have further found that flattened, unfolded proteins play a role in type 2 diabetes, cataracts, cystic fibrosis, a type of emphysema, and cardiac disease.[14]

The flattening of proteins in association with disease—the loss of molecular variability—is, from one perspective, to be expected, in light of what we now know of SuperWaves. It reflects the same wave mechanism that underlies low HRV. But at the same time, it is as striking a proof for SuperWaves as the concurrence of so-called low heart rate variability and disease is. Indeed, it reflects both the same mechanism and the same proof. Flattened waves lead to disease (i.e., a loss of proper function for the entire organism).

Similar findings are arising in other areas, as well. A recent study of people with clinical depression showed a self-similar, or fractal-type, correspondence between scales of behavior—between the behavior of the cells of the people who are depressed and the people themselves.[15] Using power laws, researchers saw a pattern: that depressed people move in a different way from healthy people. They rest for longer periods, and more often, than healthy people do. This is an expression of flattened carrier waves. Depression is, indeed, already medically associated with low HRV.

But perhaps more fascinating, and most pertinent to the current phase of our discussion, is the resemblance of this behavior to that of scales within, as described in an *Economist* article discussing this research.

When [the author of the study, Dr. Yamamoto] looked for similar power-law curves in other areas, the one which he thought most resembled that exhibited by the depressed turned out to be the pattern of electrical activity shown by nerve cells isolated in a Petri dish and unable to contact their neighbours.

It is both unnerving and intriguing that a mental disorder which isolates people from human society, and which must surely have its origins in some malfunction of the nerve cells, is reflected in the behaviour of cells that have themselves been isolated. Maybe this is just a coincidence. . . . But maybe it is not.[16]

Now we know it is the opposite of coincidence. That the behavior of cells and the behavior of the person are self-similar, across scales, is a reflection of the inherently nested, simulcausal nature of nature.*

ALL ALONG, I HAVE had to refer to scales as if they are separate, even though, as I have shown, they ultimately are not. I have had to emphasize the principles of Super-Waves separately, even though all are at play. This type of discussion has bowed to our human tendency to think in terms of parts and wholes, while honoring our ability to recognize that they point to an ultimate "one" reality. We can approach and see a greater unity, even as we think about nature one so-called part at a time.

This part-by-part discussion, while keeping the whole in mind, is like discussing a rich symphony. With an awareness of the entirety of the piece, we can discuss the individual play of each violin and flute yet not lose our sense of the indivisible whole symphony. So too we discuss the bodywave, the heartwave, the levels of the cell, the biochemistry, and even the gene. We meanwhile appreciate the role of the conductor, who shapes the play of all the musicians—a commander sculpting the performance precisely while it is being played (this is like discussing the top-down influence of the carrier wave). The symphony is the indivisible whole whose contributing aspects we can discuss individually while keeping the whole in mind.

Now we go down deeper still, to genetic mutation. When coming from the scientific perspective, people discuss mutation in terms that seem to refer to local objects, such as molecules, genes, DNA, and chromosomes. They reference stability as if it were a quality belonging to, but ultimately separate from, the material of those objects.

And yet, we can understand that our data points to a different idea. We can

---

* With an understanding of the model, one could have predicted all this cross-scale flattening of waves: the low heart rate variability and misshapen molecules present in those with chronic diseases. Those experiments were done, of course, before I discovered the phenomenon. But I predict that scientists will discover the self-similar misshaping of inner waves, including material clusters of waves such as prions, in numerous disorders. Flattened molecules, which otherwise peak and twist in healthy individuals, will be present in a host of other diseases. And power laws will further reveal self-similar patterns between the molecules involved in a disease and the behavioral patterns of the organism afflicted with it.

understand that what seems to us as instability in chromosomes is an expression of a single, inherently continuous cross-scale phenomenon. Genes are no less a part of nature than anything else is. How they work, and what they are, demand drastic reunderstanding in the context of that unity.

It turns out that the gene is a bridge where the quantum scale of the SuperWaving fractal meets with the everyday scale of the bodywaves and heartwaves—at least, from our human perspective. In truth, all is inherently connected all the time. But we glean information on our scale. We are not galaxies, and we are not electrons. So from our scale, with the information we can gather from our place in the universe, the gene pulls everything together.

We see, in the gene, all that we have understood on the micro and macro scales. We see what it is, how it works, what holds it together, and how it falls apart.

In Chapter 13, the quantum introduced us to the origin of matterspace. Matterspace, which we perceive as matter and space, emerges as waving waves sprout into clusters, that we call particles, in sharply sloping peaks. We discussed that these so-called particles cluster and sprout further as they themselves climb sharply sloping waves within waves to become atoms.

On the scale of the gene, the so-to-speak buildup from the quantum wave involves even more numerous layers of waves waving within waves. The particle clusters have grouped together and sprouted further organizational complexity. Those groups have grouped together and sprouted further organizational complexity. It goes on and on, such that the complex molecules we are familiar with, as genes, DNA, and chromosomes, come into being. This is the "what it is" of the gene.

The basic physical facts of DNA support this understanding. Scientists have found that these double helixes coil on themselves in a fractal pattern, as mentioned in the previous chapter. This intense organization, of course, exists in the center of the nucleus: the area of greatest organization and compression, such that 1.8 meters of DNA are packed into every cell nucleus, about 6×10^{-6} meters across. The DNA is not crammed in haphazardly; rather, it packs itself by coiling coil upon coil in a precise way and also unfolds, as necessary, to replicate (then refolds). This physical structure itself testifies to the SuperWaves pattern: A helix is a spiraling wave. The heart of the cell nucleus showcases spiraling (i.e., waving) waves that further spiral, swarm, and compress—in a cross-scale, self-similar pattern.

Moreover, as I said in an earlier chapter, DNA molecules never hold still. They coil and uncoil, quiver and gyrate. In their endless oscillatory motion, they are even called "dancing DNA." Scientists are aware that this motion is a significant, if not essential, feature of DNA. Helen Pearson, the features editor for the journal *Nature*, describes the scientific perspective of this motion in an article celebrating the 50th anniversary of DNA's discovery: "Some researchers believe that these mysterious movements may be just as important as the genetic sequence itself in deciding which genes are switched on and off. They even have tantalizing evidence that a failure to coordinate this sub-cellular waltz could underlie some human diseases. Half a century may have passed since the double helix made its debut, but in some ways, scientists have only just begun to understand this miraculous molecule as it twirls in time and space."[17]

The motion of the DNA is, in fact, its most essential feature. It demonstrates the SuperWaving nature of the molecule. This is "how it works."

NOW, LET'S TURN TO consider the gene, coming from the other direction, to see "what holds it together" and "how it falls apart"—the factors behind mutation.

On the way down from the bodywave and heartwave, we saw that carrier waves simulcausally shape inner waves. We discussed that carrier waves simulcausally compress inner waves in sharply sloped peaks, relative to how they disperse them in troughs. We discussed that this is the secret behind organized complexity and its partner, relative disorder and decomplexification. On and across all scales, sharply sloping peaks facilitate attraction, organization, and sprouted complexity. Relatively flattened waves facilitate repulsion, disorganization, and loss of complexity across scales.

The element critical to undesired genetic mutation is the latter point: that when the slope toward the peak of carrier waves lessens, some of their organizational and compressing power is lost. The gene, which is nested within the cell within the organ within the organism, becomes open to some degree of rearrangement. This is "how it falls apart." It is the opposite of, yet in continuum with, "what holds it together."

The point is a subtle but critical one. Genetics gets marquee billing in most people's thinking about disease. But it works in a different way than we've thought. Instead of emanating directly from a mutated gene, disease happens on many layers at once. We have seen the character of the SuperWaving fractal, and it prompts us, in all regards, to remember that our puzzle-based, well-trained instinct to search for direct, local cause and effect is not apropos. The forward progression of change is, instead, simulcausal across scales.

That is to say, disease happens on many layers at once, and not from a gene-determined direction only. As larger waves progressively flatten out, inner waves simulcausally flatten out, as well, and the sprouting and swarming and synchrony that promotes health is lost. The inner waves become unstable, spread out, and lose coherence: They are likely to mutate. Some of these changes will result in disease.

Already, scientists have discovered that the idea of one genetic change per disease is false. It turns out there are many, even hundreds, of genetic changes in people with cancer and other diseases. Different cells have different genes involved. And the function of all are nested in a greater context, as geneticist and computational biologist Aravinda Chakravarti points out in an editorial in *Science*: "The lessons from genome biology are quite clear. Genes and their products almost never act alone, but in networks with other genes and proteins and in context of the environment."[18] This is because the reality of mutation is not anchored on material objects, but on the motion of waves waving: motion that *is* the sprouted object we call a gene, motion that is responsible for organization and disorganization, and motion that is, simply, change—from the changed patterns of behavior we know as disease right down to the changes we know as unfavorable mutation.

This also explains why mutation does not equal disease. Genes for different diseases may exist in a person but may not, as science would say, "be expressed." What it means for a gene to exist but not be expressed is that a person has healthy wave patterns, on the larger scales, that keep the inner waves that are the genes in check. The overall stability and organization is maintained—even with the disorganized pattern of a much, much smaller inner wave. Only in a flattened environment would that inner wave's simulcausal influence become prominent.*

All this means that we need to think about genomic instability and disease in a different way. Not every mutation will lead to disease, but every disease involves some degree of genetic change—because the top-down organizational carrier waves are not functioning properly. The inner waves are powerfully influenced by the environmental wave patterns. Science calls it epigenetics, but it goes much further than that. Environmental waves influence the microscopic waves, on the SuperWaving fractal basis, such that what was originally stable becomes unstable and we get disorders and diseases.

A fascinating case presented in the journal *Nature* illustrates this principle in the extreme. Researchers studied a set of twins, in which only one twin had the genes for Down syndrome. The twin who was affected with Down syndrome showed "flattened gene expression" across *all* chromosomes.[19] This twin—who ostensibly differed only in having an extra copy of chromosome 21—showed a narrowed range of gene expression levels, with decreased variability, across *all* chromosomal segments. In contrast, all the healthy twin's chromosomes showed greater variability.

This finding further testifies that genetic instability occurs in the context of flattened waves—a decrease in range and variability.

The understanding of mutation as a cross-scale simulcausal phenomenon also clears up the mystery surrounding what are known as constructive, or favorable, mutations. Favorable mutations increase organization and stability on the level of the organism in relation to its environment. They play a role in the evolution of species.

---

* This is also why when a wave of radiation interacts with a gene to mutate it and disease follows, the process is not random—and the mutation is not the exclusive cause of the disease. The radiation is a wave, riding with, and therefore reshaping, other waves. Radiation is a very flat wave. It will take whatever wave it interacts with and flatten it out because waves nest with one another. But genes that nest within flattened carrier waves are more open to mutation because they are less stable. And concomitantly, the mutation will only be relevant if the carrier waves in which it nests are too flat to shepherd its motion with the collective.

Yet for science, favorable mutations present a problem. Traditional ideas about evolution through random mutation offer no pathway for information about the environment to travel to the gene and bring about those favorable mutations.

Favorable mutations happen by the same mechanism through which unfavorable changes happen, but toward greater coherence and organization instead of less. The carrier waves compress and organize inner waves toward change, instead of flattening out themselves and allowing for inner disorganization.

In other words, the environment changes, and that marks a powerful change of all the cycles in which an organism nests. Its shifting cycles coax slightly different cycles from the inner waves that climb and descend its broad span. Thus the organism's own inner waves self-similarly and simulcausally change shape. It goes all the way down scales—including to the level of the gene. We recognize these simulcausal changes in the genes and in the environment as favorable mutations.

All in all, mutation is no different from anything else whose patterns of oscillation determine its degree of cohesion and stability. The underlying pattern of stability and instability that determines whether or not a gene mutates is the same as it is throughout the rest of nature.

This broad understanding gives us what we need to go forward. As I mentioned above, I am not trying to explain the etiology of any particular disease. It is hard to say what is happening in the details of specific mutations because many nested waves are constantly changing, all together, on the very intricate level of the cell. The universal model of SuperWaves tells us the pattern through which everything in nature experiences enhanced order and organization. And it tells us that what happens in the cell is compromised when waves within waves flatten out—because when waves flatten out, all the inner waves, with their emergent complexity, flatten out, too. We see that as a loss of structural organization (i.e., instability) across scales, including in the mutated gene. This is a basic fact with regard to disease, which has been my goal to explain. Waves flatten out and bring about disease, simulcausally across scales.

I WANT TO SAY a few words about cancer. Of all diseases, it presents the most dramatic and straightforward display of the outcome of flattening waves.

Scientists have a hard time defining cancer. Different types of cancer have very different manifestations from each other. But many different physical problems are all categorized as a single disease because cancer does exhibit specific hallmarks.

One characteristic shared by all cancers is rapidly dividing cells, out of harmony with the body's needs. This is not difficult to understand. Cancer reflects a cross-scale flattening of waves. Flattened carrier waves cannot exert a powerful shaping force on the inner cells that oscillate in unison with them. Individual cells are not pressed into joining the body's rhythm. They can follow their own cycle.

That is what a cancer cell does. That is what cancer *is*. Cancer cells reproduce on their own, out of sync with the rest of the body.

Another quality is that cancer cells are immature. Cells cannot mature when the nested waves are not strong enough to sprout the characteristic complexity—the mature function—of healthy cells.

Metastasis occurs because the body's rhythms are relatively flat and allow cells to divide independently. Cells spread to other organs, rather than stay organized in their usual organ clusters. This is possible because the other organs do not create powerful waves that stop them. When a colon cancer metastasizes to the liver, the liver is not orchestrating its own cells with a powerful rhythm. The liver cells, surrounding the colon cell that drifts there, oscillate with weak, flat cycles. Nothing prevents the colon cell from reproducing. Rather than be forced into a strong top-down rhythm that would keep it in check, the colon cancer cell reproduces in relative isolation.

In my experience working with patients with metastatic cancer, I can tell you that people with cancer have very narrow ranges of heartwave variability. Instead of starting with a resting heart rate of around 60 that can go up as high as the 170s, 180s, or 190s in a series of short sprints and recoveries, they usually have a far narrower range, with a significantly dropped max heart rate. The heartwave cannot simulcausally organize and keep stable the inner biochemistry. And wayward cells cannot sprout complexity. The carrier waves are too flat to simulcause these healthy characteristics.

Cancer is a topic that deserves a book in its own right. In our current discussion, however, we can only begin to address it.

FOR TOO LONG, we have treated health as an absence of disease. In a certain sense, our approach has been terrible. Scientists and physicians home in on the gene, the chemical, the virus that makes you ill: What can that possibly tell you about *health*? How does eradicating local causes of disease help us to understand the ideal state toward which we strive? Health is too precious to be treated as a default setting.

Here, at last, is a positive definition of health. It draws from science's own findings about heart rate variability: When HRV is low, it indicates that the heartwave, as a carrier wave, is doing a poor job orchestrating the cellular, biochemical, and molecular motion necessary for good health. High HRV, then, is the key to health. But it is not because of high heart rate variability—it's because of high heart*wave* variability.

High heartwave variability indicates that the heartwave robustly shepherds our cells, biochemistry, and molecular biology to move as they should, in their proper cycles. This way they cannot stray into the errant patterns of function we call disease. High heartwave variability faithfully indicates biochemical variability, because the waves nest as a continuous fractal.

What is health? Health is the collective synchrony and organization of variable scales. A healthy individual has high-variability waves: waves that ensure, through simulcausality, that the body's chemistry and metabolism, molecular biology, proteins, DNA, and chromosomes work together like a symphony.

Health happens in this way. Highly variable waves, on all scales, simulcausally interact as they move forward. They shape one another in an exquisite interdependent dance of relatively local interactions interwoven with an over-arcing curve of top-down organization and dispersal. Such highly variable waves organize and maintain what science has called complexity.

Now we see that complexity is not complex, because there is no need to try to account for organization exclusively from local cause-and-effect interaction. The

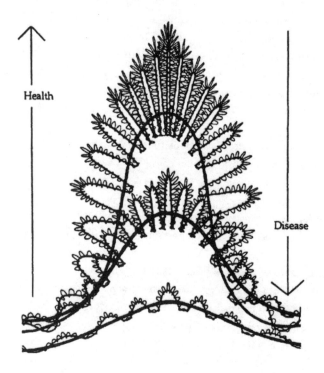

A representation of waves in the patterns that cause health and disease.

missing information—which recasts the whole picture—is simulcausality. And its most revelatory aspect is the bodywave's top-down, outside-in, collective orchestration of so-called local interactions as they unfold. This top-down orchestration, together with the way inner waves bottom-up, inside-out combine to create carrier waves, is responsible for health.

The SuperWaves pattern also shows that health, somewhat surprisingly, inherently relates with disease. Health and disease are not opposites. Rather, they reflect the continuum of waves waving.

Any living person, whether healthy or diseased, obviously exhibits extraordinarily high degrees of complexity and organization: A living person, no matter how unhealthy, is vastly more complex than a puddle or stone. The difference between a healthy and unhealthy person lies in wave patterns. The wave patterns of an unhealthy person are flattened or otherwise misshapen—and therefore do not synchronize all the internal, cross-scale cycles to work in harmonized complexity. We are healthy when we maintain the fractal-like, organized, highly variable motion of our cells, biochemistry, and molecular biology.

The obvious question becomes, how do we make waves in healthful patterns, suited to our natural cycles? I developed a program to this end, called LifeWaves®. I will discuss the program, in some detail, a little later in this chapter.

ONE WOULD THINK THAT after the thrilling resolution of the connection between low HRV and disease, things could not get any more thrilling. And yet, SuperWaves resolves another scientific mystery I mentioned at the start of this chapter: ischemic preconditioning. That resolution is itself momentous because it gives practical ideas of what we can do. It confirms, and forcefully demonstrates, the all-important inherent connection between heartwave variability and good health—and indicates how we might achieve it.

I will first describe the scientific findings about the effects of ischemic preconditioning. Then I will show how those findings testify that good health comes from properly shaped, sharp, variable waves. Last, I will introduce the program I designed to help develop good heartwave variability.

Ischemic preconditioning is a medical marvel. It is a counterintuitive procedure in which several short episodes of clamping and declamping blood vessels prior to cardiac surgery decrease damage to the heart that would occur if they were clamped straight out for an extended time period, without the preconditioning.

One would reasonably think that less clamping would be better—that the cumulative effect of episodes beforehand would only worsen the damage of the eventual prolonged clamp. But the opposite is true. Ischemic preconditioning protects the heart. It is a potent but enigmatic effect.

A group of researchers designed an experiment to measure what was believed to be a sure thing: the cumulative damage to cardiac muscles from repeated episodes of coronary angina.* First, they caused heart attacks in dogs by clamping a coronary artery in each dog for 40 minutes. They measured the degree of damage to the heart muscle in this simulated heart attack.[20]

Then, to determine how much more damage would be created by anginal ischemic episodes, they simulated repeated angina in another group of dogs. They clamped the coronary for 5 minutes, and then they declamped it for 5 minutes so that bloodflow resumed. They made four such cycles. Then the coronary was again clamped for 40 minutes.

The results were dramatically unexpected, as an editorial in the journal *Lancet* describes:

> *Several years ago it would have been incomprehensible for a rational physician or scientist to conclude that myocardial ischaemia might somehow act to protect the heart from necrosis. Then Murry et al. reported their improbable observations that an isolated 40-min occlusion in dogs resulted in an infarct whose mass could be decreased by 75% if the animals were pretreated with four cycles of 5-min coronary occlusion plus 5-min of reflow. . . . Recognition of the benefits of ischaemic preconditioning has unleashed efforts to define its characteristics, study its mechanism, and apply the principles clinically.[21]*

Since that first animal experiment, the same finding has been shown to occur in humans, as well.[22] Repeated cycles of clamping and declamping performed before a prolonged clamp not only don't damage the heart but actually protect it from the damage it would incur if simply clamped, straight out, for that prolonged period. This shows that angina episodes are, in fact, protective, as well.

SuperWaves explains what the authors of the editorial in the *Lancet* called the "incomprehensible" success of ischemic preconditioning. Clamping and declamping the coronary artery are not two discrete phenomena; rather, they constitute a powerful wave continuum of exercise-recovery (though science recognizes only the clamping aspect—calling the procedure ischemic preconditioning, where ischemia is the restricted bloodflow).

When the coronary arteries are clamped, there is a dramatic increase in survival-

---

* Two definitions for if you don't have a medical background: Ischemia is a local reduction in bloodflow caused by vasoconstriction or other obstacles; Angina is a sudden, intense pain in the chest caused by a momentary lack of adequate blood supply to the heart muscle.

oriented energy metabolism. Because the molecular biology in the heart is suddenly devoid of its oxygen, the hemoglobin works hard to release more oxygen. It is a body-wide metabolic spike of exercise. Then the coronary arteries are declamped, leading to a dramatic recovery of energy metabolism—the same as recovery from exercise.

The complex biochemical oscillations within the myocardium swing up and down cyclically through these powerful waves. The series of waves created by clamping-declamping is, in fact, a wave of waves. It therefore results in increased biochemical variability.

When the heart muscle is then exposed to prolonged clamping, those waves continue forward. The wave has, or is, a memory, which repeats forward. That is the nature of a wave: a regularity that repeats.

On the other hand, if an artery is simply clamped for a long time and then declamped (e.g., the revascularization of an acutely ischemic limb), the recovery side of the wave will swing down too far and may not pull out of the dive. The resultant crash, named "skeletal muscle reperfusion injury" by researchers,[23] will lead to decreased biochemical variability and extensive muscle death. The patient crashes, like one in a drug overdose or with post-traumatic stress disorder or in any extended cycle with inadequate recovery.

Said again because it is so important: Every time the surgeon clamps for ischemic preconditioning, the chemistry activates or, so to speak, exercises. Then declamping allows for recovery. Everything relaxes. If the surgeon cycles the on-off clamping several times, the chemistry goes up and down in a cycle of these cycles. That cycling of cycles activates a powerful forward-moving wave of waves. The surgeon simulcausally participates in shaping the wave that is protective on the cellular and molecular levels. Then, even if the surgeon clamps the coronary afterward for an extended period of time, the cycles continue to repeat, over and over, because the cycles of on-off clamping started the wave. It already set up a wave pattern that's repeating. It still oscillates as waves within waves. It is protective.

As if to add to this incredible finding, ischemic preconditioning shows what is known as the "second window of protection."[24] Twenty-four hours after the original preconditioning episode, the same protective pattern arises again. To me, this was an astounding confirmation of the pattern of the SuperWaving fractal: Here was a repeat of the wave nested in the circadian wave, 24 hours later. And the protection lasts for the duration of an ultradian wave—1½ to 2 hours. The pattern returns in an ultradian wave in the circadian wave. I was amazed to see that the continuum of waves waving, on different scales, is so powerfully organizing that its active protection repeats in waves.

TOGETHER WITH THE FINDINGS about low heart rate variability, the proof of Super-Waves by ischemic preconditioning is almost indescribably powerful.

As if they were preordained as a pair of phenomena meant to enlighten the world, the scientific account of ischemic preconditioning seamlessly complements that of heart rate variability. Together, when understood through their true identity—the heartwave—they create a total description of the causes of health and disease.

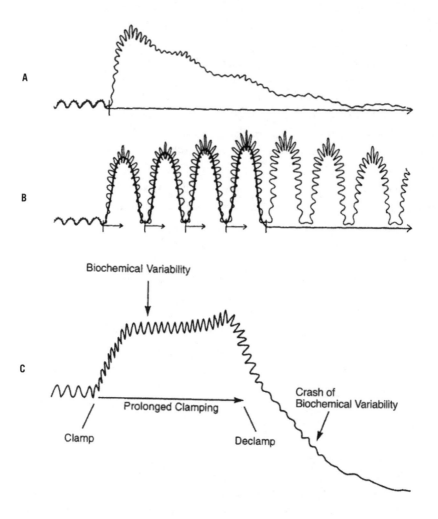

(A) Continuous coronary clamping leads to decreased biochemical variability and muscle necrosis.

(B) Successive coronary clamping and declamping leads to increased biochemical variability and maintained organization.

(C) Prolonged clamping followed by declamping results in decreased biochemical variability and muscle necrosis.

Decreased variability is the simulcause of disease; increased variability is the simul-cause of health.

That was everything in a nutshell: Decreased variability is the cause of disease; increased variability is the cause of health.

I wrote an article describing these phenomena in 1996 called "The Origin of Disease and Health, Heart Waves: The Single Solution to Heart Rate Variability and Ischemic Preconditioning." In it, I conclude that

> *both experimental mysteries—decreased HRV as a risk factor in all chronic disease and ischemic preconditioning as a protective factor against myocardial infarct— are explained and solved by the same principle of waves waving. A decrease in heart wave range (a decrease in HRV, or that which occurs with prolonged coronary clamping) is the ultimate underlying cause of chronic disorders. An increase in heart wave range (an increase in HRV, or that which occurs with cyclic preconditioning) is the ultimate underlying cause of health.*[25]

It was powerful proof: proof by Occam's razor. It could hardly be simpler. One pattern, the SuperWaves pattern of motion in motion, accounts for increased stability and organization, as well as decreased stability and organization, on and across scales. The correlation between low heart rate variability and disease/death and the success of ischemic preconditioning in preventing cardiac necrosis show without question that everything in the body is, and follows the pattern of, waves waving.*

The appeal of such a resolution can hardly be overstated: It is fundamental to our way of thinking. Human minds instinctively prefer simple explanations. Specifically, for a given set of facts, we naturally pick the explanation with the fewest assumptions over one that incorporates needless conditions. This is the principle of Occam's razor.

Occam's razor is the reason we long ago adopted the parts-and-wholes model of science. It naturally had seemed that that one assumption fit the facts best, without needing to assume anything else. Thus, the very criteria that might raise eyebrows in the realm of health and disease—that SuperWaves explains everything and simply, too—are the criteria that originally made us think nature is a puzzle.

I must emphasize, however, that as delightful and promising as simplicity is, it is

---

* This perspective offers an intriguing account of the placebo effect. It is well known in the medical establishment that the placebo effect can effectuate very real, dramatic changes in a person's health. When a person takes a sugar pill but believes it to be medication, on a regular basis, he experiences a downswing of relaxation—of relief for having treated the stressful disease. This creates a powerful wave. What had been a flat plateau of constant stress now cycles up and down between stress and relaxation. The very real wave of stress-relief simulcausally reshapes cellular, molecular, and biochemical cycles as all move forward together—sometimes organizing those inner cycles into variable, healthy ones. Thus the placebo effect can bring about a so-called spontaneous cure. It is not directly causal or always effective, but every so often, the placebo effect simulcausally works.

not a magic elixir. Occam's razor is sometimes restated as: Everything should be made as simple as possible, but not simpler.

SuperWaves offers exactly that service. It explains health and disease simply, but not simplistically. The model, in other words, is simple to understand, but the way it plays out in specific cases does involve a lot of information. Emergent complexity and its flip side, the deterioration of order, involve countless layers of constantly changing waves waving within waves, for example. Trillions of cells and molecules change together on innumerable nested scales.

The core of the simplicity lies, instead, in finally knowing how nature works and what it is. SuperWaves offers a global understanding of what is going on, as I mentioned above, and thereby eliminates scientific mystery: We no longer need to collect details for building up to a theory of everything. We have a satisfying view of nature's ultimate workings without getting mired in detail. And yet, we have much to investigate in the future. Details will still play an important role, but it will involve fleshing out the simulcausal influences of particular cycles that interest us.

Also important is that simple does not mean health can be achieved, and disease resolved, with a snap of the fingers. Health and disease are processes, not things. They are patterns of motion, visible only through time. And whether in the direction of health or disease, they take time to develop. Therefore, as much as we'd like all treatments to model antibiotics—a pill a few times a day and the disease is gone—a true resolution of chronic disease means changing and renormalizing those time-bound processes. The unfortunate rise of antibiotic-resistant infections reminds us that though drugs can be effective, a true understanding of the time-bound processes of health and disease is what we have ultimately needed. SuperWaves give us that understanding.

NOW, THE OBVIOUS QUESTION IS: What do we do about it?

The steps we can take to shape these waves are readily understandable.

We have direct control over the bodywave. We can sprint or sit, stay up late or wake early, fast for a day or nibble snacks throughout. The bodywave is unique in this way.

Now, one might have thought that the heartwave is one so-called layer away from this type of control—because we cannot mentally decide to have a rapid heartbeat the way we can decide to pick up a pen.* And one would think that cellular cycles are even more remote. Yet the reverse is true. *The inherent relationship between scales*

---

* We can bring it on with anxiety or in excited anticipation, but that wouldn't be characterized as direct control.

*means we **do** have a measure of simulcausal control over the heartbeat and over further inner layers of waves and cycles.*

The inherent relationship of waves across scales does the incredible. It spans the seemingly infinite gap we have always wanted to eliminate: the gap between that layer over which we have control and the many over which we don't.

That gap, in fact, ultimately does not exist. All layers of waves nest inherently and move forward, simulcausally, together. Therefore, when we make choices of what we do with our bodywave, we simulcausally shape the patterns of the heartwave. And in simulcausally shaping the heartwave, we simulcausally shape the waves of our internal organs as well as the cellular, molecular, and genetic waves that nest within.

Isn't that amazing?

The heartwave is a point of entry, from our scale, into the wave continuum. We simulcausally shape the forward motion of our inner waves by shaping the heartwave through the behavioral cycles of the bodywave.

There is a critical caveat to this type of control, however.

The simulcausal relationship between the bodywave, heartwave, and all levels down is simultaneously top-down (outside-in) and bottom-up (inside-out). We cannot straight-out direct the heartwave as well as deeper waves in an exclusive, one-way direction. We cannot directly control our inner workings in this way.

Simulcausality means our inner waves make up the bodywave just as much as the bodywave controls them. They are inherently intertwined as they move forward. If the cycles of our cells and biochemistry are awry, the heartwave will reflect it. It will rise and fall during cycles of exercise-recovery differently, in an aberrant pattern.

If one wants to think of exercise-recovery cycles, roughly, as steering a boatful of cellular, biochemical, and molecular cycles, simulcausality reminds us that the boat itself *is* those inner cycles and can only be steered as they will allow. The commander of the boat can decide to go this way and that. But the path the boat takes ultimately depends on the interplay between intention and the inner cycles that are the boat.

Here the beguiling relationship between the bodywave and heartwave becomes fascinating, cherishable—even priceless. It gives us an unprecedented angle of information. When we track how the heartwave responds to bodily cycles of exercise-recovery, it becomes a window through which we can peer in on the cycles of our cells, molecular biology, biochemistry, and genes.

A person can learn, in other words, about the cycles of her organs, cells, and layers below—by monitoring her heartbeat during whole-body exercise-recovery cycles. Inner cycles nest in the heartwave and bodywave. If they are misshapen, the heartwave will reflect it by not rising and falling as one would expect. The bodywave will be commensurately misshapen.

In this, I recognized the practical key for causing health. Waves wave forward, and so they can be shaped through their simulcausal nestedness. We can participate in shaping them from the scale over which we have some control, with simulcausal feedback from the scales we don't. Through the mechanism of simulcausality, this repatterning reshapes the collective rhythms of the molecular biology and genes toward the design of good health. We make stability- and organization-promoting behavioral cycles this way. Making cycles is like creating an antidote to the decreased range of variability that so endangers human health.

I had always wanted to foster and promote health, performance, and longevity. The following is the basic design of the program I developed to achieve that goal. Through this program, we can sculpt our waves into the ideal pattern for health.

THE MAIN IDEA is to increase the range of heart rate variability—what I call heartwave variability.

The popular recommendation is that your maximum target heart rate be 220 minus your age—meaning it should drop as you get older. I realized that the key to health lies in going the other direction. There's no reason to simply accept the decreasing variability that comes with age. We can stretch it, in our favor. The way to do it is not through brute force, however. Excessive exercise to try to increase your max heart rate will end up diminishing your heartwave variability.* Training in the correct pattern steers us away from the current practice of overexercising while ignoring the training of recovery.

In nature, the patterns of greatest complexity showcase dramatically increasing frequency-amplitude waves—such that the inner waves that climb that steep slope have tremendous variability—that then steeply drop into recovery. Such waves are easily created through sessions of bursts of exercise and recovery.

I developed a program called LifeWaves that delivers the program, either with a live coach or over the Internet.

To give a bit of an idea of how it's done: A person on the program wears a heart rate monitor while engaging in cycles of exercise-recovery. I call it "doing Cycles" or Cycles Exercise®, and it means doing a set of alternating exercise-recovery cycles: a spike of exercise followed by recovery, followed by another spike of exercise, followed by recovery, and so on, for several cycles.

In other words, you hit a target heart rate by doing a quick burst of exercise, and

---

* Prolonged stress creates the opposite from the desired pattern—it creates a long flat wave, not punctuated by the dips of recovery during which antioxidation, antistress processes flood the body. Repeated sharp waves, with variable peaks and troughs, create powerful, increasing amplitude waves that repeat forward (are long-lasting).

then sit down to recover *completely* before taking on the next cycle—and repeat several times. The intensity and duration of each cycle during the set varies: according to the individual and according to an analysis of his or her baseline heart rate and other factors. Each session usually takes 20 to 30 minutes.

The program uses these cycles of exercise and recovery, along with cycles of wakefulness and sleep, diet and eating, and emotional arousal and relaxation, to reshape waves into the patterns exhibited by healthy individuals. Cycles are nested in larger-scale cycles still: the monthly, seasonal, and yearly cycles in which we all live.

The heart rate numbers hit during the peak of exercise and the trough of recovery—and the speed with which a person hits the high number and how fast the numbers fall in recovery—are deeply informative. They are a way to view and measure *cellular and molecular* wave function and coordination because different layers of waves inherently, simulcausally connect.

In the program, each exercise-recovery cycle is adjusted in light of what a person's earlier heartwave patterns have been in order to help coax all of his or her waves closer to the optimal pattern.

Here's an example of how one might do Cycles. It is not a prescription for how to do it; rather, it is an example of what Cycles might look like for a specific person. The reason I am describing it at all is that this training is very different from any program done before. The description is just to give you a taste of how Cycles are done.

Person S has Parkinson's disease. Based on many factors in S's life, which will not be enumerated here, here's what one particular possible set of S's Cycles might look like on a given day.

(Again, this it is not a prescription for anyone with Parkinson's, either—every person follows a different regimen according to his or her own numbers. But it is an example of how a set of cycles is performed, in order to help you imagine it.)

First, Person S puts on a heart rate monitor—a chest strap and watch or a pulse/oxygen monitor that is worn on the fingertip, for example. He sits for 5 minutes to establish a baseline heart rate. Let's say it's 85 to 90 beats per minute.

Then he does a short burst of exercise, like pedaling on a stationary bike for at least 20 seconds. He pedals, not too hard, to reach his first target heart rate. The target is not very high: 105 for this cycle.

He dismounts from the bike and recovers fully. This means he waits until his heartbeat is as low as it will go, with several deep breaths to ensure complete recovery. It usually takes 2 to 4 minutes.

The number to which his heart rate drops, let's say 82 in this case, is also recorded, as is the pattern of its descent. Did his heart rate drop right down to 82, or did it go down slowly? Did it go straight down, or did it vary, perhaps rising a bit before descending again, on the way down? This pattern is used to help determine what his next cycle should be.

For the second cycle, Person S pedals the bike a little harder for a little longer, let's say 30 or more seconds, to reach the target heart rate: 115. He dismounts the bike and recovers fully, 2 to 4 minutes. Again, the pattern of his recovery is monitored.

And so it continues. His third cycle has a higher target that he pedals a little harder for, closer to a minute, to reach 125, and then he recovers fully. For the fourth cycle, he pedals for a shorter time but even harder, to reach a peak target, 127. He may do a fifth cycle, depending on the time of day and time of month (it differs because Cycles are nested with a person's ultradian, circadian, and monthly rhythms—sometimes only two cycles or so are done). And so he shapes his wave patterns.

During Person S's final recovery, after about 5 to 7 minutes, his resting heart rate is monitored and compared to his original baseline rate. Depending on whether a person is healthy, has a chronic disease, or has an aberrant wave pattern that can lead to a chronic disorder in the future, this comparison is meaningful for determining the targets in future Cycles.

This series of exercise-recovery cycles constitutes "doing Cycles."

For all people, Cycles are done at different times of day (morning, midmorning, or afternoon) on alternating days for 3 weeks out of a month, with a recovery week at the end. The different times correspond to our natural ultradian, circadian, and monthly rhythms and are meant to ensure a person's cycles nest properly in the natural cycles of the world we live in.

Also incorporated into the program are series of cycles that I call *breath waves*. Breath waves are a powerful tool to enhance variability that is modeled after ischemic preconditioning. After a series of deep breaths and a full exhale, a person holds his breath and continues to exercise until he must breathe again; then he sits and fully recovers. This cycle creates waves of oxygen-deprivation/oxygen-recovery across all cells of the body. It mimics ischemic preconditioning, enabling the red blood cells to release more oxygen and qualitatively enhance metabolic function across the body.

The program is tailored to the individual. Every person has their own wave pattern that is consistent with the human species but still has its own particular shape. And it is this shape, of the wave pattern, that determines the potential for disease and health. So the program is designed to focus in on the individual as unique. No two people are alike, and so no two sets of Cycles are exactly the same. The program allows a person to monitor what is happening and what needs to be done by observing and analyzing the heartwaves with a computer algorithm.

This approach allows a LifeWaves coach to determine which cycles are appropriate for each individual to help expand his range and variability. Indeed, a person can view his changes over time, and see progress that is consistent with health. (Conversely, one can see the pattern deteriorate, or flatten out, as disease progresses.) This approach is, obviously, worlds away from the standard approach of most exercise programs, where everyone runs 5 miles or does 10 pushups. With LifeWaves, you can actually see, through heartwave monitoring, increased variability corresponding with the experience of increased health, as well as flattened waves and rhythms corresponding with the experience of disease.[*][26]

For the first time, we can understand how exercise works to helps us—but also how it can hurt. Exercise helps us best through cycles of exercise-recovery. It is the opposite of prolonged exertion, like my friend Jack Kelly did; too much exercise creates one long wave of stress that can be followed by a crash. Jack, tragically, couldn't pull out of the dive. We have to train recovery right along with exercise. Cycles are most effective when nested in patterns that are ideal for us, considering our place in the SuperWaves continuum.

We can make waves in the proper pattern to increase our heartwave variability and thus increase our health, performance, and longevity. The program truly earns the name of LifeWaves.

It perhaps goes without saying that the LifeWaves program presents a plan markedly different from those aiming to control nature through direct force.

It employs no concept of causal, linear effects that one can cause as if pushing a button. It relies on no chemical or drug to kill a disease-related target. Instead, it reorients a person to understand, and to accept, the idea of waving waves that one can cultivate and sculpt toward health.

It means not having the satisfaction of overt, immediate control over nature—but surely we can recognize that there is value in our actions as participators instead of directors. We are now set up to recognize that—on the levels on which we directly participate—we have, at last, through simulcausal motion, a way to shape waves on levels that we otherwise can't directly access. Even without local, piece-by-piece control over our cellular, molecular, and biochemical cycles, our actions do reflect in and on those scales.

When you shape the levels over which you do have direct control, you conduct your inner orchestra for health or for disease. Your waves write your own prescription for the collective synchronization, or desynchronization, of all the naturally occurring drugs: the chemicals, molecules, and proteins naturally present in your body. Even without pills and injections, you can direct the activity of these natural drugs, whose activity are only mimicked, or interfered with, by artificial ones.

For the first time in fitness, we have true self-empowerment.

---

* While I am not involved with laboratory research at this time, I have advised on several studies regarding the LifeWaves program over the years. Endnote 26 includes a brief description of some of them.

A PROGRAM FOR CAUSING HEALTH: It was utterly new. In 2000, I received a patent from the United States Patent and Trademark Office for it. The patent lays out the basics of the program.

> *In this treatment, the patient in an exercise session undergoes a series of exercise-relaxation cycles in which during each cycle the pulse rate of the patient rises and falls to generate a heart wave. To enhance the efficacy of the treatment, the heart waves generated in the course of an exercise session are synchronized in time with an internal wave produced by a biological clock [i.e., ultradian, circadian, and other natural rhythms], this activity functioning to expand the range of the bio-logical wave and inducing the HRV to approach an HRV, which for the patient being treated reflects a normal condition. . . . The cure for diseases therefore is to correct, i.e., reshape the wave disorder by overriding or ablating the abnormal wave pattern, using the heart wave as the means to analyze and to bring into being over time, new heart wave patterns to prevent and reverse chronic disease or other abnormality. . . . The fractal wave pattern shape is what determines the health versus chronic disease of the organism. . . . The [program] uses the heart wave as a means of diagnosing, monitoring and modulating to recreate normal one and one-half hour ultradian wave patterns which is then used to create the appropriate 24-hour cycle, this being used to create the appropriate monthly/lunar cycle, which is then used to create the yearly cycle and ultimately the lifetime cycle.[27]*

I hope this short explanation of what the program offers and how to do it has given you a glimpse of a better, more healthy future for all of us.

FROM THAT POINT ON, I shifted my mind-set about health and disease, as I hope you will, too, now. This is not to say that relatively local cause and effect plays no role in health and disease. It surely does. But we must keep in mind that it is only relatively local—that it nests in a broader context. Carrier waves always shape the progression of all change.

The future of research lies in this recognition. There is much to be learned about how different physical cycles change depending on the carrier waves they nest in. We can research how different carrier cycles synchronize and coordinate our inner cycles for health and disease.

For example, a story in *Science News* about a surprise finding in biological syn-chronization describes that "networks of muscles, of brain cells, of airways and lungs, of heart and vessels, operate largely independently. Every couple hours though, in as little as thirty seconds, the barriers break down. Suddenly there's syn-chrony."[28] This is a mystery worth investigating.

Research on this type of topic will begin with a paradigm shift: by setting aside the idea of isolated networks that connect physically, as building blocks, to create a highway of so-called local cause-and-effect change. Studies will instead focus on how cross-scale peaks and troughs of cycles connect with each other.

This type of study will reveal the hubs: the peaks, the cores, the cycles that create so-called network organization. It will reveal how and when these peaks descend to the troughs they are in continuum with, and how that brings about dispersal, relative independence, and relative disorganization on inner scales.

Everything works in nested cycles; we will learn how those cycles that seem relatively autonomous periodically become integrated through the cycles in which they nest. We will also learn which carrier waves we can affect to help nudge errant cycles into ones that promote the organization we desire.

We will certainly continue to develop medical treatments that tap into so-to-speak local interactions. But those treatments will be in the context of larger cycles that influence them. Treatments will be vastly more powerful than ever before because they will incorporate the top-down influence of carrier waves on the inner waves that are involved in such so-called local interactions. Our new knowledge will help us understand what is going on with different diseases, and it will help us to guide inner wave motion, collectively, into the patterns we want for health.

The future of medicine will incorporate the fact that health and disease are processes, not things—that in most cases, health is not achieved, nor disease resolved, with a snap of the fingers. They are patterns of motion, visible only through time; whether in the direction of health or disease, they take time to develop.

A true resolution of chronic disease will mean changing and renormalizing those time-bound processes. We can research specific cycles for medical understanding and combine that knowledge with the LifeWaves program to create health.

THE LIVING ORGANISM REVEALS motion at its finest. Only through its organized variability can we easily see the exquisite yet simple reality that waves change together as they move forward. That one simple fact means a total reversal on many fronts, not the least of which is what simple motion really is. Our history of thought had led us to believe that we must break life apart, down to nonliving atoms and molecules, to understand it, and to use the steam engine as the model from which to understand its workings, but the living organism shows that endless facets of motion simplify into a single relationship—an inherently continuous fractal across scales. This is as simple as motion can get. It is simpler, by far, than reducing motion to a near infinite number of idealized straight lines and then trying to reconnect those idealized lines.

If simple is what we want, we have it in waves waving. We must accept its perspective. Eons ago, we ruled out the possibility of an "only one" as a choice for the nature of nature because it seemed untenable. But now we know that an "only one" phenomenon is simple and understandable. The inherent continuity of motion in motion is a powerful organizer that is arresting in just how simple it is. It brings even the most complex thing we know of, life, into the realm of the understandable. It is no less wondrous for it; it is, perhaps, even more. I had begun by recognizing it in a living organism; it, in turn, showed me the secret of life.

We have come full circle. The journey began with my discovery of the heartwave. Now we see that that heartwave is centrally involved in whether a person will experience health or disease. It is like discovering that a cherished, beautifully engraved box in our possession was all along holding, within, a jewel of inestimable worth.

We can direct the heartwave, and that opens us to a world of hope and promise. We may be growing familiar with SuperWaves as the new model of the universe, but even if we take it as a given, the way it unravels the mysteries of health and disease is nothing short of glorious. Though there is nothing new to explain about health and disease, per se—they are the same as any other example of order and disorder—SuperWaves still offers insights to take one's breath away.

CHAPTER 16

SuperWaves,
the Environment,
and the Origin and
Survival of Life

PEOPLE HAVE LONG KNOWN that environmental protection is critical for our survival. But what is the environment? The puzzle point of view had suggested it to be a soup we swim in, a cloud of ingredients we interact with piece by piece. Now we know that nature is not a puzzle; rather, it is an inherent continuum of nested waves waving within waves. Our idea of what constitutes an environment— and the true role of the environment in human health—changes in light of simulcausality.

The effect of the environment is ultimately the top-down/outside-in influence of larger carrier waves on the inner waves with which they simulcausally move forward. Good health, as well as life itself, requires that we be nested within variable environmental waves that are open and continuous across scales.

Sharp, artificial boundaries can interfere with the complexity- and stability-promoting influence of environmental carrier waves. This fact has been surfacing in research. Somewhat surprisingly, even as we have striven to conserve considerable parcels of the environment, the very act of fragmentation has endangered it. The value of land tracts set aside for conservation suffers from the boundaries around them.

This telling fact was discovered by Thomas E. Lovejoy, who went to the Amazon to determine which is best for the survival and conservation of plant and animal life: single large plots of land or multiple small ones. In Lovejoy's research, flora and fauna were counted in large areas before the areas were cleared for cattle ranching and farming. Multiple island reserves were maintained amidst the clearings, ranging in size from 1 to 100 hectares, and then tracked over the course of 32 years. Nearby forests served as control areas.

The researchers discovered that the biodiversity in parcels of rainforest land declined around the perimeters of the fragments, sometimes dramatically. Birds, butterflies, primates, other large mammals, and trees all died off or disappeared toward the boundaries. Smaller fragments lost species more rapidly than larger parcels of land did. Lovejoy and colleagues call this the "edge effect." Fragmented parcels were also more sensitive to rare weather events and to nearby human activity, such as land management practices.

Conservation biologist Jared Diamond, who conducted similar research, explained in an article for the journal *Science*, "Fragmentation causes loss of animal and plant populations by a process termed faunal relaxation." In a series of islands created by the damming and flooding of a Venezuelan valley, "the smallest islands (<1 ha) quickly lost 75% of their original species. . . . Large islands (150 ha) retained most of their original species. . . . But, within 4 years, all islands had lost their top predators: the jaguar, puma, and harpy eagle."[1]

We see, then, that what would seem to be unspoiled parcels of land lose organization and complexity near the borders, and the entire tract loses complexity and diversity over time. The parts-and-wholes perspective suggested that parts of nature can be independently curated, but even large ecosystems lose variability when boxed in. Artificial fragmentation disrupts the cross-scale continuity that enhances stability, organization, and complexity.

We impose similar artificial fragmentation on and in other areas of life and culture today. The consequences are sobering. Diseases such as cancer and cardiovascular disease are more common, as well as more aggressive, in people living in inner cities.[2] Polar bears in zoos exhibit psychotic behavior, such as repetitive pacing and self-mutilation.[3] Populations of Lyme disease–infected ticks swell as the size of forest patches shrinks.[4] Astronauts in space suffer from headaches, congestion, and vision impairment, experience significant muscle atrophy, and lose 1 percent of their bone mass a month—about 10 times the rate of bone loss in aging men and women. Violence in prisons is the norm,[5] and three-quarters of released prisoners are rearrested within 5 years.[6]

All these examples support the terrible reality that isolated controlled closed systems—in many ways, the foundation of our society—create a loss of stability, complexity, and organization.

It is a critical problem: a problem built into our civilization from early times. Intense, repetitive, confined, and low-variability behaviors, with high stress and high physical arousal, are a setup for developing a disorder. Instead of experiencing waves that rise, fall, and rise again, people fall into relatively linear, repetitive states—which means they lack strong, variable carrier waves to keep their inner cycles synchronized. These behavioral patterns, which characterize many work and school settings, facilitate disease, as discussed earlier.

It also means that when people undergo serious trauma, they may not be able to recover from their experiences. Without the variability of strong carrier waves to keep inner cycles cycling together, perturbations that would otherwise be borne and dissipate have profound, and protracted, effects. Modern-day veterans suffer from post-traumatic stress disorder (PTSD). Data shows that the soldiers fighting in the Civil War also developed what would today be called PTSD.[7] Patients who spend time in the intensive care unit—as many as 35 percent—also exhibit PTSD symptoms as long as 2 years after treatment.[8] And the effect is not limited to humans. Researchers observed that elephants from herds that have lost many members of the herd (i.e., those who have lost a large degree of their collective, nested environment) startle easily, are socially more unpredictable and solitary, practice inattentive mothering, and exhibit other behaviors that, if exhibited in humans, are associated with post-traumatic stress and similar disorders.[9]

At this point, I hope you understand the spirit in which I make these broad statements. They are meant to invite you to see how the SuperWaves fractal accounts for data already collected. It is also an invitation to investigate further—to find other unexplained phenomena that now make sense in light of SuperWaves, and more, to search for ways to solve problems unintentionally caused by our ignorance of the nature of nature. I have been working with military vets with PTSD in a rehabilitation program in Tennessee, for example, to create healthy, variable carrier waves—and have had success in easing their PTSD symptoms. Based on the work in Tennessee, I am contributing to a growing rehabilitation program in Georgia with strong initial success.

Another example of the life-sustaining simulcausal organization we get from environmental carrier waves is atmospheric stability. "The air we breathe is a potentially unstable mix of oxygen, nitrogen, water vapor and so on, and could quickly disappear in a flash of furious chemistry. Yet it has not done so,"[10] writes Mark Haw in an article for *American Scientist* magazine. Then how does the atmosphere stay stable? Whereas Haw invokes nonequilibrium thermodynamics, SuperWaves shows that inner waves are compressed, are held together, and are sprouting complexity within carrier waves. I want to drive home, for the last time, why and how that is a satisfying explanation.

The key is that we are shifting frameworks. Confusion can arise if we think of the atmosphere, for example, as a group of molecules occupying space. But that is not what it is. That is not what it *is*. The atmosphere *is* waves waving within waves: waves that compress and sprout into atoms and molecules; waves that spread out as the space of the atmosphere; waves that are the ceaseless motion of those molecules and space; waves that compress and hold the atmosphere together. Matter, motion, space, time, order, disorder—they are all expressions of the one Super-Waving continuum of waves. It's all the same stuff, ever moving and creating what we know and experience. We are here exploring, and cementing our knowledge of, its pattern.

Keeping this in mind allows us to maintain our quick pace as we visit other natural phenomena unexplained by the parts-and-wholes model. It allows us to understand, in a general way, what is going on—in line with the ideas, and described with the terms, provided by SuperWaves.

Next up is the explanation of the latitudinal diversity gradient, which is also known as the species richness gradient. This is a pattern regarding the distribution of species on our planet.

One might think that species would be equally distributed across the globe. In fact, the number of species peaks at the equator and peters out toward the poles. "In some places and some groups, hundreds of species exist, whereas in others, very few have evolved; the tropics, for example, are a complex paradise compared to higher latitudes. Biologists are striving to understand why,"[11] notes science writer Elizabeth Pennisi in an article in the 125th anniversary issue of *Science*. (Note that the author's use of the term *higher latitudes* heading toward the poles is a conventional label and is not drawn from waves; indeed, we are about to see that the poles are the lower troughs and the equator is the higher peak.) This anniversary issue poses the question, "What Determines Species Diversity?" as one of the top 25 questions facing science in the next 25 years.[12]

The species richness gradient, with its height at the equator and its fall-off toward the poles, shows a fact about Earth-related SuperWaves: that the sprouted emergent complexity of our planet reaches its peak at the equator. The greatest organization and complexity of life is present there. Thus we understand that the equator is the peak of the Earth wave, to coin a phrase.

The flip side of this pattern is seen in the fact that cancers of the breast, prostate, and colon, as well as other disorders, including depression and multiple sclerosis, are also latitudinally dependent.[13] These diseases proliferate where the organizational effects of carrier waves decrease—toward the poles—just in the same way the number of species declines.

It remains an open question how and why the equator is the peak of the cluster of waves waving within waves that is our planet. The answer will likely involve electromagnetism, the rotation of the globe, and many other layers of waves waving; the details of their nested cycles are yet to be understood. The point here is a general one—that the pattern of organization, stability, and complexity increases toward the equator and decreases toward the poles.

Evolution is faster at the equator, as well. A study published in the journal *Proceedings of the Royal Society B* expanded on findings that have shown that "faster rates of microevolution have been recorded for plants and marine foraminifera occupying warmer low latitude environments relative to those occurring at higher latitudes."* The study showed that evolution proceeds at different rates in warm-blooded animals depending on latitude and elevation, and not only in relation to time elapsed.[14]

All this comes to refute the logic and conclusions of Julius Robert von Mayer, the first person to propose science's first law of thermodynamics.†

Around the year 1840, von Mayer left Germany for a voyage in the Pacific as a ship's physician. He noticed that blood drawn from the sailors was bluer in Germany but redder by the time he reached Java.

In a creative interpretation, von Mayer supposed that because the environment at the equator was hot compared to the cold in Germany, the sailors' blood did not have to burn off as much oxygen: The environment was doing the work (i.e., creating the heat), instead of it happening inside the body. This would indicate that heat and work are the same thing in a body. It would mean, von Mayer reasoned, that energy is conserved in a closed system—in this physical process and, ultimately, in the whole universe.

Von Mayer came from life—meaning, he considered, as I did, life and its processes before all else—but, as per the Puzzle Hypothesis, he isolated human circulation as if it were a stand-alone, closed system. Though he supposed that the energy metabolism slows down when bathed in the heat of the equator, now we see that the opposite is true.

The equator, where he made his observation, is rife with nested wave motion—is rife with species, with life and health, and with evolution. Everything at the equator is full of more motion and heat, not less—just as the peak of an exercise-recovery wave is full of more motion and heat, not less. Blood, as he saw, is indeed more red there, but it's because the metabolism, complexity, heat, and organization are all higher.

The peak motion of the equator is of an utterly different nature than what von Mayer, and indeed all of science, describes with thermodynamics.

This angle of understanding illuminates the studies that show metabolism

---

* Again, the unfortunate terminology is a mismatch to reality: So-called higher latitudes approach the poles, which are troughs, and so-called lower latitudes approach the equator, which is a peak.

† The first law is often attributed to James Prescott Joule and associated with the steam engine, but von Mayer was the first to propose it.

increases, and not decreases, with caloric restriction (i.e., a dietary regime in which calorie intake is kept below what would be considered normal for health, but without causing ill health). Common sense—the common sense of the Puzzle Hypothesis— would make one think that caloric restriction would dampen a person's metabolism. That type of thinking, like von Mayer's, imagines a person as a closed system: reduce the fuel and the engine slows down.

This model is not true to nature at all, and studies indeed show that caloric restriction increases metabolism.[15] Caloric restriction creates a powerful upward swing, just as exercise does; within its arc, the inner waves it carries all increase in frequency-amplitude.

The power of waves within waves in an open system is incalculable. It is the source of life on our planet—the source that scientists have sought but missed because they searched with the tools of a closed system.

In 1953, in an attempt to simulate the conditions of our young planet in the days before life emerged, scientist Stanley Miller, under the supervision of Nobel laureate Harold C. Urey, performed what is known as the Miller-Urey experiment. Electric discharges, meant to simulate real lightning on earth, repeatedly struck heated vials of water that stood in for the primitive earth's ocean. The water was housed with an atmosphere of water vapor, methane, ammonia, and molecular hydrogen. After a week, Miller found a significant percentage of the system's carbon was locked in organic molecules such as amino acids, which are the building blocks of proteins.

This exciting experiment seemed, to scientists, to indicate that electromagnetism, in the form of sunlight or lightning, caused the molecules of life to develop. They treated the lightning as if it followed a one-way arrow—as if it embodied the second law of thermodynamics, an isolated "concentration" of energy that "dispersed."

But what was really happening, which went unnoticed, is waves waving: of lightning and *not* lightning. Lightning is electromagnetism in a powerful wave. The waves of lightning on and off took the entire experiment through cycles of light and dark, hot and cold. The strong waves (on/off) *of* strong waves (electromagnetism/lightning) caused the inner waves (carbon molecules) to compress into higher level, sprouted knots of waves waving within waves (amino acids).

It is the same pattern of light as experienced on earth, where night always accompanies day. The belief that the sun is critical for life on earth is correct. But there was never reason to assume that sunlight and heat alone drive the development of life— living creatures on earth experience a day-night *cycle*. In failing to see that, Miller and other scientists left out the cooldown, just like people who advocate the burn of exercise leave out recovery. There is another side of the wave, which is experienced by all life on our planet.

It is the same reason vents in the ocean floor have life: the hot-cold cycles, from the cold of the deep ocean and the thermal heat of the vents, create cycles that facilitate emergent organization and complexity.

These examples all show how strong, nested waves are the setting in which life occurs. It is the opposite of the randomness proposed by Darwin.

When you close a system relatively, as is often done in civilized societies, you create conditions opposite from the nested, cross-scale, continuous variability necessary for vibrant life. This is being discovered with frightening consequences today. The human-constructed environment (cities, infrastructure, and so on, and all that comes with that) is creating a shell around the earth, blocking the normal carrier waves that hold everything together.

It is not a mere theoretical concern. Researchers have found that heart rate variability narrows with ambient pollution.[16] HRV is an extraordinarily significant measure, as we have seen; it is the closest science has come to recognizing the heartwave. A narrowed HRV means the heartwave range narrows with pollution—a dangerous pattern indeed.

Being boxed in by pollution is bad for our health, in other words, and not only because breathing in chemicals is dangerous. The actual layer of pollution cages us in, cutting us off from the open nested waves that sustain our variability and complexity. It is of urgent concern that atmospheric carbon dioxide has risen over 40 percent since the start of the Industrial Revolution.[17] These emissions mean increased isolation within our otherwise open environment, which entails a narrower temperature range for us.

And, indeed, the temperature range of earth has been decreasing since the Industrial Revolution—meaning that global temperatures are growing more uniform. A study published in *Geophysical Research Letters* found that over the last hundred years, the differences in temperature between the equator and the poles have decreased, affecting the earth's climate and weather patterns.[18] People are rightly concerned about the potential effect of pollutants in the atmosphere, but it's not warming per se that is the problem. Yes, the average temperature is increasing, but it's this narrowed temperature range—the lack of variability—that simulcauses dangerous losses of order and organization. Instead of global warming, I call this threat global storming. The recent increase of wild or extreme weather is evidence that the narrowed range of variability is affecting us severely.

Yet there is hope for significant and almost immediate reversal. Following the crises of September 11, 2001, in the United States, commercial flights were grounded for 3 days. Data collected from 4,000 weather stations showed that the range of the average daily temperature increased by about 2°C while flights were suspended.[19] Without the shroud of airplane emissions, the variability of the earth's temperature range expanded.

What does this all mean? It means we have to understand the top-down effects of carrier waves and act upon that understanding. We saw that the health of the individual depends upon high heartwave variability: variable, nested, open waves. The health of the environment depends on, as Lovejoy discovered, retaining its open nestedness. And the health of the entire planet depends on keeping it open. We cannot let the earth become an increasingly closed system, or it will all fall apart. We must look to the survival not only of the individual but also of the planet itself.

Much like the Gaia hypothesis proposed by chemist James Lovelock and biologist Lynn Margulis, which proposes that we view our planet as a unified self-regulating organism, there is a continuity of the earth as a whole. The biosphere, atmosphere, oceans, and soil are not Mother Earth per se, but it is not wrong to see the whole earth as alive. SuperWaves shows that life itself sprouts when everything is cycling and rhythmic and all is in harmonious motion.

Stability, organization, and complexity resulting from high-variability nested waves—we see it on the scale of the individual, on the scale of the environment, and, now, on the scale of the planet itself.

In 1954, Maurice Allais, who would later win a Nobel Prize in economics, observed a Foucault pendulum swinging during a solar eclipse. As the moon blocked the sun, the pendulum changed its course to a more rapid and chaotic one; it returned to normal only as the eclipse waned. Allais went on to observe the same pronounced shift during a 1959 eclipse.

The Allais effect has been studied on numerous occasions, often confirmed, though sometimes not. Research conducted during the 1970 eclipse in Boston, for example, supported Allais's findings with a Foucault pendulum, including "variations [that] are too great to be explained, on the basis of classical gravitational theory, by the relative change of position of the moon with respect to the earth and sun."[20] The inconsistent findings continue to be puzzling, as science writer Phil McKenna explains in *New Scientist*: "If the anomaly exists, it would challenge our ideas about how gravity works. Neither Newtonian physics nor general relativity can explain it."[21]

The Allais effect has, obviously, been observed when the rays of the sun are temporarily blocked. We can understand this too without getting into specifics, which are of interest but not our current concern. We can understand that nested inner waves are being cut off from their normal carrier waves. Normally, they nest in the electromagnetic waves emanating from the sun. The eclipse blocks these carrier waves and thus interferes with their capacity to create stability and organization. The inner waves temporarily flatten out, and chaos—the relative chaos of the pendulum's irregularity—ensues.

THIS HAS BEEN a brief excursion into the natural environment and life on earth, from the perspective of SuperWaves. But I think you will agree that the inherent continuity of nested waves shows that the natural environment is even more important for sustaining human health than we'd ever realized—and highlights which aspects of the environment simulcause health and disease. We can see, too, how potentially fragile our system is and how it is in danger of disastrous disruption by the products of human activity. The concerns go beyond the simple fact of rising global temperatures. The open, nested character of the environment is extraordinarily complex, organized, and variable, and we must take care to dial back our interference with it.

CHAPTER 17

The Solar System, the Galaxies, and the Nature of Gravity

IN 1885, KING OSCAR II of Sweden and Norway offered a prize to anyone who could discover what is responsible for the stability of the solar system. He wanted to know why it didn't just fly apart. French mathematician and physicist Henri Poincaré won the prize, not by answering the king's question but by demonstrating that the stability cannot be mathematically established. He showed that we can say things about bodies rotating about one another and describe and predict their motion, but that the mathematics ultimately suggests chaotic behavior, or a sensitivity to initial conditions.

This so-called sensitivity to initial conditions was publicized in the 1960s and '70s by meteorologist and mathematician Edward Lorenz, who called it the butterfly effect. It is also known as chaos theory.

To Lorenz, Poincaré, and others who study similar questions, stability is mathematically unsolvable—an incarnation of the n-body problem. But Poincaré's and Lorenz's arguments shed their weightiness in the SuperWaves framework. Theirs were arguments about local cause-and-effect solutions. Theirs were attempts to describe natural processes as interactions between parts.

Treating nature as a puzzle, as a sum of wholes and parts, is the source of the n-body problem's thorniness. As we have repeatedly seen, the nature of nature allows that the whole is often greater than the sum of the parts.

The solar system stays together in the same way as everything else in the universe does—in a way not yet modeled with mathematics.* Powerful, high-variability waves waving within waves—in the nested design of organization, which I have repeatedly

* I say modeled with, and not captured by, mathematics, because ultimately all systems are never truly closed.

described—foster maintained stability, as well as relative instability. The earth, as well as the solar system in which it nests, as well as the galaxy in which it nests, moves in such waves within waves. This waves-within-waves motion is the key to the earth's relative stability, as well as the stability of the solar system and the galaxy.

It is the key to all stability in nature—the key that eluded Poincaré. SuperWaves answers the question by reversing our order of understanding. Not only do you not need mathematics to explain the motion of the solar system, but its stability comes from the motion itself—right along with the relative instability with which it is in continuum.

A moment's thought confirms this reality. Our solar system lies within a galaxy, and that galaxy lies within a galaxy cluster. The term *lies*, as it were, however, is a poor descriptor of how the solar system and galaxies exist. *Lies* denotes a stillness of being, and nothing could be further from the truth.

The planets are ever in wave motion. Think about how the planets actually move. The earth revolves around the sun. The sun revolves around the center of the galaxy. The galaxy is in motion, as well. But none travel true elliptical paths. They *spiral* through space.

Every orbit that we see as an ellipse is really a spiral. Picture the earth spinning on its axis while spinning around the sun, and then picture the whole solar system in motion, such that the spinning motions are spinning, too: That is the spiral.

These are helix shapes, but not plain helixes. Each helix is itself coiled as a helix, on and on. It is waves waving—the natural, inherently continuous motion of the Super-Wave fractal on an exceedingly grand scale. The pattern is self-similar to the one we've seen on the scale of DNA: coiled helixes upon helixes, which cluster and disperse.

This is why and how the solar system maintains its stability.

When nested waves flatten out, on the other hand, chaos ensues. It allows for the so-called butterfly effect. With decreased variability, everything becomes more even, more uniform, and closer to equilibrium. Then, in such cases, small perturbations can make a big, chaotic effect. Nothing is stably held together through the simul-causal, top-down effects of carrier waves, so it does not take much of a disturbance to upend what was otherwise ordered and organized.

The galaxy, however, nested as it is in the galaxy cluster, and that, nested in a supercluster, prevents such instability in the solar system. SuperWaves shows that if the galaxy were somehow removed, the solar system would indeed fall apart. But the top-down effect of simulcausal carrier waves prevents it.

NOW WE COME at last to gravity: gravity, one of the great laws of physics, taught to the youngest of children and easily invoked, in casual conversation, as the reason things fall down. Gravity is due its reexplanation.

Let us recall that physicists have faced problems with gravity. Newton's laws made it possible to calculate gravity's pull, but they were vexed by its apparent action at a distance; Newton himself deplored it. Einstein addressed that problem with his proposal of curved spacetime, but others remain.

Why don't galaxies fly apart? After all, as I noted in an earlier chapter, observable matter amounts to less than 5 percent of the mass required to hold the galaxy together, if indeed gravity is what is holding it together. What, then, is missing mass, or dark matter, which supposedly holds galaxies together—and why can't we detect it? What is dark energy, that which seems to be making the universe fly apart?* How do the push and pull of dark energy and dark matter relate with one another?

What is a black hole? Why are stars born around black holes, such that scientists call them "the odd couple"? In an article in the journal *Science*, astrophysicist Philip J. Armitage says that "The observation of stars close to the galactic center requires a rethink of the star formation process. . . . A new mode of star formation may be needed to explain stars observed in the immediate vicinity of the supermassive black hole at our own galactic center." [1] The way stars form is unclear.

The outer edge of the Milky Way, on the other hand, is encircled by several giant, doughnut-shaped rings of surprisingly old stars. The rings feature wavelike, concentric ripples. [2] Why do old stars ring the galaxy in waves?

Finally, how does gravity link up with the realm of the quantum, to connect the macro universe with the micro universe?

These basic questions have to do with the attraction and repulsion that we observe throughout the universe. And the concurrence of black holes and young stars indicates that these forces, somehow, relate with the creation of matter and space.

The scientific questions about gravity fall away in the context of the SuperWaves understanding. As I have said, the unsolved mysteries derive from the assumption that nature takes the design of a puzzle. But nature is not a whole made of parts, and it *is* waves waving within waves. SuperWaves is both a model for and, at the same time, the stuff of nature.

As in life, as in complexity, as in quantum physics, so with the galaxies. We can plug in our facts to the new model and everything makes sense.

ATTRACTION-REPULSION takes a starring role in explaining gravity. Its role in what we have been calling gravity is so direct, it is almost obvious. Let's just spell it out.

---

* Recall that modern physicists posit that dark energy makes up 68 percent of the universe (though we know nothing about it, other than that it supposedly is slowing the expansion of the universe); that dark matter makes up 27 percent of the universe (though we know nothing about it other than that it stops the galaxies from flying apart); and that everything we detect makes up a mere 5 percent of the universe.

What has always seemed like a "force" pulling things together is in fact the compression in the peaks of large-scale carrier waves on inner scales. This is gravity.

Such compression naturally pairs with sprouting. This is why matter exists in the middle of high-gravity regions.

This means that matter results from SuperWaves compression: It emergently sprouts in regions of densely peaking peaks. Those are the very regions that seem to exert strong gravity. Matter and so-called gravity go hand in hand.

This understanding immediately and elegantly reverses our understanding of the matter-gravity relationship. The idea that matter exerts gravity is false. Gravity is not secondary to matter. Nor is it its own force. It is the compressing effect of large-scale carrier wave peaks curving from all directions—the same peaks that compress matter into being. They happen together. Because there are no linear dimensions, the compression results in spherical shapes.

Gravity is neither a unique nor a stand-alone phenomenon. It is a necessary aspect of the natural universe. Matter, which was thought to exert gravity, is the same—an equally necessary aspect of the natural universe, neither unique nor standing alone.

They both come about because of SuperWaves. It is the nature of nature—the nature of the universe—that nested peaks simultaneously compress and sprout stable inner complexity. Gravity is the compressing aspect as identified, by humans, on the scale of massive objects; matter is the sprouted compressed waves. They come about together.

This is gravity's true identity. It does not stem from a new idea. Rather, it stands as confirmation that the nature of nature is an inherently nested fractal of waves waving within waves. Its seamless explanation provides further evidence that SuperWaves—and not a model of parts and wholes—is the correct way to understand nature.

FOR THE REST of the chapter, we will see attraction-repulsion again and again, whirling slowly around as on a great carousel, carrying forward different features of gravity to be reunderstood through SuperWaves. They will not be unique or independent resolutions. Every idea to follow relates with all the rest. This is to be expected, as all facets of gravity are expressions of the "one" SuperWaves continuum.

To begin: Many questions dissipate on the fact that matter is beholden to waving waves for its existence, and not the other way around. The question of why galaxies don't fly apart, for example, becomes a nonquestion. It is the nature of nature that galaxies hold together. There is no need to contrive the idea of dark matter to account for anything. Matter does not effectuate gravity; matter and gravity are both natural aspects of SuperWaves.

The equations correlating mass and gravity don't add up because there is no equality. The mass is huge, condensed SuperWaves, and the gravity is an even *higher* layer of SuperWaves that compresses it and everything around it.

Like the other problems we have seen before it, the problem of dark matter has been a trick outcome of the Puzzle Hypothesis. Physicists posited it exists because the whole of gravity, as they calculate it, is greater than the sum of the parts that seem to exert it. But nature is not a whole made of parts. So there is no need to posit parts that might exert more gravity. No parts exert any gravity. Carrier waves hold everything together, not material substance; they meanwhile compress in peaks to exhibit that material substance. This means that stars, just like other forms of sprouted complexity, inherently, simulcausally, coexist with the compression we thought was gravity coming from their physical matter.

The concept of dark matter, like the ether of yesteryear, becomes superfluous.

The patterning of stars throughout the universe confirms this conception. Stars exist in galaxies—which are clusters of stars. Those galaxies exist in galaxy clusters. Galaxy clusters group as superclusters. This is the pattern of the SuperWaves universe.

One might say that galaxy clusters are the same type of cluster as human beings are. Each star is a sprouted cluster of tightly peaking waves waving within waves, just as each cell in the body is. Galaxies are clusters of stars held together by compressing carrier waves, just as organs are clusters of cells held together by compressing carrier waves. The intensified grouping of these groups—galaxy clusters and human organisms, respectively—are peaks of peaks of peaks clustering together. The wave-shaped arms swirling around the heart of spiral galaxies offer a particularly striking visual illustration of their nested wave nature, but all galaxies showcase a group of stars clustered together. These then further cluster into superclusters.*

The so-called expansion of the universe demonstrates the flip side of these processes.

Space, as we have seen, is spreading waves. This is why what we have been calling intergalactic space is full of waves always, from light to microwave background radiation. (It is well established that space and light, or other so-called electromagnetic waves, are, in practice, inseparable. By tradition, we refer to them separately, as if one is a medium that the other travels through, but in reality they always go

---

* Superclusters are as large a wave peak as we can detect. According to the SuperWaves Principle, they indicate the compressed peaks of enormous waves, which then, by their nature, must dip into troughs—but never exhibit an edge. This means that there is no boundary that would stand as the edge of the universe. What goes on beyond is unknown. There may be other universes with other clusters within, just beyond the troughs that seem to demarcate our own. This is only speculation, of course, but it is something to think about.

together.) The fact is that space is never empty, but always full of waves, because it *is* waves by its nature. People have given the name sunlight to the lovely emanations from our nearest star, for example, but those emanations actually comprise the so-called space surrounding the sun.

Nested as they are in the SuperWaves continuum, these waves naturally spread out in a nonlinear way. It is the nature of waves to spread before they again compress, just as it is their nature to compress before they spread. It all is ultimately the same stuff: When compressed, we call it matter and gravity; when spreading, we call it electromagnetic waves and space. But as they spread, they seem, from the human perspective, to be expanding.*

Thus, when we invert our understanding of where space "comes from," as we did with matter and gravity, the mystery of the universe's expansion falls away. There is no need to contrive dark energy. The so-called force that is said to cause the universe to expand is in fact waves spreading out in troughs, as per their nature. What seems to be expansion of space is that natural spreading.

If we were to draw yet another representation of the SuperWaves continuum, we would show the inseparable team of space and light spreading in the troughs of giant carrier waves and the inseparable team of gravity and mass clustering in their peaks.

It may seem like we are reeling off so many points about gravity. The fact is, there isn't anything unique to say about it. Gravity is to be expected in the framework of SuperWaves. The equations physicists use to calculate compression are very fine indeed, but what they are measuring and predicting is a normal feature of the Super-Waving universe. There is no disrespect intended in redefining gravity or dismissing dark matter. This new framework simply recasts the identity, as aspects of Super-Waves, of what we had thought were independent players in nature. And it doesn't take much to do that.

THERE IS, OF COURSE, much to be discovered about how particular realms of nested waves progress, and this will allow for better predictions, technological progress, and regional understanding. But the ultimate secret—of what it all points to—is already discovered. The universe is SuperWaves, and that needs to be shown. It is what I am showing here. And it allows the gravity-related mysteries created by the Puzzle Hypothesis to be easily finished off.

The scientific idea of fields, for one, sheds its opacity in this framework. A field has been conceived, by scientists, as a region surrounding an object where a force occurs. It is measured mathematically. Its value has been that it eliminates action at

---

* They compress again into another star or galaxy or galaxy cluster, depending on the scale. The space between seems to be spreading because that's what trough waves do.

a distance while offering a means to calculate the effects of a force. But its identity has been a mystery.

SuperWaves clarifies what the so-called field has been all along. The "region surrounding an object" is spreading waves surrounding compressed waves (the object). The "force" felt in this region is the effect of carrier waves—either compressing or dispersing (i.e., pulling or pushing).

That is to say, a field has represented a heroic scientific effort to incorporate more than one scale of waves into a single phenomenon. It is an idea science uses to identify relatively local cross-scale waves working together. The so-called field refers, all at once, to spreading waves—which we experience as space—and also to carrier waves whose effects seem to be forces exerted in that space.

If it weren't tucked in with so many other astounding ideas, this alone would be a mind-blowing insight. But so it is. Like the elegant reversal of the relationship between matter and gravity, this elegant reversal of the relationship between so-called space and so-called forces redefines what we thought we knew. It's not that space carries forces. SuperWaves in motion generate, and are, the spreading troughs that we perceive as space, right along with the carrier waves whose effects we perceive as forces felt in that space.

This is how and why we perceive seemingly instantaneous effects of forces through space.* Scales change simultaneously, so carrier waves collectively influence many inner waves at once.

This, in itself, is a phenomenally striking reality. Historically, and troublingly for scientists, instantaneous effects across space were perceived as action at a distance. Fields were a way for science to address that problem. Now we see how they work. In a moment, we'll return to the apparent action at a distance of gravity. Einstein addressed it with the idea of curved spacetime; we will see how it too ultimately works.

But first we will spell out the relationship between what science calls the four fundamental forces of nature. These are gravity, electromagnetism, the strong force, and the weak force.

Each of these "forces" is waves waving within waves—specifically, the effect of carrier waves on inner, nested waves. By now, it is almost awkward to refer to them by separate names, but since science has done so until now, we must use this nomenclature to deliver to each force its true identity.

We have seen, from several different angles, that electromagnetism nests in gravi-

---

* It is also the natural explanation for why the force of gravity seems greatest near the object and why it dissipates with distance. Near the peak is where material substance sprouts and where gravity seems greatest. The distance away from the object is the trough between wave peaks, in which waves are inherently less compressed. Therefore, we perceive less gravity. It's also why high-mass objects seem to exert "high gravity"—because they are invariably nested in enormous, powerful carrier waves. Indeed, that is how they sprouted, or materialized into being, in the first place. They are in enormous, powerful waves.

tational waves, though it has not been made explicit until now. The effect of carrier waves is apparent in the way electromagnetism progresses. For, despite the way they are popularly described, electromagnetic waves are not truly linear. They still feature peaks and troughs. Indeed, even the most regular of all electromagnetic waves, so-called cosmic microwave background radiation, show small variations. This is because they nest in larger waves (which are themselves nested in larger waves).

These carrier waves are flatter than the peaks of waves that we perceive as gravity and mass. This is why they feature relative spreading instead of relative compression. Whereas gravity and mass are how we perceive highly compressed, sharply sloping peaks, electromagnetism and space are how we perceive relatively spreading, relatively flattened troughs.

Meanwhile, on scales within, in what we identify as the atomic nucleus, the so-called strong force and weak force exhibit the same relationship. The strong force is compressing like gravity; the weak force is radiating outward like electromagnetism.

These effects are self-similar to those of the gravitational waves and electromagnetic waves within which they nest.

It is the same principle of attraction-repulsion we have seen again and again.

It is, indeed, all the same stuff.

Now we come back to Einstein's attempt to unify forces. Einstein tried to unify electromagnetism and gravity, not realizing that the star itself and light and space are a continuum.

I am going to move a little quickly now, to get through the basic points. There is really nothing new here; I am just showing how a number of fragmented scientific ideas were, all along, aspects of the SuperWaving IC fractal. A few moments' reflection will make clear anything that doesn't jump out on first reading.

In his general theory of relativity, Einstein described spacetime as a continuum and said that matter bends spacetime. Now we know what is going on: Highly peaking waves waving within waves materialize as the matter that is the star and also compress in the way we call gravity, and the troughs, which are continuous with the compressions, spread out as light, or electromagnetism, and space. Earlier, I said that I have given the name matterspacetime to this continuum.

The matterspacetime continuum means that Einstein's view that matter bends spacetime approaches the truth even if it is, ultimately, incorrect. The bend is a natural part of the inherent relationship between the three. Of course space bends near matter; matter exists at the peak, and that's where the space emanates and spreads out from (or in the reverse, the space condenses toward the center peak where the matter is).

Here, an idea that has the potential to be very confusing instead makes a lot of sense. Einstein's idea that spacetime is curved is not at all radical once one understands

SuperWaves. Spacetime *must* curve because it *is* waving waves nested in waves. Matter naturally exists, as a compression and sprouting of waves waving within waves, right where spacetime curves most intensely. This makes perfect sense, for that is where the larger nested waves come to a peak.

This takes us to a topic that has long eluded scientists: the connection between the quantum and gravity. The SuperWaves continuum here offers another beautiful insight into how nature works. The realm in which we see the effects of gravity is inherently continuous with, and self-similar to, the world of the quantum.

Think of it: On smaller levels, waving waves peak and compress into what we have called a particle. Then carrier waves, in the guise of the strong force, compress what we have called quarks into larger particles. They compress protons and neutrons into atomic nuclei. Then on larger levels still, grander scales of waving waves peak and compress and sprout regions of what we have called mass, under the guise of what we have called gravity.

Likewise, waves in troughs disperse on small scales in the same way they do on larger scales. On quantum scales, dispersal takes the form of there being no particle detected, or of electromagnetism emitted as light on the scale of the electron and the atom, or of decay attributed to the weak nuclear force; on larger levels, the same, self-similar dispersal is identified as electromagnetism and space.

It is the same, self-similar process and the same stuff on different so-called scales.

One could even say that when a quantum particle appears in the peak of a wave— a real wave climbing a wave, accelerating to the peak, such that a so-called particle appears—it is the same as a star existing in a gravitational field. The peak of the quantum wave is an IC fractal of the gravity wave with and within which it nests.

They're IC fractals of each other, nested across scales.

It's a beautiful thing.

LET ME REITERATE ONE of the most important effects on how we must think about nature after the discovery of SuperWaves. This is not an attempt to solve science's mysteries. It is a placement of established facts into a new model. With this placement, old questions, about how everything fits together, vanish.

This is the answer: an all-at-once shift to a far simpler model that dissolves the need to unify fractured bits of knowledge. There is surely much that needs to be calculated, but as to what these things *are* and how they connect: They are all Super-Waves, and they have never been disconnected. The phenomena we see are all easily understandable as aspects of nature as an inherent continuum of waves waving within waves.

And this brings us back around to what science calls black holes.

The so-called curvature of spacetime, as proposed by Einstein, is the reason, scientists say, a black hole is formed. The gravitational pull of an extremely dense object supposedly deforms spacetime such that gravity overwhelms all other forces.

Fascinating as black holes may seem, SuperWaves shows that they don't exist in the way scientists have described them. What scientists have been describing as the extreme compression of black holes does not emanate from a condensed star. Gravity doesn't come from matter, so it cannot come from a so-called black hole. The black hole must be, and is, supercondensed peaks of peaks—compressed as can be, from our perspective. The compressing effect of these incredibly dense waves is what astrophysicists have been calculating as the gravity of a black hole.

This explains why black holes—long thought to be spent stars, about as dead as anything in the universe—are a site where stars are born. It is naturally where stars will be born. In such an incredibly intense compression, you will find the sprouting of new matter.* Instead of calling whatever exists in the peak a "black hole," something along the lines of "SuperWaving emergent fusion" is more appropriate. In contrast, old stars ring the edge of the galaxy because that is where it all spreads out.

THESE TERRIFICALLY IMPORTANT POINTS have tremendous practical implications. They create a future field of study that I call dynothermatics—the opposite of what science has called thermodynamics. Dynothermatics takes our information about heat, motion, mass, and work and reverses the processes as scientists have understood them.

The science of thermodynamics has posited that heat creates motion. This understanding directs the development of every engine today, as well as every attempt at energy production. Create heat, says science, and motion will become available for use. Some will go to waste, but some can be harnessed for work.

This is why scientists, trying to mimic the process of stellar fusion, apply enormous heat to accelerating particles. It is an attempt to fuse them and release further heat. They try to create fusion in a lab with intense heat and high-speed particle collisions because that is their understanding of how fusion comes about.

The grand master of this approach is none other than the Puzzle Hypothesis, wearing a lab coat and creating a plan based on its vision of the universe. Add up certain parts—high heat, high speed, intense collisions—and you may, hopefully, end up with fusion.

SuperWaves inverts the entire process. In nature, things get together through carrier wave compression. Heat rises as motion increases. It comes about naturally

---

* As I described above, this is the same event, on a different scale, as what happens in quantum physics. In the quantum realm, the sharper the slope, the greater the probability of finding a particle in the peak. In this realm of astrophysics and gravity, the sharper the slope of spacetime, the greater the likelihood of finding a star. This once again, and dramatically, demonstrates the IC fractal between quantum scales and the scales on which we perceive gravity.

in peaks of peaks. We see this with human exercise: At the peak of exercise-recovery waves, with all the peaking inner waves, higher temperatures naturally happen.

It's the same relationship we saw when we put the quantum and thermodynamics together. It's the same thing that happens in a so-called black hole, where stars are born and atoms are created.

I realized that because SuperWaves is an IC fractal, there is no reason that we cannot mimic the way waves wave to compress and make a star—and do it on our scale—to create clean energy. We simply need to replicate the process. We can create waves within waves within waves. That will, top down, coax inner waves together such that they appear to fuse—that is, compress and sprout emergent inner complexity.

It's not cold fusion, per se, though that is what science has called it. It just doesn't require intense heat as a starter. It requires waves waving within waves, because it is that motion that makes heat. SuperWaves shows that waves can be squeezed together and actually create melded nuclei to release an incredible flow of heat. It taps into the same process we see in gravity and the creation of stars.

It will happen if we pulse electromagnetic waves in a fractally nested pattern to create compression—and with it, heat, organization, and matter. The compression of inner waves in the peak creates both the heat and the matter. This is fusion.

My colleagues and I have made promising strides in this field, as I will show in the final chapter.

THIS CHAPTER BRINGS US to the end of a basic discussion of SuperWaves. We have covered the highlights of what science knows about nature and have put it all in the SuperWaves context.

All along, I have been establishing that waves waving within waves is both the true model of nature and also the stuff of nature itself. Whereas science incorporates an artificial separation—a separation between the model (wholes and parts) and the stuff of nature (matter, motion, space, and time)—waves waving within waves is "one." It *is* what nature is, and it is also how it works, how it holds together, and how it falls apart. I have shown how this idea eliminates many scientific problems created with the parts-and-wholes model, and I have brought multiple proofs to attest to it.

You have seen that this reality hands us a new approach. Instead of solving science's problems, SuperWaves eliminates the framework in which they were considered to be valid. Instead of setting up questions that call for big, climactic answers, it allows us to list former questions and then wipe them away. The singular aha! moment is all in one go: an understanding of the true nature of nature. There is no longer a call to piecemeal solve scientific mysteries in search of an ultimate truth.

Going back to the beginning of my journey, I had had no choice but to begin with life and death: with life as I had worked with it—in school, as a surgeon, and with the Olympics—and from the death of my friend Jack Kelly. It is a dazzling truth that a person can come from the most complex existence—from life—to see a simple commonality. And, no less, a commonality that is logical, natural, and sensible and that we have all experienced. We all know motion in motion. It is woven in the fabric of every life.

Through life, we see the hidden secret of the universe. Nature, and the nature of nature, is SuperWaves.

PART 5

The Future

CHAPTER 18

A New Beginning

MANY SCIENTISTS HAVE CONJECTURED that even if they found a theory of everything, it would have no use. It would just be a mathematical formula; it would be something you could wear on a T-shirt. Stephen Hawking, in *A Brief History of Time*, muses that "the discovery of a complete unified theory . . . may not aid the survival of our species. It may not even affect our life-style."[1]

While SuperWaves was not a culmination of scientific research in the conventional sense, it is a discovery that gives us a theory of everything. It sets us in an entirely different direction, both in the realm of understanding the nature of nature *and* in the ability to harness that knowledge for immense practical good.

In terms of helping us understanding nature, SuperWaves has no parallel in the history of knowledge. I don't say this lightly, but it is true. SuperWaves tells us the nature of nature. We now have the whole picture: what nature is *and* how it works, including how it holds together and falls apart. One discovery—of simultaneous heart- and bodywaves—pointed to the reality that nature is an inherently continuous fractal of waves waving within waves. The facts of science supported it. Now the rest of the details science has accumulated can, at last, be appreciated in context.

And a new beginning awaits us. We can use our knowledge of SuperWaves.

To me, one of the most extraordinary insights is the new understanding of causality. The reality is that there is a forward-moving mechanism of coordinated change across scales we'd never before recognized. There is a large degree of what we had thought was local cause and effect, but at the same time, a top-down/outside-in organizing influence naturally unfolds together with it—shaping it and being shaped by it. This natural, inherently cross-scale mechanism creates organization going forward, through what I have called simulcausality. We can embrace the recognition—and we can tap into it.

It is time to stop idealizing direct control through domination. It is time to realize there is more to nature than Darwinian materialist thinking would have us believe. The way to survive, to be fit to live with nature, and to genuinely thrive is to participate in the wondrous wave phenomenon happening all around and in us: to make rhythms *in* rhythm with nature's natural cycles.

There exists, as it were, an intelligent design to nature, not in the sense of a supernatural guiding hand but, instead, inherent in the nature of nature. Our knowledge of it gives us power in a surprising new way. Cooperation and conflict follow a pattern we can understand and influence. We can participate in nature's rhythms to cultivate and nurture the developments we want in the world. We can, with our knowledge of nature's nature, bring about organizational order at last.

The changes we must make are not small, but neither will they have a small impact. They will change the way we live in the world and will foster our own health and longevity. They will change the world for the better.

OUR NEW UNDERSTANDING of nature directs us to cultivate health through collective synchronization. The approach is not one that regards parts of the body as independent, like pieces of a puzzle, and evaluates targets for destruction, nor does it take a holistic approach, as integrative medicine does. The approach draws its strength from the common-denominator language of all health and disease: the underlying phenomenon of nested wave motion. It focuses on changing patterns of behavior, on influencing motion, rather than trying to target a physical gene, molecule, or chemical. Chronic diseases are rhythmic patterns of decreased variability and organization, which repeat forward in patterns over and over. We can work to change those patterns over time.

We can calm the immune system from overly active waves of behavior to obliterate autoimmune diseases, for example. We can do this by creating powerful carrier waves of exercise-recovery, hunger-eating, emotional arousal-relaxation, wakefulness-sleep, and other cycles nested within the powerful circadian, ultradian, monthly, and yearly cycles with and within which our cycles naturally cycle. The same can be done to synchronize the wayward rhythmic changes that characterize other systemic problems, such as cancer, cardiovascular disease, neurological disorders, and behavioral disorders. Through cycles, we can orchestrate our own health.

Remember the clocks on Christiaan Huygens's wall? Imagine that your body is that wall, made up of all the clocks within—the clocks being your organs and cells. You can make powerful rhythms that knock all those inner clocks into synchrony. The clocks can't act *out* of synchrony with the rest of the wall when the wall makes powerful rhythms.

If your cells can't act out of synchrony with the rest of your body's healthy rhythms, you essentially don't have one of the health problems mentioned above.

Organized, synchronized motion across all your cells and organs: If you have that, you are essentially healthy.

We can create masterful full-body rhythms if we coordinate powerful body-waves, which are carrier waves for the inner cycles of our bodies, with the natural environmental waves we all live in. Those carrier waves, from the environment down through our bodies, down through our organs and cells, will simulcausally synchronize our inner waves as an orchestrated collective as those inner waves synchronously participate in shaping those larger waves. Immune system cycles will naturally come to harmonize with all our bodies' other cycles. Cancer cells will stop dividing independently. Cardiovascular, neurological, and behavioral cycles will renormalize.

We can exercise and recover and shape other cycles for health. We can exercise and recover to shape our waves for longevity.

All the medical details held at bay in our discussion will become useful in unprecedented ways. We will adjust the use of our many good and effective medications, which sometimes are more effective on some people than others, sometimes lose effectiveness over time, and sometimes cause terrible side effects. We will learn how to administer them in synchrony with the oscillatory processes of the body so the medicines hit their mark but do not otherwise affect what they ideally are not supposed to. We will medicate more wisely—at certain times of day, in the context of different behaviors, so the medications spur our cells and organs to function when it is most beneficial for health.

But our health is not in isolation; to stay healthy, we must care for the health of the environment and the health of the planet. The word *environment* may come from *environ*, which means to circle or surround, but the connection is deeper—and inherent. We must safeguard the environment as much as we safeguard our own health—for our personal stability and the stability of the environment and the planet are a continuum.

This is not idealism. Variability is essential for our survival. We need variability on all scales if we are, as a species, to survive and flourish. A flattened Earth wave, with decreased variability, cannot adequately shape human inner waves.

Some practices that will yield immediate benefits are those that open scales. I described some in earlier chapters, such as cleaning the atmosphere to prevent the global storming that happens as we box ourselves in with pollution; I noted that pollution decreases the heartwave variability that we need for health. I mentioned the work of Thomas E. Lovejoy, which shows that when we preserve land for parks, we must keep the parcels large and in open continuum with one another. We must reunderstand and design our living spaces this way—our homes, our workplaces, our schools, and our cities. We must incorporate this knowledge into how to build roads and otherwise work with the land, to prevent the "edge effect" described by

Lovejoy and others, through which flora and fauna die off or disappear. We need to preserve the variability of life on earth: When a species becomes extinct, literally one more inner peak of the entire Earth wave is lost, and it flattens.

We must also stop isolating different crops and instead allow them to grow together—an idea championed by Wes Jackson, founder and president of The Land Institute. Jackson showed me, when I met with him, that crops grown together thrive without pesticides. This is one way we can draw extraordinarily practical benefits from variability: Nested, polyculture developments promote order and organization, whereas fields that grow individual crops in isolation promote an overgrowth of pests. This open pattern of crops growing together, endemic in complex ecosystems but rarely used in agriculture, offers a promising avenue for feeding all people organically.

We need to make these changes to take advantage of the organizational aspects of SuperWaves.

WE WILL ALSO, in the future, use SuperWaves to develop practical technologies. One dramatic technological advance we will soon make involves what is known as cold fusion, or low-energy nuclear reaction. This clean source of heat for our engines and other technologies would seem like a dream come true, but it is a natural step to take now that we know the nature of nature. I have worked with physicists in Israel, the United States, and Italy on how to achieve it, and I will use it as an example of how to apply SuperWaves to technology.

Our new knowledge of nature gives us two ideas that open up the area: what so-called fusion really is, and why and how scientists haven't been able to do what they have been trying to do (i.e., get excess heat from their experiments). These ideas take us directly to a practical technological application—what I call Emergent Fusion.

First, the fusion itself. We already know what happens, and it's not separate hydrogen atoms fusing to become a helium atom. So-called fusion is, simply, another example of compression and sprouting. We have seen it on and across scales—in the human body, in quantum particles, in so-called emergent complexity, and in the sun and the galaxies. All exhibit this emergent pattern of complexity. We have called this process "fusion" when it happens in the heart of the sun.* But the term *fusion* wrongly puts the focus on one aspect of SuperWaves—as if it were two atoms smashing together on a single scale. The reality is that nested waves peaking across scales bring on a symphony of compression, organization, emergent complexity, synchrony, and heat.

---

* Said another way: The sun, as we have already discussed, is the compressed peaks of peaks of waves waving within waves. Sunlight is relatively flattened trough waves, dispersing from the peak compression that is the sun. And heat is how we experience heightened wave motion.

That is the key. The way nature works is that compression and heat naturally happen together, when, across scales, high frequency-amplitude waves climb the high frequency-amplitude waves within which they nest. Nested wave motion is the secret: It is everything, the stuff of the matter as well as the accompanying heat.

Scientists have never been able to bring out more heat than they put into their experiments because they have not known that this is how it works. They have tried to draw excess heat based on their idea of thermodynamics. They begin with heat and try, in an environment of tremendous heat, to smash and fuse highly accelerated atoms. But fusing particles to make heat is backward. Heat does not cause motion, and in any case, pushing nuclei together is not the goal.

I call the correct approach dynothermatics, a new field I described in the previous chapter: using the motion of waves waving to create heat and matter. It gives us all we need to know to achieve Emergent Fusion. Heat is not an *effect* of so-called fusion; heat occurs naturally, simultaneously, as inner nested waves sprout complexity, which both occur naturally and simultaneously as upwardly spiking carrier waves come to a peak. We can entice open systems to become both more compressed and hotter by creating highly peaking nested wave patterns. We can create the heat we desire by starting with the motion.

This is how to do it: Pulse waves in the form of so-called electricity, in large, arcing waves. Then pulse waves into the peaks of those waves. Then pulse smaller waves yet into the inner peaks of those waves. Unlike Thomas Edison's direct current, which is a straight line, or Nikola Tesla's alternating current, which alternates around an axis of a straight line, this is a "current" pulsed in the pattern of the SuperWaves IC fractal.

With this carrier-wave pulse, the inner clusters of SuperWaves (the hydrogen atoms) will themselves pulse and pulse in ever-increasing frequency-amplitudes, until they compress—until they sprout a new level of complexity and become what we recognize as a helium atom. It is no different from the amoebas that oscillate and compress into a slime mold. Heat—the heat we are trying to produce—will naturally happen at the same time.

Years ago, I worked with a company called Energetics Technologies in Israel to investigate SuperWaves in Emergent Fusion (which was still called low-energy nuclear reactions in our work at that time).* For my contribution of SuperWaves Theory to the design of these experiments, I received the Giuliano Preparata Medal at the International Conference on Cold Fusion in Washington, DC, in August 2008. The scientists who awarded the medal—including Michael McKubre, at the

---

* Using four experimental approaches—electrolysis, glow discharge, gas loading in catalyst cells, and high-pressure high-temperature cells with ultrasonic wave excitation—we obtained preliminary results. A significant amount of excess heat was measured in the first glow-discharge experiment—up to as much as 2,500 percent, or 25 times the input power. SuperWaves drove the excess heat higher than direct heat.

time director of the Energy Research Center at SRI International—are all true visionaries. They have risked scorn from the scientific community at large for pursuing cold fusion, which has a bad reputation in the United States, in hopes of a greener, cleaner, and oil-politics-free world. As it happens, Lord Kelvin said that heavier-than-air machines were impossible, a mere 10 years or so before the Wright brothers flew. Just because it seems something is impossible, does not mean that we have understood nature well enough to pass verdict.

Emergent Fusion is an area in which I was very involved, but SuperWaves promises new prospects for every technology. For example, magnetohydrodynamics, the study of the magnetic properties of electrically conducting fluids, has applications in the mixing of molten metals. One of the scientists with whom I collaborated on cold fusion, Herman Branover, decided to try applying SuperWaves to this field.

Magnetohydrodynamics is essential to building strong metals. The better mixed the metal, the stronger it is. Magnets are used for stirring because the molten metals eat up propellers or other mixers.* Magnetic fields race around the vat, and as the steel swirls after the magnetism, it gets mixed. However, the turbulence from this process eats away at the vats that hold the molten metal.

By pulsing the waves in the SuperWaves pattern, the metallurgists got better mixing as well as lesser drag on the vat walls—thereby extending the vat's life. The energy of the turbulence was several hundred times stronger than the usual energy that is induced with a sine wave, which yielded more homogeneous, stronger metals. It also reduced impurities, and the speed of the mixing process was dramatically increased. The economic boost of extending vat usage (so that they need replacing less frequently) was a bonus.

This one experiment suggests we may be on the road to stronger metals—imagine more robust car frames or jet planes—that are cheaper, as well. This from one person's inventive application of SuperWaves.

I HAVE BROUGHT UP these technological examples to drive home the idea that our approach to nature must change. Emergent Fusion will have profound effects; metallurgy will, too. Who can say what incredible applications are waiting for SuperWaves?

Physicist David Bohm pointed out, as I have, that we treat nature in a fragmentary way, and so we think of it that way. It is an amazing and beautiful reality that human beings can transcend our limits to see a world unbound by our limitations. Even though we think in wholes and parts, we can now understand that nature itself

---

* Molten steel, for example, can reach over 1,500°C.

is not wholes and parts. We can understand that we have projected onto nature a quality of our minds but that nature itself has a nature all its own—a nature that science's own findings point to. In his book *Wholeness and the Implicate Order*, Bohm states that "science itself is demanding a new, non-fragmentary world view, in the sense that the present approach of analysis into independently existent parts does not work very well in modern physics."[2] His call for inherent continuity could not be more timely. He was looking for a hidden essence, the essence being an invariant, something that is "one." The "one" he sought is here: motion in a pattern, with a design to it, rather than a puzzle.

In this book, I have attempted to establish, above all, with a number of basic facts, the discovery that nature is one inherently continuous fractal of waves waving within waves. The puzzle-derived model of parts and wholes is not true to nature, and that is the reason that science has come up with the mysteries it has. The details of science support this conclusion. Idealizing lines from natural reality, and then trying to piece them back together, will never bring us to the ultimate truth, but once you accept the model of SuperWaves motion, all the scientific facts make sense. The discovery, and the model it generates, offers humanity an angle of influence over and with nature that can change our future.

Now SuperWaves is in your hands. I hope that there are those of you who will take this idea and run with it. It is inspiring to see so many possibilities before us, and it will take many able minds, working alone and together, to develop Super-Waves into workable technologies.

What if we had heart monitors that track more truly the changes of the heart while we exercise and recover in cycles, so that we can see the heartwave with all its nuances? What if we had cars that ran almost endlessly, without pollution and with no need for gas? We need to develop an Emergent Fusion cell, a contained, portable unit that generates power, so that our homes, our offices, our libraries, and our schools can run on this clean fuel technology. Think about statistics. Imagine what we'll be able to calculate to much greater certainty now that we know the role of car-rier waves. Nutrition, education, architecture, city planning, weather forecasting—what expertise *doesn't* benefit from knowing the nature of nature? I was pleasantly surprised by the potential applications to metallurgy. We need people to work on technologies and pathways to knowledge.

And we have ahead of us the wonderful work of reinterpreting our scientific facts. We can crack open their piece-by-piece, invariant shells and weave them together—as they wave in waves—under the illuminating guidance of the SuperWaves perspective. There is a new world ahead of us—truly a new world, one which we can understand as exactly as we can given the ever-changing wonder that is SuperWaves.

The natural constraints of a book—and of the human mind of the author—

mean that *The Nature of Nature* can only introduce this topic. There are concepts in nature that I could not touch upon but that are poised for understanding and development in light of SuperWaves. There are untold details that await explanation. Whole fields of study are ready for renovation. If you know one, carry the torch.

It is up to you, and it is up to us all to change how we live with and within nature.

WHAT LIES BEFORE US is to see how what we know to be SuperWaves *is* SuperWaves, on different levels, and to figure out how to work with it. Science has forever tried to know what "it" is. Here, we just know what *is*.

Nature itself is "one." It is a single phenomenon. It is an incredible phenomenon that is beyond comprehension. Why there is something rather than nothing, as a question, hardly touches on the marvel of it—it's *ordered*. How incredible is that? Obviously it cannot be an accident. It is profound.

One would think that knowing "the answer" might decrease our sense of wonder, but SuperWaves reveals a grand, sparkling design that increases it beyond words. The perfection of SuperWaves, ever moving forward, is awe-inspiring. It is an invariance that varies, a symmetry that is asymmetrical, an unchanging phenomenon of change.

How it came to be is beyond anyone's guess. That it exists is beyond question. It is reality. It is SuperWaves.

ACKNOWLEDGMENTS

MY DEEPEST THANKS TO the wonderful and visionary Maria Rodale, without whom this book would not be possible, and to the entire team at Rodale including Leah Miller and Anna Cooperberg. Thank you Roger Lewin for your outstanding editing.

Thank you Sidney Kimmel for all your years of support. You played a special role with your powerful commitment to the work we are doing.

Alison Godfrey: Your creativity, dedication and hard work are unequaled. Thank you. Trevor and Whitney, thanks for always being there to discuss SuperWaves.

Over the years I have discussed SuperWaves with many people and I thank you all for your support. There have been many ups and downs along the way, and it takes courage to stand by a revolutionary idea that challenges the status quo (and the people entrenched in the status quo).

Special thanks to Mike McKubre, as well as (in alphabetical order) Gideon Ariel, Kiira Benzing, Mike Catalano, Dick Fox, Ryan Freilino, Marshall Gisser, Lana Israel, Duncan Kennedy, Susan Love, Marianne Macy, Mike Melich, Herman Rush, David Saloff, Stan Smith, Morty Wolkowitz, Kevin White, and Bert Zarins. You have all been important to me in different ways and I appreciate it. And thank you to those who have performed experiments and are continuing to conduct research with SuperWaves—you know who you are. Keep up the great work!

Thank you to my family, including my children Belle, Shimona, Estee and Judah, and Trevor and Whitney; my brother Herbert Dardik and my sister Ray Josell; and my family members who have passed on: my brother Elliot Dardik, my sisters, Sylvia Plutchok and Gloria Schoenholtz, and my parents, Moshe and Sarah Dardik.

For this book itself: Thank you Joey, and Tamar, Isaac, Dovid, and Naomi, for your patience and participation as Estee wrote this book with me. And thank you Estee. You are amazing and your contribution was fantastic. We did it!

SOURCE NOTES

INTRODUCTION

1 Stephen Hawking, *A Brief History of Time*, 10th anniversary ed. (New York: Bantam Books, 1998), 11–12.

PART 1 PRELUDE

1 Louis de Broglie, *New Perspectives in Physics* (New York: Basic Books, 1962).

2 Albert Einstein, "Physics and Reality," in *Ideas and Opinions* (New York: Crown, 1954), 290–323.

CHAPTER 1

1 Hugo Lagercrantz and Theodore A. Slotkin, "The 'Stress' of Being Born," *Scientific American*, June 1986, 100.

2 Elizabeth Pennisi, "In Nature, Animals That Stop and Start Win the Race," *Science* 288, no. 5463 (2000): 83–85.

3 J. O'Brien, H. Browman, and B. Evans, "Search Strategies of Foraging Animals," *American Scientist* 78 (March–April 1990), 154, http://fishlarvae.com/common /SiteMedia/AmericanScientist_1990_Reduced.pdf.

4 Pennisi, "In Nature."

5 T. M. Williams, R. W. Davis, L. A. Fuiman, et al., "Sink or Swim: Strategies for Cost-Efficient Diving by Marine Mammals," *Science* 288, no. 5463 (2000): 133–36.

CHAPTER 2

1 David Green and Terry Bossomaier, *Patterns in the Sand: Computers, Complexity, and Everyday Life* (New York: Basic Books, 1998).

2 Jacob Bronowski, *The Origins of Knowledge and Imagination,* rev. ed. (New Haven, CT: Yale University Press, 1979), 16–18.

3 F. David Peat, *From Certainty to Uncertainty: The Story of Science and Ideas in the Twentieth Century* (Washington, DC: Joseph Henry Press, 2002), 230.

4 Françoise Balibar, *Einstein: Decoding the Universe* (New York: Harry N. Abrams, 2001), 143.

5 John D. Barrow, *Pi in the Sky: Counting, Thinking, and Being* (New York: Back Bay Books, 1992), 295.

6 E. T. Bell, *Men of Mathematics: The Lives and Achievements of the Great Mathematicians from Zeno to Poincaré* (New York: Simon & Schuster, 1965), 13. Reprinted with permission of Touchsone, a division of Simon & Schuster, Inc. from *Men of Mathematics* by E. T. Bell. Copyright © 1937 by E. T. Bell. Copyright renewed © 1965 by Taine T. Bell. All rights reserved.

7 Stephen Hawking, *A Brief History of Time,* 10th anniversary ed. (New York: Bantam Books, 1998), 11–12.

8 John D. Barrow, *The World within the World* (Oxford, England: Oxford University Press, 1991), 398. By permission of Oxford University Press.

9 Albert Einstein, "Physics and Reality," in *Ideas and Opinions* (New York: Crown, 1954), 290–323.

10 David Darling, *Equations of Eternity: Speculations on Consciousness Meaning and Mathematical Rules That Orchestrate the Cosmos* (New York: Hyperion, 1993), 34.

11 Sigmund Freud, "The Future of an Illusion," in *The Freud Reader,* ed. Peter Gay (New York: W. W. Norton and Company, 1989), 693.

12 Darling, *Equations of Eternity,* 77.

13 Richard P. Feynman, with Robert B. Leighton and Matthew Sands, *Six Easy Pieces: Essentials of Physics Explained by Its Most Brilliant Teacher* (New York: Basic Books, 1995), 2.

PART 2 PRELUDE

1 Brian L. Silver, *The Ascent of Science* (New York: Oxford University Press, 1998), xvi.

2 John D. Barrow, *New Theories of Everything: The Quest for Ultimate Explanation* (New York: Oxford University Press, 2007), 110.

3 Albert Einstein, "Physics and Reality," in *Ideas and Opinions* (New York: Crown Publishers, 1982), 290.

4 Richard P. Feynman, with Robert B. Leighton and Matthew Sands, *Six Easy Pieces: Essentials of Physics Explained by Its Most Brilliant Teacher* (New York: Basic Books, 1995), 2.

5 Richard Feynman, *The Character of Physical Law* (Cambridge, MA: MIT Press, 1965), 147. Excerpt reprinted with permission from the publisher.

CHAPTER 3

1 F. David Peat, *From Certainty to Uncertainty: The Story of Science and Ideas in the Twentieth Century* (Washington, DC: Joseph Henry Press, 2002), 29.

2 Mark Haw, *Middle World: The Restless Heart of Matter and Life* (New York: Palgrave Macmillan, 2007), 115.

3 Leon Lederman with Dick Teresi, *The God Particle: If the Universe Is the Answer, What Is the Question?* (Boston: Houghton Mifflin, 1993), 278.

4 Hermann Weyl, *Symmetry* (Princeton, NJ: Princeton University Press, 1952), 21.

5 John S. Rigden, "Einstein's Revolutionary Paper," *Physics World*, April 2005, 18; see also, Albrecht Folsing, *Albert Einstein: A Biography* (New York: Penguin Books, 1998),143.

6 John Gribbin, *Almost Everyone's Guide to Science* (New Haven, CT: Yale University Press, 2000), 6–7.

7 Ibid., 62.

8 Robert Wilson, *Astronomy through the Ages: The Story of the Human Attempt to Understand the Universe* (Princeton, NJ: Princeton University Press, 1998), 275.

9 Richard Feynman, *The Character of Physical Law* (New York: Random House, 1994), 143–44.

10 K. C. Cole, *Mind over Matter: Conversations with the Cosmos* (Orlando: Harcourt, 2003), 101.

11 James D. Watson with Andrew Berry, *DNA: The Secret of Life* (New York: Alfred A. Knopf, 2003): xii–xiii. Excerpt reprinted with permission from the publisher.

12 Richard P. Feynman with Robert B. Leighton and Matthew Sands, *Six Easy Pieces: Essentials of Physics Explained by Its Most Brilliant Teacher* (New York: Helix Books, 1995), 4. Excerpt reprinted with permission from the publisher.

CHAPTER 4

1 Roger Penrose, "Must Mathematical Physics Be Reductionist?" in *Nature's Imagination: The Frontiers of Scientific Vision*, ed. John Cornwell (Oxford, England: Oxford University Press, 1995), 12–13. By permission of Oxford University Press.

2 Lawrence M. Krauss, *Fear of Physics: A Guide for the Perplexed* (New York: Basic Books, 1993), 7–8.

3 Roger G. Newton, *Galileo's Pendulum: From the Rhythm of Time to the Making of Matter* (Cambridge, MA: Harvard University Press, 2004), 1–3, 137.

4   James S. Perlman, *Science without Limits: Toward a Theory of Interaction between Nature and Knowledge* (Amherst, NY: Prometheus Books, 1995), 126.

5   Krauss, *Fear of Physics*, 9.

6   Albert Einstein, foreword to *Dialogue Concerning the Two Chief World Systems,* by Galileo Galilei, trans. Stillman Drake (Berkeley: University of California Press, 1967). Excerpt reprinted with permission from the publisher.

7   Ibid., 8.

8   "General Scholium," in *The Principia: Mathematical Principles of Natural Philosophy,* by Isaac Newton, trans. I. Bernard Cohen, Anne Whitman, and Julia Budenz (Berkeley: University of California Press, 1999), 943.

9   Quote taken from http://plato.stanford.edu/entries/qm-action-distance/.

10 Sadi Carnot, *Reflections on the Motive Power of Fire* (New York: Dover, 1960), 7.

11 Steven Strogatz, *Sync: How Order Emerges from Chaos in the Universe, Nature, and Daily Life* (New York: Hyperion, 2003), 69.

12 Sir Arthur S. Eddington, *The Nature of the Physical World* (Cambridge, England: Cambridge University Press, 1948), 37, http://henry.pha.jhu.edu/Eddington.2008.pdf.

13 John P. Briggs and F. David Peat, *The Looking Glass Universe: The Emerging Science of Wholeness* (New York: Cornerstone Library, 1984), 108.

14 John Maddox, *What Remains to Be Discovered: Mapping the Secrets of the Universe, the Origins of Life, and the Future of the Human Race* (New York: Free Press, 1998), 177.

CHAPTER 5

1   Robert H. MacArthur, *Geographical Ecology: Patterns in the Distribution of Species* (Princeton, NJ: Princeton University Press, 1984), 1.

2   Richard Feynman, *The Character of Physical Law* (Cambridge, MA: MIT Press, 2001), 13.

3   K. C. Cole, *The Universe and the Teacup: The Mathematics of Truth and Beauty* (New York: Harcourt, Brace, 1998), 72.

4   Ibid., 73.

5   Ibid., 42.

6   Keith Devlin, *The Language of Mathematics: Making the Invisible Visible* (New York: Holt Paperbacks, 2000), n.p.

7   Galileo Galilei, *Discoveries and Opinions of Galileo,* trans. Stillman Drake (Garden City, NY: Doubleday, 1957), 237–38.

8   Paul Davies, "What Are the Laws of Nature," in *Doing Science: The Reality Club,* ed. John Brockman (New York: Prentice Hall, 1991), 57.

9   Lawrence M. Krauss, *Fear of Physics: A Guide for the Perplexed* (New York: Basic Books, 1993), 26.

10 In Richard Panek, "Has the Einstein Revolution Gone Too Far?" *Discover*, March 2008, http://discovermagazine.com/2008/mar/01-has-the-einstein-revolution-gone -too-far; and in a letter to Arnold Sommerfeld in "The Collected Papers of Albert Einstein," vol. 5 (October–November 1912), p. 324, http://einsteinpapers.press. princeton.edu/vol5-trans/346?highlightText=%22enormous%20respect%20for%20 mathematics%22.

11 Albert Einstein and Alan Harris, *Einstein's Essays in Science* (Mineola, NY: Dover Publications, 2009), 17.

12 Davies, "What Are the Laws of Nature," 47.

13 Paul Davies, "Taking Science on Faith," *New York Times*, November 24, 2007, nytimes.com/2007/11/24/opinion/24davies.html.

14 David Darling, *Equations of Eternity: Speculations on Consciousness Meaning and Mathematical Rules That Orchestrate the Cosmos* (New York: Hyperion, 2003), 75.

15 Keith Devlin, *Life by the Numbers* (New York: Wiley, 1998), 18.

16 Jeremy Rifkin with Ted Howard, *Entropy: A New World View* (New York: Viking, 1980), 29.

17 Stephen Hawking, *A Brief History of Time*, 10th anniversary ed. (New York: Bantam Books, 1998), 11.

PART 3 PRELUDE

1   Heinz R. Pagels, *The Dreams of Reason: The Computer and the Rise of the Sciences of Complexity* (New York: Bantam Books, 1989), 308.

2   Ibid.

CHAPTER 6

1   "The Nobel Prize in Physics 1926," www.nobelprize.org/nobel_prizes/physics /laureates/1926/.

2   "Presentation Speech by Professor C. W. Oseen, member of the Nobel Committee for Physics of the Royal Swedish Academy of Sciences, on December 10, 1926," www.nobelprize.org/nobel_prizes/physics/laureates/1926/press.html.

3   Paul Davies, *Cosmic Jackpot: Why Our Universe Is Just Right for Life* (Boston: Houghton Mifflin, 2007), 43.

4   Heinz R. Pagels, *The Cosmic Code: Quantum Physics as the Language of Nature* (New York: Simon & Schuster, 1982), xiii.

5   P. C. W. Davies and J. R. Brown, eds., *The Ghost in the Atom: A Discussion of the*

Mysteries of Quantum Physics (New York: Cambridge University Press, 1993), 25–26. Excerpt reprinted with permission from the publisher.

6 Leon Lederman with Dick Teresi, *The God Particle: If the Universe Is the Answer, What Is the Question?* (Boston: Houghton Mifflin, 1993), 23–34.

7 Aage Petersen, "The Philosophy of Niels Bohr," *Bulletin of the Atomic Scientists* 19, no. 7 (September 1963): 12.

8 © The Nobel Foundation, 1954.

9 Max Born, "Statistical Interpretation of Quantum Mechanics," *Science* 122, no. 3172 (October 14, 1955): 675–79.

CHAPTER 7

1 Albert Einstein, "On the Electrodynamics of Moving Bodies," in Alan Lightman, *The Discoveries* (New York: Vintage Books, 2005), 72–83.

2 Richard Feynman, *The Character of Physical Law* (New York: Random House, 1994), 129.

3 Quoted in Werner Heisenberg, *Physics and Beyond: Encounters and Conversations,* trans. Arnold J. Pomerans (New York: Harper & Row, 1971), 206.

4 Richard P. Feynman, with Robert B. Leighton and Matthew Sands, *Six Easy Pieces: Essentials of Physics Explained by Its Most Brilliant Teacher* (New York: Basic Books, 1995), 2.

5 Francis Galton, *Natural Inheritance* (New York: Macmillan, 1889), 66.

6 Ian Stewart, *Does God Play Dice? The New Mathematics of Chaos* (Malden, MA: Blackwell, 2002), 47–48. Excerpt reprinted with permission from the publisher.

7 Hans Christian von Baeyer, *Maxwell's Demon: Why Warmth Disperses and Time Passes* (New York: Random House, 1998), 89–91.

CHAPTER 8

1 Richard Feynman, *The Character of Physical Law* (New York: Random House, 1994), 39.

2 Ibid., 172.

3 *Tom Stoppard: Plays 5: Arcadia, The Real Thing, Night & Day, Indian Ink, Hapgood* (London: Faber and Faber), 545.

4 John Holland, *Emergence: From Chaos to Order* (Reading, MA: Addison-Wesley, 1998), 1.

5 Erwin Schrödinger, *What Is Life?* (New York: Cambridge University Press, 2012), 71.

6 Paul Davies, *The Fifth Miracle: The Search for the Origin and Meaning of Life* (New York: Simon & Schuster, 1999), 98.

7 Paul Davies, *The Cosmic Blueprint: New Discoveries in Nature's Creative Ability to Order the Universe* (New York: Simon and Schuster, 1988), 20.

8 H. Tsuji, F. J. Venditti, E. S. Manders, et al., "Reduced Heart Rate Variability and Mortality Risk in an Elderly Cohort: The Framingham Heart Study," *Circulation* 90 (1994): 878–83.

9 Ary Goldberger, "Chaos Theory and Creativity: The Biological Basis of Innovation," *Journal of Innovative Management* 4 (1999): 15–23.

10 Per Bak, *How Nature Works: The Science of Self-Organized Criticality* (New York: Copernicus, 1996), 21.

11 Davies, *The Fifth Miracle,* 257.

12 G. S. Engel, T. R. Calhoun, E. L. Read, et al., "Evidence for Wavelike Energy Transfer through Quantum Coherence in Photosynthetic Systems," *Nature* 446 (April 12, 2007): 782–86, explained in Davide Castelvecchi, "Quantum Capture: Photosynthesis Tries Many Paths at Once," *Science News* 171, no. 15 (April 11, 2007): 229. Quote is from *Science News*; research article is in *Nature.*

13 Steven Strogatz, *Sync: How Order Emerges from Chaos in the Universe, Nature, and Daily Life* (New York: Hyperion, 2003), 289.

14 Dennis Overbye, "A Quantum of Solace: Timeless Questions about the Universe," *New York Times,* July 1, 2013, nytimes.com/2013/07/02/science/space/timeless-questions -about-the-universe.html?_r=0. Excerpt reprinted with permission from the publisher.

15 Hans Christian von Baeyer, *The Fermi Solution: Essays on Science* (New York: Random House, 1993), 152.

16 Max Tegmark, *Our Mathematical Universe: My Quest for the Ultimate Nature of Reality* (New York: Vintage, 2015), 6, 267.

17 Eugene Wigner, "The Unreasonable Effectiveness of Mathematics in the Natural Sciences," *Communications on Pure and Applied Mathematics* 13 (February 1960): 1–14. The full paper can be found at https://www.dartmouth.edu/~matc /MathDrama/reading/Wigner.html.

18 Marjorie Sun, "How Do You Measure the Lovejoy Effect," *Science* 247, no. 4947 (March 9, 1990): 1174–76.

19 See discussion in Ros Clubb and Georgia Mason, "Animal Welfare: Captivity Effects on Wide-Ranging Carnivores," *Nature* 425 (October 2, 2003): 473–74.

PART 4 PRELUDE

1 Heinz R. Pagels, *The Cosmic Code: Quantum Physics as the Language of Nature* (Mineola, NY: Dover Publications, 2011), 321–22.

2 Timothy Ferris, *Coming of Age in the Milky Way* (New York: Morrow, 1988), 346.

3 Richard P. Feynman, *"What Do You Care What Other People Think?" Further Adventures of a Curious Character*, Ralph Leighton, ed. (New York: Norton, 1988), 244.

4 K. C. Cole, *First You Build a Cloud: And Other Reflections on Physics as a Way of Life* (San Diego: Harcourt, Brace, 1999), 153–54.

5 Ibid., 156–57.

6 Stephen Hawking, *A Brief History of Time: From the Big Bang to Black Holes* (New York: Bantam Books, 1988), 189.

CHAPTER 10

1 J. Dimsdale, L. H. Hartley, T. Guiney, J. N. Ruskin, and D. Greenblatt, "Postexercise Peril. Plasma Catecholamines and Exercise," *JAMA* 251, no. 5 (February 3, 1984): 630–32.

2 Irving Dardik, "Research Proposals for Studying Mind-Body Interactions: Cardiocybernetics, Relaxation through Exercise," *Advances* 3, no. 3 (summer 1986): 56–69.

3 See, for example, the finding that "after intense long-term exercise, the immune system is characterized by concomitant inflammation and temporary suppression of the cellular immune system," in B. K. Pedersen, T. Rohde, and K. Ostrowski, "Recovery of the Immune System after Exercise," *Acta Physiologica Scandinavica* 162, no. 3 (March 1998): 325–32. Also see the research of B. K. Pederson and L. Hoffman-Goetz, "Exercise and the Immune System: Regulation, Integration, and Adaptation," *Physiological Reviews* 80, no. 3 (2000): 1055–81.

4 C. M. Masi, L. C. Hawkley, E. M. Rickett, and J. T. Cacioppo, "Respiratory Sinus Arrhythmia and Diseases of Aging: Obesity, Diabetes Mellitus, and Hypertension," *Biological Psychology* 74, no. 2 (February 2007): 212–23.

5 Simon A. Levin, "The Problem of Pattern and Scale in Ecology: The Robert H. MacArthur Award Lecture," *Ecology* 73, no. 6 (1992): 1943–67.

CHAPTER 11

1 Steven Strogatz, *Sync: The Emerging Science of Spontaneous Order* (New York: Hyperion, 2003), 182.

2 Amanda Gefter, "Don't Mention the F Word," *New Scientist* 193, no. 2594 (March 10, 2007): 30–33.

CHAPTER 12

1 Helen Pearson, "DNA: Beyond the Double Helix," *Nature* 421 (January 23, 2003): 310–12. Excerpt reprinted with permission from the publisher.

2 J. H. Levine, Y. Lin, and M. B. Elowitz, "Functional Roles of Pulsing Genes in Genetic Circuits," *Science* 342, no. 6163 (December 6, 2013): 1193–200.

3 "Hormone Cycling Found to Affect Gene Activity," National Institutes of Health, August 16, 2009, nih.gov/news-events/news-releases/ hormone-cycling-found-affect-gene-activity.

4 Franklin M. Harold, *The Way of the Cell: Molecules, Organisms, and the Order of Life,* p. 5, quoting Erwin Chargaff, *Heraclitean Fire: Sketches from a Life before Nature* (New York: The Rockefeller University Press, 1978), 87.

5 Zoltán N. Oltvai and Albert-László Barabási, "Life's Complexity Pyramid," *Science* 298, no. 5594 (October 25, 2002): 763.

6 Paul Davies, *The Fifth Miracle: The Search for the Origin and Meaning of Life* (New York: Simon & Schuster, 1999), 257.

7 Ibid., 98.

8 F. David Peat, *Synchronicity: The Bridge between Matter and Mind* (New York: Bantam Books, 1987), 87.

CHAPTER 13

1 John P. Briggs and F. David Peat, *The Looking Glass Universe: The Emerging Science of Wholeness* (New York: Cornerstone Library, 1984), 115–16.

2 Hans Christian von Baeyer, *Maxwell's Demon: Why Warmth Disperses and Time Passes* (New York: Random House, 1998), 89–90.

3 Mark Haw, "Einstein's Random Walk," *Physics World*, January 2005, 19–22.

4 F. David Peat, *Synchronicity: The Bridge between Matter and Mind* (New York: Bantam Books, 1987), 88.

5 Heinz R. Pagels, *The Cosmic Code*: *Quantum Physics as the Language of Nature* (New York: Simon & Schuster, 1982), xiii.

6 Richard P. Feynman with Robert B. Leighton and Matthew Sands, *Six Easy Pieces: Essentials of Physics Explained by Its Most Brilliant Teacher* (New York: Basic Books, 1995), 4.

CHAPTER 14

1 Sir Arthur Eddington, *The Nature of the Physical World* (Cambridge, UK: Cambridge University Press, 1948), 37, http://henry.pha.jhu.edu/Eddington.2008.pdf.

2 Paul A. Schlipp, ed., *Albert Einstein: Autobiographical Notes* (Chicago: Open Court Publishing Company, 1999), 31.

3 Paul Davies, "Searching for the Fourth Law," *New Scientist* 188, no. 2523 (October 29, 2005): 51.

4 Paul Davies, *The Fifth Miracle: The Search for the Origin and Meaning of Life* (New York: Simon & Schuster, 1999), 257.

5 David Robson, "Disorderly Genius: How Chaos Drives the Brain," *New Scientist* (June 27, 2009): 34.

6 Jerry A. Coyne, "Genetics and Speciation," *Nature* 355 (February 6, 1992): 511.

7 Gregory S. Engel, et al., "Evidence for Wavelike Energy Transfer through Quantum Coherence in Photosynthetic Systems," *Nature* 446 (April 12, 2007): 782–86. doi:10.1038/nature05678.

CHAPTER 15

1 T. Hisako, M. Larson, F. Venditti, E. Manders, C. Feldman, and D. Levy, "Impact of Reduced Heart Rate Variability on Risk for Cardiac Events, the Framingham Heart Study," *Circulation* 94, no. 11 (December 1, 1996): 2850–55.

2 J. Hayano, Y. Sakakabara, M. Yamada, et al., "Decreased Magnitude of Heart Rate Spectral Components in Coronary Artery Disease, Its Relation to Angiographic Severity," *Circulation* 81, no. 4 (April 1990): 1217–24.

3 L. Temoshok, A. O'Leary, and S. Jenkins, "Survival Time in Men with AIDS: Relationships with Psychological Coping and Autonomic Arousal," Abstract from Sixth International Conference on AIDS, San Francisco, CA, June 20–24, 1990; A. O'Leary, L. Temoshok, and S. Jenkins, "Autonomic Reactivity, Immune Function and Survival in Men with AIDS," 1992.

4 B. Neubauer and H. J. G. Gundersen, "Analysis of Heart Rate Variations in Patients with Multiple Sclerosis; A Simple Measure of Autonomic Nervous Disturbances Using an Ordinary ECG," *Journal of Neurology, Neurosurgery, and Psychiatry* 41, no. 5 (May 1978): 417–19.

5 A. Murry, D. Ewing, I. Campbell, M. Neilson, and B. Clarke, "RR Interval Variations in Young Male Diabetics," *British Heart Journal* 37 (1975): 882–85; H. J. G. Gunderson and B. Neubauer, "A Long-Term Diabetic Autonomic Nervous Abnormality," *Diabetologia* 13 (1977): 137–40.

6 J. Hirsh, A. Aguirre, and R. Leibel, "A Non-Invasive Tool to Assess Energy Metabolism: Measurement of Heart Rate Variability," Fifth International Congress on Obesity, 1989.

7 A. Garfinkel, S. Raetz, and R. Harper, "Heart Rate Dynamics after Acute Cocaine Administration," *Journal of Cardiovascular Pharmacology* 19, no. 3 (1992): 453–59.

8 A. Raine, P. Venables, and M. Williams, "High Autonomic Arousal and Electrodermal Orienting at Age 15 Years as Protective Factors against Criminal Behavior at Age 29 Years," *American Journal of Psychiatry* 152 (November 1995): 1595–600; A. Raine, P. Venables, and M. Williams, "Relationships between Central and Autonomic Measures of Arousal at Age 15 Years and Criminality at Age 24 Years," *Archives of General Psychiatry* 47 (November 1990): 1003–7.

9 R. Lowensohn, M. Weiss, and E. Hon, "Heart-Rate Variability in Brain-Damaged Adults," *The Lancet* 1, no. 8012 (March 19, 1977): 626–28.

10 H. Tsuji, F. Venditti, E. Manders, et al., "Reduced Heart Rate Variability and Mortality Risk in an Elderly Cohort, the Framingham Heart Study," *Circulation* 90, no. 2 (August 1994): 878–83.

11 Ibid.

12 Ary Goldberger, "Chaos Theory and Creativity: The Biological Basis of Innovation," *Journal of Innovative Management* 4 (1999): 15–23.

13 Ines Moreno-Gonzales and Claudio Soto, "Misfolded Protein Aggregates: Mechanisms, Structures and Potential for Disease Transmission," *Seminars in Cell & Developmental Biology* 22, no. 5 (July 2011): 482–87; Laura Sanders, "Alzheimer's May Be Handiwork of 'Prion' Proteins: A-Beta Moves from Cell to Cell, Spreading Destruction," *Science News* 182 (July 2012): 5–6.

14 Amy Dockser Marcus, "Mad-Cow Disease May Hold Clues to Other Neurological Disorders," *Wall Street Journal*, December 3, 2012, wsj.com/articles/SB1000142412 7887324020804578151291509136144.

15 T. Nakamura, K. Kiyono, K. Yoshiuchi, R. Nakahara, Z. R. Struzik, and Y. Yamamoto, "Universal Scaling Law in Human Behavioral Organization," *Physical Review Letters* 99, no. 13 (September 28, 2007): 138103.

16 "Something in the Way He Moves," *The Economist* (September 27, 2007): page 85. Excerpt reprinted with permission from the publisher.

17 Helen Pearson, "DNA: Beyond the Double Helix," *Nature* 421 (January 23, 2003): 310.

18 Aravinda Chakravarti, "Genomics Is Not Enough," *Science* 334, no. 6052 (October 7, 2011): 15.

19 A. Letourneau, F. A. Santoni, X. Bonilla, et al., "Domains of Genome-Wide Gene Expression Dysregulation in Down's Syndrome," *Nature* 508 (April 17, 2014): 345–50.

20 C. E. Murry, R. B. Jennings, and K. A. Reimer, "Preconditioning with Ischemia: A Delay of Lethal Cell Injury in Ischemic Myocardium," *Circulation* 74, no. 5 (November 1986): 1124–36.

21 M. V. Cohen and J. M. Downey, "Ischaemic Preconditioning: Can the Protection Be Bottled?" *Lancet* 342, no. 8862 (July 3, 1993): 6. Excerpt reprinted with permission from the publisher.

22 See, for example, Peter Moore, "Science: Absence of Blood Makes the Heart Grow Stronger," *New Scientist*, December 11, 1993, 19.

23 J. Crinnion, S. Homer-Vanniasinkam, and M. Gough, "Skeletal Muscle Reperfusion Injury: Pathophysiology and Clinical Considerations," *Cardiovascular Surgery* 1, no. 4 (August 1993): 317–24.

24 N. Yamashita, S. Hoshida, N. Taniguchi, T. Kuzuya, and M. Hori, "A 'Second Window of Protection' Occurs 24 h after Ischemic Preconditioning in the Rat Heart," *Journal of Molecular and Cellular Cardiology* 30, no. 6 (June 1998): 1181–89.

25 Irving I. Dardik, "The Origin of Disease and Health, Heart Waves: The Single Solution to Heart Rate Variability and Ischemic Preconditioning," *Cycles* 46, no. 3 (1996): 77

26 In 2003, several researchers and I conducted a feasibility study at the Wistar Institute of The University of Pennsylvania in conjunction with Philadelphia FIGHT, an AIDS service organization. The protocol involved 52 HIV-positive people who were taking medications to suppress the virus but suffering poor quality of life from the side effects of the medications. Over the course of 8 weeks, participants did Cycles, aerobics, or no supervised exercise. Compared to aerobics or the control group who did not exercise, doing Cycles improved quality of life. Patients reported at Week 8 (compared to Week 1) an increase in happiness, improved sleep, less effort required for daily activities, a decrease in thinking that one's life is failed, and an increase in verbal interaction with others.

In another study, my colleagues and I tracked the markers of overall health of healthy women doing Cycles. This study was designed to evaluate the Cycles program compared to aerobics and interval training to establish the feasibility and efficacy of the program, as well as to encourage further research. This small study established that even with only 10 women participating in the 8-week-long program, there were significant changes in cardiovascular fitness, neuroautonomic dynamics, and psychological well-being, warranting further investigation. We found that in this short time, subjects showed a 15.5 percent increase in the maximum rate at which the heart, lungs, and muscles burn oxygen; a 13 percent improvement in the efficiency of oxygen delivery; a 7.5 percent decrease in diastolic blood pressure; a 9 percent increase in heart rate variability; heightened positive moods; and decreased anxiety. R. L. Goldsmith, I. Dardik, D. M. Bloomfield, et al., "Implementation of a Novel Cyclic Exercise Protocol in Healthy Women," *American Journal of Sports Medicine* 4, no. 2 (2002): 135–51.

In a 2003 study several researchers and I did on Parkinson's disease, patients doing Cycles experienced significantly improved quality of life and an increase in anti-inflammatory signal molecules, among other benefits, in comparison with a sample group. ACTH levels determined in plasma by radioimmunoassay as well as cytokine (IL-10) levels determined in plasma by enzyme-linked immunosorbent assay both showed significant improvement. These improvements, together with the fact that tremors and spasticity were improving, all showed the process of Parkinson's beginning to reverse. P. Cadet, W. Zhu, K. Mantione, et al., "Cyclic Exercise Induces Anti-Inflammatory Signal Molecule Increases in the Plasma of Parkinson's Patients," *International Journal of Molecular Medicine* 12 (2003): 485–92.

W. Stanton Smith, a client of mine, and his wife, Rosalind Lewis-Smith, wrote about his dramatic experience reversing Parkinson's with LifeWaves Cycles in *Waving Goodbye to Parkinson's Disease: A Journey of Hope* (available at wstantonsmith.com/waving-goodbye-to-parkinsons-disease.php).

In 2001, researchers and I performed a study on individuals doing Cycles, technically tracing their levels of nitric oxide, and found they experienced enhanced pulsations of blood vessels and reduced inflammation. G. B. Stefano, V. Prevot, P.

Cadet, and I. Dardik, "Vascular Pulsations Stimulating Nitric Oxide Release during Cyclic Exercise May Benefit Health: A Molecular Approach (Review)," *International Journal of Molecular Medicine* 7, no. 2 (2001): 119–29.

A 2016 study (in which I was not involved) showed that passive recovery led to improved performance and reduced stress. A. T. Scanlan and M. C. Madueno, "Passive Recovery Promotes Superior Performance and Reduced Physiological Stress across Different Phases of Short-Distance Repeated Sprints," *Journal of Strength and Conditioning Research* 30, no. 9 (September 2016): 2540–49.

27 Irving I. Dardik, Patent USRE38749 E1, filed September 15, 2000, and issued June 28, 2005.

28 Elizabeth Quill, "When Networks Network: Once Studied Solo, Systems Display Surprising Behavior When They Interact," *Science News* 182, no. 6 (September 22, 2012): 18.

CHAPTER 16

1 Jared Diamond, "Dammed Experiments!" *Science* 294, no. 5548 (November 30, 2001): 1847.

2 Peter Aldhous, "Neighborhoods That Can Kill: The Strain of Inner-City Life May Send Tumors into Overdrive," *New Scientist*, January 16, 2010, 6.

3 "Captive Polar Bears Are 'Psychotic,'" *New Scientist*, May 21, 1987, 29.

4 Nicola Jones, "Diseases Are Running Rife in Forest Remnants," *New Scientist*, February 8, 2003, citing Richard S. Ostfeld and Felicia Keesing, "Biodiversity and Disease Risk: The Case of Lyme Disease," *Conservation Biology* 14, no. 3 (June 2000): 722–28.

5 Michael Schwirtz, "Riker's Island Struggles with a Surge in Violence and Mental Illness," *New York Times*, March 18, 2014.

6 "Recidivism," National Institute of Justice, nij.gov/topics/corrections/recidivism/Pages/welcome.aspx, last modified June 17, 2014.

7 Bruce Bower, "Combat Trauma from the Past: Data Portray Civil War's Mental, Physical Fallout," *Science News* 169, no. 6 (February 8, 2006): 84.

8 Jan Hoffman, "Nightmares after the I.C.U.," *New York Times*, July 22, 2013, http://well.blogs.nytimes.com/2013/07/22/nightmares-after-the-i-c-u/?_r=0.

9 Charles Siebert, "An Elephant Crackup?" *New York Times Magazine*, October 8, 2006, nytimes.com/2006/10/08/magazine/08elephant.html.

10 Mark Haw, "From Steam Engines to Life?" *American Scientist*, November–December 2007, 472.

11 Elizabeth Pennisi, "What Determines Species Diversity?" *Science* 309, no. 5731 (July 1, 2005): 90.

12 Description of the 125th anniversary issue of *Science*, July 1, 2005, accessed October 25, 2017, sciencemag.org/site/feature/misc/webfeat/125th/.

13 See, for example, C. L. Hanchette and G. G. Schwartz, "Geographic Patterns of Prostate Cancer Mortality. Evidence for a Protective Effect of Ultraviolet Radiation," *Cancer* 70, no. 12 (December 15, 1992): 2861–69: "Prostate cancer mortality exhibits a significant north-south trend, with lower rates in the South. These geographic patterns are not readily explicable by other known risk factors for prostate cancer."

   Also, William B. Grant, "An Ecological Study of Cancer Incidence and Mortality Rates in France with Respect to Latitude, an Index for Vitamin D Production," *Dermato-Endocrinology* 2, no. 2 (April–December 2010): 62–67; M. F. Borisenkov, "Latitude of Residence and Position in Time Zone Are Predictors of Cancer Incidence, Cancer Mortality, and Life Expectancy at Birth," *Chronobiology International* 28, no. 2 (March 2011): 155–62; and S. Simpson Jr., L. Blizzard, P. Otahal, I. Van der Mei, and B. Taylor, "Latitude Is Significantly Associated with the Prevalence of Multiple Sclerosis: A Meta-Analysis," *Journal of Neurology, Neurosurgery & Psychiatry* 82, no. 10 (October 2011): 1132–41.

14 L. N. Gillman, D. J. Keeling, H. A. Ross, and S. D. Wright, "Latitude, Elevation and the Tempo of Molecular Evolution in Mammals," *Proceedings of the Royal Society B* 276, no. 1671 (September 22, 2009): 3353–59.

15 See, for example, "All You Can't Eat," *Economist*, March 31, 2005, economist.com/node/3809652, citing Elaine A. Hsieh, Christine M. Chai, and Marc K. Hellerstein, "Effects of Caloric Restriction on Cell Proliferation in Several Tissues in Mice: Role of Intermittent Feeding," *American Journal of Physiology—Endocrinology and Metabolism* 288, no. 5 (May 2005): E965–72.

   Also M. P. Mattson, D. B. Allison, L. Fontana, et al., "Meal Frequency and Timing in Health and Disease," *Proceedings of the National Academy of Sciences of the United States of America* 111, no. 47 (November 25, 2014): 16647–53.

16 D. R. Gold, A. Litonjua, J. Schwartz, et al., "Ambient Pollution and Heart Rate Variability," *Circulation* 101, no. 11 (March 21, 2000): 1267–73.

17 Intergovernmental Panel on Climate Change, Climate Change 2013: *The Physical Science Basis* (New York: Cambridge University Press, 2013), 9, https://www.ipcc.ch/pdf/assessment-report/ar5/wg1/WGIAR5_SPM_brochure_en.pdf.

18 A. I. Gitelman, J. S. Risbey, R. E. Kass, and R. D. Rosen, "Trends in the Surface Meridional Temperature Gradient," *Geophysical Research Letters* 24, no. 10 (May 15, 1997): 1243–46., http://onlinelibrary.wiley.com/doi/10.1029/97GL01154/epdf.

19 D. J. Travis, A. M. Carleton, and R. G. Lauritsen, "Climatology: Contrails Reduce Daily Temperature Range," *Nature* 418, no. 6898 (August 8, 2002): 601.

20 E. J. Saxl and M. Allen, "1970 Solar Eclipse as 'Seen' by a Torsion Pendulum," *Physical Review D* 3, no. 4 (February 1971): 825.

21 Phil McKenna, "Eclipse Sparks Hunt for Gravity Oddity," *New Scientist*, July 25, 2009, page 101.

CHAPTER 17

1   Philip J. Armitage, "Stars in the Making," *Science* 321, no. 5892 (August 22, 2008): 1047.

2   Y. Xu, H. J. Newberg, J. L. Carlin, et al., "Rings and Radial Waves in the Disk of the Milky Way," Cornell University Library, March 1, 2015, https://arxiv.org/pdf /1503.00257v1.pdf.

CHAPTER 18

1   Stephen Hawking, *A Brief History of Time: From the Big Bang to Black Holes* (New York: Bantam Books, 1988), 13.

2   David Bohm, *Wholeness and the Implicate Order* (London, New York: Routledge, 2002), xiii.

INDEX

Boldface page references indicate illustrations.

A

Abstraction, 77–78, 108, 117
Acceleration, 56, 58, 215
Action at a distance, 64–65, 73–74, 106–7, 119, 173, 182
    Einstein and, 107, 119, 199, 209, 284–85
    Newton and, 64–65, 106–7
Allais, Maurice, 276
Almost Everyone's Guide to Science (Gribbin), 43–44
Amplitude, 54, 66–67, 112, 143, 173–77, **175**
Anti-inflammatory responses, 150–51, 153
Antioxidant responses, 150–51, 153
Anxiety, 14, 150, 155–56
Areas under a curve, 80
Aristotle, 50
Armitage, Philip J., 280
Arrhythmia, 149, 151, 163
Arrow of time, 72, 124, 213
Ary Paradox, 125, 244
The Ascent of Science (Silver), 29
Astronomy through the Ages (Wilson), 45
Atmospheric stability, 271–72
Atomic hypothesis, 36–38, 47–48, 71, 92, 94, 97–98, 130, 197, 222
Atoms, 36–38, 45, 68, 92–93, 104, 113–14, 130–31
    dance of, 139
    mindless atoms, 194
    sprouting complexity, 214
    validated by Einstein, 38, 92, 113–14, 132
Attraction and repulsion, 204, 210, 222, 280
    central nucleation, 216
    four forces of nature and, 96
    gravity and, 280–81
    in peaks and troughs of waves, 172, 177–79, **178**, 184, 201, 216
Attractor waves, 193, 198

Author's story, 142–47
    education, 142–45
    medicine, 144–45, 156, 186
    sports, 142–43, 145–47

B

Bacon, Roger, 79
Bak, Per, 126
Balibar, Françoise, 17
Barabási, Albert-László, 188–89
Barrow, John D., 17, 19, 32
Bateson, Gregory, 78
Bell, E. T., 17
Bell, John Stewart, 107
Bell curve, 109–13, 131–32, 219
Beta-sheets, flat, 247
Big bang, 97
Biodiversity, 270
Blackbody radiation, 40–41, 70, 94, 108, 114, 127, 199, 229, **231**
Black holes, 96–97, 119, 286–87
Blood pooling, 149, 151–53
Bodywave, 190–92, 194, 248–50, 260–62
    as carrier wave, 246, 295
    control over, 260–61
    coordination for health, 295
    exercise and recovery, **159,** 159–60, 163–64, 174, 180, **183,** 190, **191**
    failure to organize, 246
    heartwave and, **159,** 159–60, 163–64, 167, 170–71, 179, 182–83, **183,** 186, 190, 192, 198, 206, 229, 237, 244, 260–62
    inherently continuous fractal pattern, 184
    inner cycles nested in, 261–62
    self-similarity, **183**
    top-down orchestration, 255
Bohm, David, 124, 219, 298–99

Bohr, Niels, 107, 128, 197, 199
  complementarity principle, 43, 94, 203,
    219
  hydrogen emission spectrum, 105
  on quantum world, 98–99, 106
Boltzmann, Ludwig, 111, 122
Born, Max, 42, 99, 199, 206
Bosons, 39–40, 96, 215
Bossomaier, Terry, 10
Boundaries, 9–15, 60
  idea of absolute, 36
  lacking for waves, 138, 154, 157–58,
    160–61
  language reinforing sense of, 20
  numbers, 37
  our perception of, 9–11
  tendency to observe, 22
  world without artificial, 4
Brain waves, 236
Branover, Herman, 298
Breath waves, 264
A Brief History of Time (Hawking), 293
Briggs, John P., 72, 211
Bronowski, Jacob, 10
Brown, Julian R., 93
Brown, Robert, 194
Brownian motion, 38, 113, 115, 131, 193–94
Butterfly effect, 278

C

Calculus, 80, 111–12, 218
Caloric restriction, 274
Cancer, 252–53, 295
Carnot, Sadi, 68–71, 111, 228
Carrier waves, 174–79, **175, 178,** 189
  change shaped by, 266
  compression of waves, 198, 201–5, **205,**
    208, 281–82, 286, 287
  control of health with, 294–95
  disease and, 250, 252–53
  electromagnetism and, 285
  environment, 269, 271
  flattening of, 245, 247, 252–53
  force and, 210
  heartwave as, 244
  influenced by inner waves, 209
  molecular motion and, 193
  mutation and, 252

  nested, 266
  organization of complexity, 217
  particle formation, 205
  phase transitions, 237
  power scaling laws, 236
  quantum jumps and, 201, **202**
  in a relatively closed system, 230
  simulcausality, 228, 250, 269, 279
  speciation, 237–38
  synchronicity, 238–40, 266
  top-down effects, 269, 271, 276, 279
  variability, 271
Causality, 180–81
  direction of, 227
  forward-moving mechanism of change
    across scales, 293
  simulcausality (see Simulcausality)
Cause and effect
  in closed system, 85
  in health and disease, 266
  local, linear, 19, 63–65, 67, 73–74, 108,
    126, 180, 209, 217–18
  simulcausality (see Simulcausality)
Cell cycles, 186–96
Cells, 45–46, 217, 294–95
Cellular and molecular motion, 170–71,
  186–96, **191**
Central dogma of molecular biology, 73, 124
Certainty of number, 37
Certainty versus uncertainty, 119
Chakravarti, Aravinda, 250–51
Change
  description versus understanding of, 82
  order of change, 222
  shaped by carrier waves, 266
  simulcausality, 180
  time as progression of, 213
  waves waving, 220–21
Chaos, 29–30, 76, 82, 122
  flattening of waves, 279
  lack of absolute, 237
  order and, 13–14, 16, 18, 30, 32, 82–83,
    231
  self-organized criticality, 235–37
  systems on the edge of chaos, 235–36
The Character of Physical Law (Feynman), 45,
  76, 106, 117–18
Chargaff, Erwin, 188–89
Chronic disease, 133, 245–46, 260, 267, 294

Civilization, growth of, 19–22
Clausius, Rudolph, 71
Clocks, 55, 181–82, 294
Clockwork universe, 63–65, 81
Closed system, 71–72, 102–4, 123–24
    atoms as, 36
    Brownian motion, 194
    cause and effect in, 85
    disorder promoted in, 232
    energy and, 134, 221
    idealized, 38, 85, 230, 233
    sprouting hampered by, 234
    steam engine, 69, 71, 111, 123
    in string theory, 44
    thermodynamics and, 85, 120, 127, 230–
        33, **231**
Cold fusion, 288, 296–98
Cole, K. C., 46, 76–78, 139
Commonsense experience, 9, 11, 22
Complementarity principle, 43, 94, 203, 219
The Complete Book of Running, 148
Complexity, 130, 182
    of data, 131
    decomplexification, 243, 245–46, 250
    effectiveness of mathematics in, 130,
        219–20
    emergent (*see* Emergent complexity)
    energy contribution to, 221
    explanations for, 122–23
    gravity and, 125
    irreducible, 124
    latitudinal dependence, 273
    loss in environments, 270
    organization by carrier waves, 217
    pathology correlation with loss, 243
    sprouting, 214–15, **215**
    synchronicity, 127
    thermodynamics and, 119–24, 127
Continuity, 15–18, 156–57, 165. *See also*
        Discontinuity
    motion and, 103
    stability due to, 193
    waves and, 153–57, 163
    waves waving, 169–70, 176
    words ill-suited to address issues of, 20
Cooper, Kenneth, 149
Copernicus, Nicolaus, 51
The Cosmic Blueprint (Davies), 124
The Cosmic Code (Pagels), 93, 220

Cosmic Jackpot (Davies), 93
Cosmic microwave background radiation,
    115, 285
Counting, 37, 78
Coupled oscillations, 181
Coyne, Jerry A., 237
Crick, Francis H. C., 46
Crystallization, molecular, 217
Curie, Marie and Pierre, 38–39
Current, 297
Curved spacetime, 107, 115, 118–19, 280,
    284–86
Curves, 101, 109–16
    bell curve, 109–13, 131–32
    calculus and, 80
    carrier wave, 175
    combining multiple, 112
    force fields, 72, 115
    Galileo and, 57–59, 69, 73
Cybernetics, 73, 115, 122
Cycles, 4–8, 76
    of athlete training, 174
    cell, 186–96
    in human behaviors, 155–56
    lack of fixed boundaries, 5–7
    metabolic, 187
    nested, 6, 261–66, 267, 273
    oscillating, 55
    simultaneous, 5–6
Cycles Exercise, 262–63

D

Dalton, John, 38
Dardik Biograft, 145
Dark matter/energy, 97, 125, 280, 282–83
Darling, David, 20, 23, 83–84
Darwin, Charles, 46, 74, 85–86
Davies, Paul, 194, 233
    The Cosmic Blueprint, 124
    Cosmic Jackpot, 93
    The Fifth Miracle, 123, 126, 188–89, 235
    on mathematics, 79, 81, 83
Day-night cycle, 274
de Broglie, Louis, 199–200
    guide waves, 124, 219
    matter-wave duality, 42, 95, 114, 203
    quantum weirdness, 200
    on tyranny of preconceptions, 2

Decomplexification, 243, 245–46, 250

Democritus, 35–36, 38, 43, 98, 104, 131, 212

Descartes, René, 62, 72, 77, 212

Design, 2, 9, 28

Desynchronization, 239–40

Devlin, Keith, 79

Diamond, Jared, 270

Diastole, 175–76, 190, **191**

Dichotomies, 118–28, 131

Differential equation, 80

Discontinuity, 11–19, 24, 101, 163

Discrete, 17, 42–44, 47, 108, 198–99

Disease, 133, 245–46, 260, 267, 294. *See also*
    Health and disease

Divisibility, 23, 211
    continuous, 17, 41, 163
    indivisibility of nature, 28, 34

DNA, 46, 73, 124, 249
    change guided by other waves of motion, 195
    motion of, 186–87, 249–50

DNA: The Secret of Life (Watson), 46–47

Does God Play Dice? (Stewart), 112

Double-slit experiment, 94, 104, 114

The Dreams of Reason (Pagels), 90–91

Dynothermatics, 287

E

Earthquakes, Richter scale of, 235

Earth's hum, 115

Eddington, Sir Arthur, 72, 232

Edge effect, 133, 270, 295–96

Edges, 9–10, 15, 22, 44, 60

Edison, Thomas, 297

Einstein, Albert, 41–43, 96
    action at a distance, 107, 119, 199, 209,
        284–85
    atoms validated by, 38, 92, 113–14, 132
    Brownian motion, 38, 113, 115, 131, 194
    curved spacetime, 107, 280, 284–86
    Einstein-Podolsky-Rosen (EPR) paradox,
        107
    electromagnetism and, 285
    e=mc², 95, 114, 221
    on Galileo, 60–61
    general relativity, 118, 182, 285
    gravity and, 115, 118–19, 182, 280, 284–85
    light and, 41–42, 95, 114, 199, 203
    on mathematics, 79–80, 84

nature as a puzzle, 2, 19, 22

photoelectric effect, 41

quantum weirdness, 200

science as everyday thinking, 33

search for Theory of Everything, 90, 118

space-time continuum, 115, 118–19

special relativity, 85, 86, 102

statistical interpretation of quantum
    mechanics, 219

time and, 213

Einstein: Decoding the Universe (Balibar), 17

Einstein-Podolsky-Rosen paradox, 107

Einthoven, Willem, 54, 72–73, 163, 165, 244

Eldredge, Niles, 238

Electricity, 118, 297

Electrocardiogram, 72–73, 163–66, 242, 244

Electromagnetic waves, 40–41, 66, 70, 195,
    209, 282–83, 285

Electromagnetism, 65–66, 96, 210, 274,
    284–86

Electrons, 38–39, 42
    motion of, 104–5, 114
    quantum jumps, 236
    as waves, 114, 205, 207

Elements, periodic table of, 215

Elowitz, Michael B., 187

Emergence: From Chaos to Order (Holland), 120

Emergent complexity, 122, 126–27, 134,
    214–17
    described, 120–21
    energy and, 221
    fusion as, 296
    hampered by isolation/closed systems,
        232–34
    of organized matter, 233
    slime molds, 239
    species richness gradient, 272
    sprouting and, 232–34

Emergent Fusion, 296–98

Empedocles, 35

Endurance, 149, 151–52

Energy, 220–21. *See also* Thermodynamics
    availability of free energy, 122
    cellular, 187
    clean, 288, 296
    closed system and, 134, 221
    conservation, 85, 127
    dark, 97, 125, 280, 283
    defined, 102–3

dispersal of, 230
e=mc², 95, 114, 221
emergent complexity and, 221
exchange of, 102
heat, 229
heat loss in exchange of, 120, 122
kinetic and potential, 102
light, 40–42, 199
low-energy nuclear reaction, 296–97
mass relationship to, 93, 221
matter and, 129, 140
motion and, 102–4, 129, 220
operational definition, 220
particles as abstract mathematical
    quantities of, 93
quantum, 40–42, 201, **202**
waves and, 95, 115, 139–40, 143, 155, 162
waves waving, 220–21
what it is, 129
Energy metabolism, 257, 273
Entropy, 122, 126, 230, 232–33
Entropy (Rifkin), 86
Environment, 269–71
    atmospheric stability, 271–72
    carrier waves, 269, 271
    edge effect, 133, 270, 295–96
    fragmentation, 269–70
    latitudinal gradient, 272–73
    pollution, 275, 295, 299
    temperature range of earth, decreased
        variability in, 275
    treatment of, 133–34
Epicurus, 38
Epigenetics, 124, 251
Equations, 78
Equations of Eternity (Darling), 20, 23, 83–84
Euclid, 62
Evolution, 46, 74, 85–86, 194–95
    favorable mutations, 251–52
    punctuated equilibrium, 238
    rate of, 273
    speciation, 237–38
Exercise
    heart rate after, 150–52, 157–59, 162, 167
    LifeWaves program, 262–66
    as pro-oxidation, proinflammatory, 153
    recovery and, 149–60, 163, 229, 261–66,
        274, 295
Experiment, 19–20, 33–34, 56, 79, 90, 107, 171

F

Facts, reunderstanding of, 170
Faraday, Michael, 65
Fear of Physics (Krauss), 52, 59, 79
Feedback loops, 115, 126
Fermions, 39
Feynman, Richard P., 24, 34, 47
    on atoms and dance, 139
    The Character of Physical Law, 45, 76, 106,
        117–18
    on scientific truth, 107
    Six Easy Pieces, 24
Fields, 65, 72, 115, 199, 210, 283–84
The Fifth Miracle (Davies), 123, 126, 188–89,
    235
Fire, mastering, 24
First You Build a Cloud (Cole), 139
Fixx, Jim, 148
Flattening of waves, 245–48, 251
Fölsing, Albrecht, 42
Force(s), 39, 56–57
    attraction and repulsion in peaks and
        troughs of waves, 172, 177–79, **178,**
        184
    carrier waves and, 210
    Faraday and, 65
    fields and, 115, 283–84
    four forces of nature, 95–96
    lines of, 65
    Newton and, 62–64
    operational definition of, 177
    as particles, 95–96
    waves waving, 284
Fourier, Joseph, 66–67, 109
Fractals, 126. *See also* Inherently continuous
    fractal
    health and, 243–44
    life as, 189–91
    in nature, 184
    waves waving and, 183–85
Fragmentation, 269–70
Framingham Heart Study, 125, 242
Frequency, 54, 66–67, 112, 143, 173–77,
    175
Friction, 57
From Certainty to Uncertainty (Peat),
    10, 37
Frost, Robert, 141
Fusion, 287–88, 296–98

G

Gaia hypothesis, 276
Galaxies, 97, 279–82
Galileo Galilei
  acceleration, 56, 58
  curves, 57–59, 69, 73
  gravity, 56–58
  inclined-plane experiments, 57
  linearity, 52–74, 77, 83, 100–101, 115–16,
    165–67
  mathematics as language of God, 79
  motion, 51–62
  pendulum motion, 54–55
  projectile motion, 57–58
  treatment of reality by, 90–91, 165–66
Galton, Francis, 110
Gefter, Amanda, 184
General relativity, 118, 182, 285
Genes, 46–47, 73, 98, 124, 133, 249–52
Geographical Ecology (MacArthur), 76
The Ghost in the Atom, 93
Global storming, 134, 275
Glucocorticoids, 187–88
The God Particle (Lederman), 96
Goldberger, Ary L., 125, 243–44
Gould, Stephen Jay, 238
Graviton, 39–40, 96
Gravity, 56–58, 63–65, 96–97, 279–87
  attraction-repulsion and, 280–81
  black holes, 119, 287
  complexity and, 125
  connection to quantum, 286
  Einstein and, 115, 118–19, 182, 280, 284–85
  general relativity and, 118, 182
  matter and, 281–82
  Newton and, 63–65, 81, 106–7, 182, 280
  strangeness of, 119
  SuperWaves and, 280–83, 286
Green, David, 10
Gribbin, John, 43–44
Guide waves, 124, 219

H

Haw, Mark, 38, 219, 271
Hawking, Stephen, 18, 34, 87, 140, 293
Health and disease, 241–68
  cultivation of health, 133
  flattening of waves, 245–48, 251

health as an absence of disease, 253
heart rate variability, 241–47, 253, 257–58
ischemic preconditioning, 241, 255–59, **258**
latitudinal dependence, 272
as processes not things, 267
synchrony and organization, 254, 294–95
Heart
  cellular and molecular waves, 189–90, **191**
  cycles of, 6
  ischemic preconditioning, 241, 255–59, **258**
  systole and diastole, 175–76, 189–90, **191**
Heart attack/disease, 148–49, 151–52, 241
Heart rate
  after exercise, 150–52, 157–59, 162, 167
  on electrocardiogram (ECG), 163–66
  LifeWaves program and, 263–64
  sinus arrhythmia, 163
Heart rate variability, 125, 241–47, 253,
    257–58, 262–66, 275
Heartwave, 179–81, 190–92, 245–50, 268
  bodywave and, **159,** 159–60, 163–64, 167,
    170–71, 179, 182–83, **183,** 186, 190,
    192, 194, 198, 206, 229, 237, 244,
    260–62
  as carrier wave, 244
  exercise and recovery, **159,** 159–60, 163–64,
    166–67, 174, 177, 180, 190, **191, 261**
  flattening of, 245–46
  health and disease, 245–50, 260–65, 268
  inherently continuous fractal pattern, 184
  inner cycles nested in, 261–62
  in ischemic preconditioning, 259
  self-similarity, **183**
  shaping of, 261, 264
  variability, 245, 253–59, 262, 265, 275–76,
    295
Heat, 68–72, 120–22
  cold fusion, 296–98
  energy, 229
  motion and, 287
Heat death, 120–21
Heisenberg, Werner, 105–6
Heisenberg uncertainty principle, 105, 197,
    199, 207–8
Helmholtz, Herman von, 120
Higgs boson, 40, 96, 215
Holland, John Henry, 120
Hooke, Robert, 45
Hormones, 187–88, 193

How does it work?, 28–29, 31, 49–74, 91,
    108–12, 155
How Nature Works (Bak), 126
Hunger, 5, 14, 155–56, 294
Huygens, Christiaan, 181–82, 294
Hypothesis, 15, 19, 56, 170–71

I

Inclined-plane experiments, 57
Independent velocities, law of, 58, 228
Industrial Revolution, 67, 71, 103, 111, 275
Infinitesimals, 80
Inherent, use of term, 176
Inherently continuous fractal, 197, 200–201,
    220, 235–38
    cross-scale continuity of, 236
    heartwave organization, 184
    motion, 192–93, 218
    power scaling laws, 235–36
    quantum and gravity, 286
    self-organized criticality, 235–37
    synchronicity, 238–40
    waves waving as, 173, 203, 210, 221
Inherent nonlinearity, 172–77
Instability, 13, 30–31
Interference pattern, 104–6, 114
Interval training, 144
Invariance
    Einstein's law of special relativity, 86
    facts about nature, 131
    lack of existence of true, 228
    laws as, 75–76, 83, 86
Invariants
    atom as, 36
    genes as, 47
    idealized, 59
    linearity, 52–54, 57–61, 72, 74
    mathematics and, 80, 117
    matter, 50–51
    measurement of, 78
    search for, 31–36, 44–45, 84
Ischemic preconditioning, 241, 255–59, **258**
Isochronism, 55
Isolationism, experimentation through, 232

J

Jackson, Wes, 296

K

Kaufman, Stuart, 122
Kaufmann, Walter, 215
Kelly, Jack, 148, 163, 166, 265, 289
Kelvin, Lord. *See* Thomson, William (Lord
    Kelvin)
Krauss, Lawrence M., 52, 59, 79

L

Language, 19–23, 79, 108
Laplace, Pierre-Simon, marquis de, 63
Latitudinal density gradient, 272–73
Laughlin, Robert B., 90
Laws, 75–87, 128–29, 134–35
    conservation of energy, 85, 127
    disparity in, 117–18
    first law of thermodynamics, 85, 120, 127,
        129
    of independent velocities, 58, 228
    invariance and, 75–76, 86
    lack of existence in nature, 82–83
    mathematical, 81–82
    predictive power of, 77, 81–82
    search for, 117
    second law of thermodynamics, 71–72,
        120–24, 129
    special relativity, 86, 102
    Theory of Everything, 90
    unification of, 86–87
    universal gravitation, 106
Lederman, Leon, 90, 96
Leibnitz, Gottfried Wilhelm, 80, 218
Leptons, 39
Leucippus, 35–36, 43
Levin, Simon A, 167
Levine, Joe H., 187
LifeWaves, 193, 262–67
Light
    Einstein and, 41–42, 95, 114, 199, 203
    as electromagnetic wave, 66
    energy, 40–42, 199
Lightening, 274
Lin, Yihan, 187
Linear frequencies, 108
Linearity, 52–74
    calculus, 80
    cause and effect, 63–65, 67, 73–74
    Galileo and, 52–74, 77, 83, 165–67

Linearity *(cont.)*
  of heat flow, 69–71
  motion, 52–74, 77, 100–101, 165
  time, 55, 77
Linear superposition, 67
Lines, 52–74
  curves treated as, 80
  invariant, 72, 74
  in space, 62, 84
  waves as, 66–67
Locality, 36
The Looking Glass Universe (Briggs and Peat),
    72, 211
Lorenz, Edward, 278
Lovejoy, Thomas E., 133, 269–70, 276,
    295–96
Lovelock, James, 276
Lucretius, 38

M

MacArthur, Robert H., 76
Maddox, John, 74
Magnetism, 118
Magnetohydrodynamics, 298
Mandelbrot, Benoit, 126
Margulis, Lynn, 276
Mass
  dark matter, 97
  electrons, 215
  energy relationship to, 93, 221
  missing from universe, 125, 280
  sprouting, 215, 286
Mathematical laws, 81–82
Mathematics, 17
  applied to motion, 108
  calculus, 80, 111–12
  certainty, 37
  formalization of nature as puzzle, 22–25
  invariants, 80
  as language of discreteness, 108
  limitations in describing nature, 218–20
  matter and, 93
  measurements, 78–79
  nature understood through, 79–87
  power scaling laws, 235–36
  predictive power of, 109, 129
  probability, 105–6
  as the reality, 199

Theory of Everything and, 90
  as ultimate reality, 129
  unreasonable effectiveness of in natural
      sciences, 130–31, 218–19
  waves and, 143–44, 162, 165
Matter, 16, 35–48
  created by consciousness, 199
  dark, 97, 125, 280, 282
  dematerialization, 92–97
  as discrete, 47, 94
  distribution in space, 184
  duality with waves, 203
  energy, 129, 140
  gravity and, 281–82
  invariant, 50–51
  matterspace, 234, 249
  matterspacetime, 213, 221, 223, 285
  motion and, 12–13, 29, 31, 48–52, 100–103,
      222
  as the "one," 16
  quantum mechanics, 41
  space and, 12–13
  stability and, 13, 29–32
  wave-particle duality, 94–95
  waves waving compared, 162
  waves waving compressing to become,
      203–4, 209, 212, 221, 281
  What is it? question, 28–29
  where we stand today with regard to, 92–99
Matterspace, 210, 234, 249
Matterspacetime, 213, 221, 223, 285
Maxwell, James Clerk, 66, 109, 111, 118, 121,
    165
Maxwell's Demon (von Baeyer), 113
McKenna, Phil, 276
McKubre, Michael, 297
Measurement, 78–79, 208
Mechanical universe, 81, 84, 86
Mechanistic change, 63–64
Medicine
  author's experience in, 144–45, 156, 186
  divide and conquer strategy, 133
  health and disease, 241–68
  implications for control over waves, 179
  integrative, 294
Meditation, 149–50, 152, 155
Men of Mathematics (Bell), 17
Metabolism, 146, 187, 257, 273–74
Metallurgy, 298

Metastasis, 253
Microwave background radiation, 115, 285
Middle World: The Restless Heart of Matter and Life (Haw), 38
Miller, Stanley, 274
Mind-body connection, 150
Mind over Matter: Conversations with the Cosmos (Cole), 46
Molecular movements, 186–96
Molecules, flattening of, 247
Momentum, 208
Mortality, low heart rate variability linked to, 241–47
Motion, 12–13, 16, 49–74
   Aristotle and, 50
   bell curve as definitive pattern of group motion, 110
   Brownian, 38, 113, 115, 131, 193–94
   calculus applied to, 80–81
   cellular and molecular, 170–71, 186–96, **191**
   curved, 80
   of DNA molecule, 186–87, 192
   of electrons, 104–5, 114
   energy and, 102–4, 114, 129, 140, 221
   Galileo and, 51–62
   gravity, 56–58
   heat and, 287
   invariant, 50–52
   lack of boundaries, 160–61, 211
   linearity, 52–74, 77, 100–101, 165
   mathematics applied to, 108
   matter and, 12–13, 29, 31, 48–52, 100–103, 222
   in motion, 161, 169, 172, 182, 200, 222, 268
   Newton's laws of motion, 62–64, 81, 101, 108–9, 165
   as the "one," 16
   patterns of, 76–79, 157
   perpetual motion of particles, 204, 222
   quantum, 104–6
   reductionism, 50–51, 54, 58, 60
   space and time, 13
   stability and, 13, 29–32
   steam engine, 67–71
   time and, 55
   top-down influences, 179–82
   ubiquity of, 172
   vibratory, 188
   waves, 66–67, 138–40, 160, 279
   where we stand today with regard to, 100–116
   Zeno's paradoxes, 100–101
Mutations, 193, 195, 248, 250–52

N

Natural Inheritance (Galton), 110
Natural selection, 74, 86, 194–95
Nature
   existence of, 3
   indivisibility of, 28, 34
   matter, space, motion, and time as parts of, 12–17
   as "one," 15–16, 18, 31, 34, 38, 59–60, 64, 84, 90–91
   as puzzle, 11–25, 28–34, 38, 43, 46, 48–51, 59, 61
   separation from, 21–22
   stratification into different scales, 14
   as whole made of parts, 2–3, 9, 11, 13, 18, 28, 32, 35, 48, 61
Nature's Imagination: The Frontiers of Scientific Vision, 50
N-body problems, 111, 218, 278
Negation, fallacy of, 18
Negative entropy, 122, 126
Neutrinos, 39
New Theories of Everything (Barrow), 32
Newton, Isaac, 55–56, 62–65
   action at a distance, 64–65, 106–7
   calculus, 80, 218
   gravity and, 63–65, 81, 106–7, 182, 280
   motion and, 62–64, 81, 101, 108–9, 165
   Philosophiae Naturalis Principia Mathematica, 65
Nonlinearity of waves waving, 172–77, 219
Nucleation, central, 214, 216
Nucleus, atomic, 39, 285, 286
Numbers, 79, 82
   atoms and, 37
   boundaries, 37
   as distinct, 22–23
   measurement, 78

O

Observation, 12, 56
Occam's razor, 198, 259–60

Oltvai, Zoltán N., 188–89
Oncogenes, 73
"Only one"
  light energy as, 42
  motion in motion and, 161, 172, 268
  nature as, 34, 43, 59–60, 84, 134, 168, 222, 268
  waves waving as, 170, 198, 223
On the Origin of Species (Darwin), 237
Open system, 103, 131, 274
Operational definition, 102
Oppenheimer, Frank, 77
Opposites, 118–28, 134
Optimistic arrow, 124
Orbits, 279
Order, 76–77, 79–80
  of change, 222
  chaos and, 13–14, 16, 18, 30, 32, 82–83, 231
  inherently continuous fractal, 236
  religious origin of, 33
  synchronization and desynchronization, 240
  universe as ordered, 218
Organization, 13, 15
  emergent complexity of organized matter, 233
  hampered by isolation/closed systems, 232–34
  latitudinal dependence, 273
  network, 267
  wave waving, 234–35
Original Theory of Everything, 24, 35, 48, 51, 113–14, 125, 127
The Origins of Knowledge and Imagination (Bronowski), 10
Oscar II (King of Sweden and Norway), 278
Oscillators, 55
Our Mathematical Universe (Tegmark), 128
Overbye, Dennis, 128

P

Pagels, Heinz R., 90–91, 93, 137, 220
Parabola, 57, 69, 80
Particles, 38–44, 92–99, 132
  appearance from mathematical probability, 206
  forces as, 95–96
  lack of boundaries, 204
  light as, 41–42, 95

perpetual motion of, 204, 222
quantum motion/jumps, 104–6, 199, 201, 202
uncertainty principle, 105
virtual, 199
wave-particle duality, 94–95, 104, 114, 119, 170, 197
waves waving, 197, 203–4, 205, 207–9, 286
Particle zoo, 95–96, 98, 129
Patterns, 76–79, 139, 157
Patterns in the Sand (Green), 10
Pearson, Helen, 249
Peat, F. David, 10, 37, 72, 195–96, 211, 219–20
Pendulums, 54–55, 77, 181, 276
Pennisi, Elizabeth, 7, 272
Penrose, Roger, 50
Performance, 146–47
Perrin, Jean Baptiste, 92
Petersen, Aage, 99
Phase transitions, 237
Pheidippides, 148, 166
Philosophiae Naturalis Principia Mathematica (Newton), 65
Photoelectric effect, 41–42
Photons, 39, 42–43, 199, 203, 215
Photosynthesis, 239–40
Pietronero, Luciano, 184
Pi in the Sky (Barrow), 17
Pilot waves, 124, 219
Placebo effect, 259
Planck, Max, 40–43, 70, 94–95, 108, 114, 197–203, 228–29
Planets, motion of, 279
Podolsky, Boris, 107
Poincaré, Henri, 278–79
Points, 18, 62, 77, 80, 164, 183, 218
Pollution, 275, 295, 299
Post-traumatic stress disorder, 271
Power scaling laws, 115, 235–36, 247–48
Preconceptions/preconditions, 2, 34, 37, 101
Prigogine, Ilya, 122
Prions, 247, 248
Probability wave, 105–6, 206
Projectile, motion of, 57–58
Proteins, 73, 247, 287, 293
Pulsilogium, 54
Punctuated equilibrium, 238
Puzzle Hypothesis, 15–24, 30–32

atomic hypothesis and, 36–38
failure to evaluate it, 140–41
inherent shortcoming of, 30

Q

Quantum, 72, 127, 170, 198, 286
Quantum energy, 40–42, 201, **202**
Quantum entanglement, 240
Quantum field theory, 199
Quantum jumps, 199, 201, **202**, 204, 211–12, **231**, 236
Quantum mechanics, 108, 118, 219
Quantum motion, 104–6
Quantum physics, 39–44, 82, 94–95, 106–7
   basic problems of, 198–201
   creation of, 40–71, 70, 114, 201
   gravity and, 118
   thermodynamics and, 70, 127–28, 228–29
Quantum possibilities, 199
Quantum theory, 107, 128
Quantum tunneling, 115
Quantum weirdness, 94, 198, 200, 207–8
Quantum world, 98–99, 171, 198
Quarks, 39, 44, 286

R

Reality, 90–91, 101, 128, 165–66
Recovery
   anti-inflammatory and antioxidant responses, 150–51, 153
   dying during, 148–49, 151
   exercise and, 149–60, 163, 229, 274, 295
   training, 151, 262
Reductionism, 33, 50–51, 54, 58, 60, 124
Reflections on the Motive Power of Fire (Carnot), 69
Regularities, 76–81, 128, 130
   in disease, 245
   regularity of irregularity, 222
   search for, 134–35
Relational reality, 128
Relativity, 72, 86, 102, 118, 182, 285
Relaxation, 14, 150
Repetition, 270–71
Rhythms
   assumption of discontinuity, 14–15
   author's experience with, 142, 146–47

creating full-body, 295
disturbances in, 7
DNA movement, 187
examples, 4–5
in humans, 72–73, 146
lack of fixed boundaries, 5–7
motion, universality of rhythmic, 222
participation in nature's, 7, 294
physical laws, 76
treated as pieces of a puzzle, 21
universality of, 3–8
world as experienced by early humans, 4–8
Rifkin, Jeremy, 86
RNA, 73, 187
Rosen, Nathan, 107
Running, 148–49
Running without Fear (Cooper), 149
Rutherford, Ernest, 39

S

Scales, simultaneous change across, 169
Schrödinger, Erwin, 42, 122
Schrödinger wave function, 199, 205–6, 208
Science
   creation of, 33
   experiment, 33–34, 56, 90, 107, 171
   hypothesis, 15, 19, 170–71
   material progress, 90
Scientific laws, 75–87
Scientific method, 33–34, 232
Scott, David, 56
Self-organization, spontaneous, 122, 126
Self-organized criticality, 122, 235–37
Self-similar waves, **183**, 183–84, 201, 214–15, 246
Senses, 223
Sensitivity to initial conditions, 278
Silver, Brian L., 29
Simulcausality
   carrier waves, 279
   disease and, 246
   environment and, 269
   forward-moving mechanism of change across scales, 293
   guidance from carrier waves, 228
   heart rate variability, 244
   shaping of waves, 261–62
   SuperWaves and, 227–28

Simulcausality *(cont.)*
    synchronicity, 238–40
    as waves waving principle, 173, 180–82,
        184, 201, 209, 217–18
Singularity, 96–97, 119
Six Easy Pieces (Feynman), 24
Skeletal muscle reperfusion injury, 257
Sleep, 156
Slime molds, 239
Solar system, stability of, 278–79
Space
    cosmic microwave background radiation,
        115, 285
    forces in, 65
    lack of boundaries, 211
    linearity, 62, 84
    matter and, 12–13, 209, **210**
    matter distribution in, 184
    matterspace, 210, 234, 249
    matterspacetime, 213, 221, 223, 285
    motion and, 102–3
    solar system stability, 278–79
    waves spread to form, 204, 209–12, 221,
        282–83
Spacetime, curved, 107, 115, 118–19, 280,
    284–86
Space-time continuum, 115
Special relativity, 85, 86, 102
Speciation, 237–38
Species diversity, 272
Species richness gradient, 272
Spontaneous self-organization, 122, 126
Sprinting, 152
Sprouting, 213–19, **215,** 232–34, **234**
Stability, 13, 29–33
    atmospheric, 271–72
    atoms, 36
    complexity and, 120–21, 122
    continuity and, 193
    definition of, 124–25
    latitudinal dependence, 273
    mathematical laws and, 81–82
    matter, 50
    scientific laws, 87
    solar system, 278–79
    variability as, 125
Standard Model of particle physics, 39–40,
    95–96
Stars, 119, 280, 282, 287–88

Statistics, 109–12, 131–32
Steam engine, 67–71, 103, 111, 120, 123
Stewart, Ian, 112
Stoppard, Tom, 118
Stress, reduction with meditation, 149–50
String theory, 44, 115
Strogatz, Steven, 69–70, 127, 176–77
Strong force, 96, 210, 285
Subatomic particles, 95, 97, 115
Superclusters, 282
SuperWaves, 170, 223–38
    applications, 293–99
    brain waves, 236
    cold fusion and, 296–98
    discovery of, 171, 244
    DNA and, 249–50
    environment and, 171–72
    epigenetics and, 251
    fields and, 283–84
    flattening of molecules, 247
    gravity and, 280–83, 286
    health and disease, 241, 246–47, 252,
        255
    inherently continuous fractal nature of,
        235–40
    ischemic preconditioning, 255–59
    Occam's razor, 259–60
    organizational aspects, 296
    punctuated equilibrium, 238
    resistance to discovery of, 225
    significance of, 293, 300
    simulcausality and, 227–28
    synchronicity, 238–40
Sympathetic resonance, 181
Sync (Strogatz), 69–70, 127, 176–77
Synchronicity, 127, 181, 238–40
    carrier waves and, 266
    health and, 294–95
    medications administered in, 294
    synchronization and desynchronization,
        240, 265
Synchronicity (Peat), 195–96, 219–20
Systems on the edge of chaos, 235–36
Systole, 175–76, 189–90, **191**

T

Tegmark, Max, 128
Temperature range of earth, 275

Tesla, Nikola, 297
Thales of Miletus, 35
Thermodynamics, 70–72, 85
    complexity theory and, 119–24, 127
    first law of thermodynamics, 85, 120, 127,
        129, 233–34, 273
    fourth law of, 233
    irreversible processes of, 111
    nonequilibrium, 271
    quantum physics and, 127–28, 228–29
    second law of thermodynamics, 71–72,
        120–24, 129, 213, 221, 229–32, **231**
Thomson, J. J., 38–39
Thomson, William (Lord Kelvin), 71, 120,
    298
Thought, development of human, 28
Time
    curved spacetime, 107
    irreversibility, 213
    as line, 55, 77
    matterspacetime, 213, 221, 223, 285
    motion and, 102–3
    as progression of change, 213
    sense of, 12–13, 16
    space-time continuum, 115
    straight line of, 71–72
    waves waving, 212–13
Trial and error, 19
Tumor-suppressor genes, 73

U

Ultraviolet catastrophe, 40
Uncertainty principle, 105, 140, 197, 199,
    207–8
Understanding, 3, 11–12
Unification, 33
Unity, words ill-suited to address issues of, 20
Universality, 5
Universe
    expansion of, 44, 282–83
    heat death of, 120–21
    missing mass, 125, 280
    multiple universes, 199
    as ordered, 218
The Universe and the Teacup, 76–78
The Unnatural Nature of Science (Wolpert),
    98
Urey, Harold C., 274

V

Vacuum, 114
Variability
    decreased in disease, 294
    as essential for survival, 295
    heart rate, 125, 241–47, 253, 257–58,
        262–66, 275
    heartwave, 245, 262, 295
    preservation of, 296
Vibratory motion, 188
Virtual particles, 199
von Baeyer, Hans Christian, 113, 128, 219
von Mayer, Julius Robert, 273–74

W

Wakefulness, 156
Water, 35
Watson, James D., 46–47
Wave function, 105–6, 199
Wavelength, 54, 66
Wavenergy, 221
Wave packets, 202, 206
Wave-particle duality, 94–95, 104, 114, 119,
    170, 197, 199, 203
Waves. *See also* Waves waving
    attraction and repulsion in peaks and
        troughs of, 172, 177–79, **178,** 184
    attractor, 193, 198
    brain, 236
    breath, 264
    carrier (*see* Carrier waves)
    cellular and molecular motion, 170–71,
        186–96, **191**
    continuity, 153–57, 163
    definitions of, 66, 138–40, 143, 154
    dispersal of, 230, **231,** 232
    duality with matter, 203
    electromagnetic, 40–41, 66, 70, 195, 209,
        282–83, 285
    energy and, 95, 115, 139–40, 143, 155, 162
    exercise and recovery, 153–60, 163
    flattening of, 245–48, 251, 276, 279
    guide, 124, 219
    helix as spiraling wave, 249
    in humans, 72–73
    inherently continuous fractal, 173, 184,
        192–93
    interference pattern, 104–6, 114

Waves *(cont.)*
  lack of boundaries, 138, 154, 157–58,
    160–61
  as lines, 66–67
  mathematical analysis, 143–44, 162, 165
  motion, 66–67, 138–40, 160
  as nonmaterial, 65–66, 139, 143
  ocean, 138, 140, 142–43
  particles, 42–43, 94–95, 104, 114, 119, 170
  as pattern of repetition, 154
  in patterns that cause disease and health,
    254
  perceived as unnecessary side effect of
    nature, 21
  pilot, 124, 219
  probability, 105–6, 206
  relationship across scales, 261
  self-similar, **183,** 183–84, 201, 214–15, 246
  shaping of, 261–62, 264–65
  space formed by spread, 204, 209–12, 221
  sprouting, 213–15, **215**
  as sum of parts, 67
  within waves, 163–64, 166, 197–98,
    201–5, 209–10, **210,** 213, **216,** 217
Waves waving, 168, 189, 192–98, 201–17,
  215–16
  atmosphere as, 272
  in biological organism, 167, 195, 196
  characteristics of, 171–73
  compressing to become matter, 203–4,
    209, 212, 221
  continuity, 161, 169–70, 176, 197, 208, 210
  energy, 220–21
  forces, 284
  fractals, 183–85
  genes and, 249
  health and disease, 241, 255, 259
  heartwave and bodywave, 158, **191**
  in ischemic preconditioning, 257
  latitudinal dependence, 272
  matter and, 162, 209
  Miller-Urey experiment and, 274

nested scales, **191**
ocean waves, 138
particles, 197, 203–4, **205,** 207–9, 286
power scaling laws, 236
principles of, 169–85
  attraction and repulsion in peaks and
    troughs of waves, 172, 177–79, **178,**
    184, 201, 216
  nonlinearity of continuum of frequencies
    and amplitude, 172–77, **177,** 184, 219
  simulcausality, 173, 180–82, 184, 201,
    209, 217–18
quantum realm and, 197–223
Schrödinger wave equation, 199, 205–6
simultaneity, 161–62, 165–66, 209
spacetime, 286
sprouting, 213–15, **215**
time and, 212–13
top-down influences, 179–82
within waves, 197–98, 201–5, 209–10,
  210, 213, **216,** 217
Wavicles, 202
Weak force, 96, 210, 285–86
What holds things together?, 28–32, 76, 91
What is it?, 28–29, 31, 35–48, 91, 155
What is Life? (Schrödinger), 122
Wheeler, John Archibald, 98, 137, 198
Wholeness and the Implicate Order (Bohm),
  299
Why do things fall apart?, 28–32, 76, 91
Wiener, Norbert, 73
Wigner, Eugene, 130–31, 218
Wilson, Robert, 45
Wolpert, Lewis, 98
Work, 68–69, 71, 102–3, 220, 229–30, 273
The World within the World (Barrow), 19

Z

Zeno of Elea, 100–101, 211
Zeno's paradoxes, 100–101, 211–12
Zero-vacuum fluctuations, 114–15, 209